ENHANCING DECENT WORK IN THE AGRICULTURE SECTOR

ENHANCING DECENT WORK IN THE AGRICULTURE SECTOR

Dr. Hazem Shawky Fouda

Delve
PUBLISHING

www.delvepublishing.com

Enhancing Decent Work in the Agriculture Sector

Dr. Hazem Shawky Fouda

Delve Publishing
4164 Lakeshore Road
Burlington, ON L7L 1A4
Canada
www.delvepublishing.com

Email: orders@arclereducation.com

ISBN: 978-1-77956-476-4 (Hardcover)

ABOUT THE AUTHOR

Dr. Hazem Shawky Fouda has a PhD. In Agriculture Sciences, obtained his PhD. from the Faculty of Agriculture, Alexandria University, 2008, MSc. In Agriculture Sciences from the Faculty of Agriculture, Alexandria University in 2004, Post-Grade Diploma in Cotton, 2001, BSc. in Agriculture Sciences, from the Faculty of Agriculture, Alexandria University, 1997, worked in Cotton Arbitration & Testing General Organization (CATGO) from 1999 till 2018 was working in the International Cotton Training Center (ICTC) – Cotton Arbitration & Testing General Organization (CATGO) from 2000 till 2015, as a Lecturer & Classer's Trainer for Egyptian and foreigner classers, technicians, ginners, spinners & traders in all cotton aspects. Besides that he was an editor and active member in the Research & Translation Committee, participating in issuing weekly, monthly and annually issues about the international & local cotton market including price trends and direction, recent developments & researches concerning cotton production, protection, harvesting, ginning, fiber testing, spinning & weaving since its foundation in 2000 till 2014 and from 2015 till 2018 he worked as an inspector, since 2018 till present works as a consultant.

TABLE OF CONTENTS

LIST OF FIGURES

LIST OF TABLES

PREFACE

The agriculture sector is central to the livelihoods of millions across the globe, providing essential food and resources. However, those who work within this industry often face severe challenges related to fair wages, working conditions, and basic human rights. It is against this backdrop that the concept of *decent work*—as outlined by the International Labour Organization (ILO)—emerges as a critical framework for ensuring that agricultural workers are afforded opportunities for fair treatment, safe environments, and job security.

This book, *Enhancing Decent Work in the Agriculture Sector*, aims to explore the multifaceted dimensions of labor in agriculture, focusing on the implementation of decent work principles. It seeks to identify and analyze the key issues facing agricultural workers today, ranging from child labor and gender inequality to occupational health and safety.

Chapter 1 provides an overview of the concept of decent work, rooted in the ILO's framework. It outlines the challenges associated with its implementation, particularly in the agricultural sector, where child labor, occupational safety and health (OSH), and gender disparities persist. Special attention is given to the "Decent Work, Decent Life for Women" campaign, which advocates for equal rights and opportunities for women in agriculture. This chapter emphasizes the importance of addressing these issues in agricultural supply chains and improving the rights and conditions of waged agricultural workers.

Chapter 2 offers a detailed examination of the working conditions faced by agricultural laborers across the globe. It discusses the prevalence of seasonal and migrant labor, characterized by irregular work hours, low wages, and job insecurity. Agricultural labor often involves long hours under difficult conditions, and this chapter highlights the critical need for fair compensation and improved labor relations. The struggles faced by agricultural workers are explored, particularly in light of state interventions, labor mobility, and the economic realities of agricultural economies. The role of consumers in advocating for better labor conditions is also discussed, offering insights into how collective action can drive positive change.

Chapter 3 tackles the sensitive and pervasive issue of child labor within the sector. Despite international efforts to eradicate child labor, it remains prevalent in many agricultural communities, driven by poverty, economic pressures, and cultural norms. This chapter examines the drivers of child labor, including the role of global supply chains and the seasonal nature of agricultural work. It highlights the FAO's guiding principles on addressing child labor and presents strategies for promoting safe agricultural practices, integrating agriculture and nutrition into rural education curricula, and empowering youth.

Chapter 4 addresses the persistent gender inequalities that characterize agricultural labor. Women play a crucial role in agriculture, yet they face numerous barriers, including unequal access to land, financial resources, and decision-making power. This chapter examines gender wage gaps, the contribution of women to food security, and the barriers that prevent gender equality in agricultural labor. Strategies for promoting gender equality and improving working conditions for women are discussed, with a focus on policy frameworks, community-based initiatives, and the importance of addressing cultural norms that perpetuate discrimination.

Chapter 5 focuses on the health hazards associated with agricultural work, particularly the dangers posed by pesticide exposure and the lack of adequate health and safety measures. This chapter examines the risks faced by farmworkers, including exposure to harmful chemicals and the long-term health effects associated with pesticide use. It also explores strategies for reducing health risks, such as implementing safer pesticide application methods and promoting better health risk assessments. Improving occupational safety and health in agriculture is vital to ensuring that workers can carry out their duties in safe and healthy environments.

Chapter 6 focuses on the livelihoods of agricultural workers, discussing the various categories of workers—full-time, temporary, seasonal, migrant, and indigenous—and the complexities of wage structures, job insecurity, and labor organization. It emphasizes the importance of trade unions and collective bargaining in improving working conditions and securing sustainable livelihoods. The chapter also addresses challenges like irregular employment, poverty, and inadequate labor standards, while highlighting the role of technology and labor markets in transforming agricultural work.

Chapter 7 shifts attention to the stresses faced by rural communities, particularly the impacts of economic instability and limited social protection policies. It reviews existing policies and strategies aimed at mitigating these stresses, such as cash transfers, vouchers, and targeting mechanisms, and emphasizes the importance of integrating social protection with agricultural growth.

Chapter 8 examines the role of labor unions and worker organizations in advocating for agricultural workers' rights, exploring the historical context, union structures at national and international levels, and their impact on sustainable agricultural production. Collective bargaining is presented as a key tool for improving labor conditions and securing fair wages.

Chapter 9 links sustainability with decent work, examining how practices like fair trade, organic farming, and responsible sourcing can enhance labor conditions. The chapter also highlights the potential of sustainable agriculture to provide green jobs and secure livelihoods, emphasizing the role of ethical practices and policy frameworks in ensuring the welfare of agricultural workers.

The book is aimed at policymakers, labor organizations, researchers, and advocates working to improve labor conditions in agriculture. It provides insights into the challenges faced by agricultural workers and offers strategies to promote fair wages, safe working conditions, and gender equality. It is essential reading for those committed to advancing the rights and well-being of agricultural laborers globally.

CHAPTER 1

INTRODUCTION TO DECENT WORK IN AGRICULTURE

Contents

INTRODUCTION

Decent work is defined by the International Labour Organization (ILO) as "productive work for women and men in conditions of freedom, equity, security, and human dignity." Creating jobs and making sure everyone has a good job are crucial components of sustainable development. Good jobs and decent working conditions reduce poverty and inequality while empowering people, particularly women, young people, and the most vulnerable, like those with disabilities. As a result, we assist our partner nations in establishing and advancing gainful employment, guaranteeing secure working environments, offering social security, and defending the rights of workers.

In this era of worldwide transformation, the ILO has expressed its commitment to advocating for the objective of ensuring fair employment opportunities for women and men worldwide. Fair employment is a prevailing need globally, and the ILO has transformed this need into four interconnected foundations, collectively forming the Decent Work Agenda. These fundamental elements of decent work are: fundamental principles and rights at work; employment; social protection; and social dialogue.

The comprehensive notion of decent work encompasses a multitude of issues, such as: the right to secure, non-discriminatory employment; income commensurate with basic economic, social, and family needs and obligations; equal and just treatment, regardless of national origin, race, ethnicity, gender, or age; adequate social protection; access to training and skill development; and the freedom to partake in decision-making processes regarding work-related matters, either directly or through representatives chosen freely.

These qualities offer an equity framework since they are essential components of any program aimed at protecting human rights and guaranteeing social justice. But just as significant is the fact that they are also required for the long-term growth of productivity, which means they offer an efficient framework for socioeconomic development.

A more favorable international environment for bringing up important issues in agriculture is being created by the rise of the rights-based approach to development, the increased focus on reducing poverty, and awareness in some quarters of the connections between agriculture and sustainable development. The justification for decent work in agriculture thus stems from two different viewpoints: the first is that of fundamental and universal rights; the second is that of the role of decent work in poverty alleviation and sustainable agriculture, and consequently, of its contribution to sustainable development, which is defined as development that is socially just, ecologically balanced, and economically viable.

1.1. CONCEPT OF DECENT WORK (ILO FRAMEWORK)

Decent work is defined as employment that upholds both the workers' rights to safety and compensation at work and the fundamental rights of every human being respect for an employee's mental and physical health while they perform their job duties. Both the formal and informal sectors are subject to the concept of decent work. It needs to cover all types of occupations, individuals, and households. Decent work, according to the ILO, includes opportunities for productive work that pays fairly, job security and social protection for families, improved opportunities for social integration and personal growth, the freedom to organize and voice concerns, the ability to participate in decisions that impact one's life, and equal treatment and opportunity for men and women.

To create those opportunities and contribute to the reduction and eradication of poverty, the ILO is formulating an agenda for the community of work, represented by its tripartite constituents, to mobilize their significant resources. The goals of full and productive employment and decent work for all are pursued at the international, regional, national, sectoral, and local levels through the balanced and integrated programmatic approach known as the ILO Decent Work Agenda. Its four pillars are social protection, enterprise development and employment creation, workplace standards and rights, and social dialogue (FAO, 2017; FAO and ILO, 2018; Kerr and Camfield, 2018).

1.1.1. Challenges in Implementation

While most people agree with the Decent Work Agenda in theory, there are disagreements and difficulties in implementing decent work in practice. For instance, in Africa, well-paying jobs with social protection benefits are the exception rather than the rule. This has been linked to the growing pressure of globalization making it harder to find employment in the formal sector. However, there are disagreements over whether social welfare would result from shrinking the informal economy or forming trade unions for its participants. The ILO brought the scope and diversity of decent work deficits to the attention of states attending a meeting of the UN Economic and Social Council (ECOSOC) in 2006. These deficits included gender inequality, unsafe work and insecure income, poor quality and unproductive jobs, unemployment, and underemployment, and rights violations.

National and international organizations must dedicate themselves to the goal of creating high-quality jobs and address its obstacles if they are to fulfill the Decent Work Agenda. One challenge is that it can be challenging to persuade a nation's citizens that supporting growth and job creation overseas has advantages at home as well. Governments are tempted to restrict markets and reduce labor standards to stay competitive in the global economy, which is thought to result in poor working conditions and wages. Although current incentives and conditions do not always support the Decent Work Agenda, a variety of actors can influence the provision of decent work. To illustrate:

National governments use industrial and economic policies to establish decent work. Globalization-related factors, like declining wages and less flexibility in macroeconomic policy, have made it harder for national governments to accomplish this objective alone.

- The European Union committed itself to "contribute" to the decent work agenda in its 2006 *Communication, Promoting decent work for all: The EU contribution to the implementation of the decent work agenda in the world.*

1.1.2. Eradicating Child Labor in Agricultural Supply Chains

Child labor is still a major problem in agriculture, particularly in developing nations where children are more susceptible to exploitation due to poverty, lack of access to education, and unstable economies. Millions of children still labor in fields, plantations, and farms around the world, frequently in dangerous and taxing conditions, in spite of global efforts and rising awareness. A comprehensive strategy including legislative frameworks, educational programs, awareness campaigns, and businesses' active involvement to guarantee ethical supply chains is needed to address this issue. Agriculture serves as the foundation of the economies of many developing nations. A sizable section of the population depends on it for their livelihoods, but families frequently have to use child labor to make ends meet due to financial difficulties. Children are hired to perform agricultural work like planting, weeding, harvesting, and processing. These jobs can be hazardous and physically taxing, exposing kids to long workdays, harsh weather, and pesticides. In addition, a lot of kids work without the right safety gear, which raises the possibility of accidents and health problems. In addition to denying kids their right to an education and a safe upbringing, child labor in agriculture feeds the poverty cycle because kids who don't go to school are more likely to work in low-wage jobs for the rest of their lives.

Governments, corporations, non-governmental organizations (NGOs), and international organizations have led the charge to end child labor in agriculture. The creation and implementation of regulatory frameworks is one of the main tactics. To end child labor, the ILO has created conventions on a global scale. The Convention on the Worst Forms of Child Labor of the ILO (No. 182), as well as the Convention on Minimum Age (No. 138), are two significant pieces of legislation that establish guidelines for preventing children from engaging in dangerous labor and making sure they don't work before the legal working age.

Although many nations have incorporated these conventions into their domestic legislation, enforcement is frequently a problem. It can be challenging to monitor and regulate in rural and isolated areas where agricultural work is common. The efficient application of child labor laws is further hampered by weak legal systems, corruption, and a lack of funding. To address this and develop systems that can more effectively monitor and enforce compliance, governments, law enforcement, and local communities must work together more closely.

A key factor in stopping child labor is education. Because schools are inaccessible in rural areas or because their families cannot afford to send them to school, many children work in agriculture. Children can be kept in the classroom instead of working in the fields by making investments in educational infrastructure, such as constructing schools near rural areas, offering scholarships, and putting in place initiatives that promote attendance. Families may be encouraged to place a higher priority on education by providing flexible educational programs that take into account the seasonality of agricultural labor. Children can learn the fundamentals of reading and numeracy through non-formal education programs, which also accommodate their need to help support the family financially. Another essential element in the fight against child labor in agriculture is awareness campaigns. Increasing public awareness of the detrimental effects child labor has on children's development, health, and prospects for the future can aid in altering cultural perceptions that frequently normalize the practice. Employers, parents, and communities can all be educated by campaigns about the long-term negative effects of child labor and the advantages of keeping kids in school. Additionally, these campaigns can educate families about government assistance programs or other sources of income that can lessen their dependency on child labor. Additionally, increasing global awareness can put pressure on businesses and governments to enact stricter laws against child labor and increase supply chain transparency.

When it comes to combating child labor, businesses involved in agricultural supply chains are crucial. Numerous international corporations purchase agricultural goods from areas where child labor is common, including cocoa, coffee, cotton, and tobacco. They must thus make sure that their supply chains are morally sound and devoid of exploitation. Initiatives related to corporate social responsibility (CSR) may serve as a catalyst in this respect. Codes of conduct that forbid the use of child labor and mandate that suppliers follow moral labor standards can be put into place by businesses. Third-party inspections, certification programs, and routine audits can all aid in keeping an eye on compliance. Additionally, businesses can collaborate with neighborhood groups to fund community development initiatives that give families access to healthcare, education, and other sources of income, thereby lowering the need for children to work. Child labor can be eradicated in a sustainable setting when businesses make proactive investments to raise the standard of living in farming communities.

Addressing the underlying socioeconomic factors that contribute to the prevalence of child labor is also crucial for an effective campaign against it. Many families are forced to rely on their children's incomes due to poverty, a lack of social safety nets, and limited adult economic opportunities. Thus, measures to end child labor must be combined with more comprehensive development programs that seek to raise economic stability, encourage fair wages, and give adults employment opportunities. Communities can be improved and become less reliant on child labor by supporting smallholder farmers, offering microfinance, and encouraging sustainable farming methods. Families are less likely to send their kids to work when they are financially stable and have access to resources (FAO, 2017; FAO and ILO, 2018; Kerr and Camfield, 2018).

1.1.3. Improving Occupational Safety and Health (OSH)

It is commonly acknowledged that one of the most dangerous industries for employees is agriculture. It includes a variety of tasks that expose people to hazardous equipment, poisonous substances, and challenging environmental circumstances. Risks for farmers, farmworkers, and other agricultural workers range from operating large machinery to extended exposure to pesticides and inclement weather. Implementing thorough safety plans that put employees' health first is necessary to address these risks. Improving safety regulations, offering suitable protective gear, and putting in place efficient training initiatives are essential measures to lower mishaps and health hazards in the agriculture industry.

The use of large machinery, including harvesters, tractors, and other mechanical devices, is one of the main risks in agriculture. Although these devices are necessary for increasing productivity, improper operation or maintenance can make them extremely dangerous. Serious injuries, such as fractures, amputations, and even death, can result from accidents involving machinery. Strict safety regulations must be established, and all equipment must be routinely inspected and maintained, to improve safety in this area. Clear rules for machinery safety, such as procedures for routine inspections, emergency stop mechanisms, and the use of protective guards, can be established by governments and agricultural associations. Additionally, manufacturers ought to be urged to create devices with safety features that reduce the possibility of accidents, like preventing unintentional starts and stops.

Another significant issue in agriculture is pesticides and other agrochemicals. Although these chemicals are necessary for maintaining crop yields and managing pests, improper handling of them can be harmful to human health. In addition to long-term illnesses like cancer and neurological disorders, pesticide exposure can cause acute health effects like skin irritations, respiratory problems, and poisoning. Enforcing laws that restrict the use of dangerous chemicals and encourage the use of safer substitutes is essential to reducing these risks. To guarantee that only authorized chemicals are used and that employees are aware of the possible risks, governments can impose stringent regulations on pesticide registration, use, and labeling. Additionally, farmers ought to be urged to embrace integrated pest management (IPM) strategies, which use biological control and other sustainable farming methods to lessen dependency on chemical pesticides.

Ensuring agricultural workers are protected requires the provision of suitable personal protective equipment (PPE). PPE may consist of gloves, goggles, masks, helmets, and protective clothes, depending on the nature of the work. For example, to prevent skin exposure and inhalation, workers who handle pesticides must wear masks and gloves that are resistant to chemicals. To lower the chance of harm, people who operate machinery should also wear helmets, safety goggles, and earplugs. PPE alone, however, is insufficient; employees also need to receive training on proper usage and the significance of wearing it on a regular basis. Employers are responsible for making sure that the equipment is available, well-maintained, and appropriate for the particular tasks that the employees perform. Additionally, educating employees about the value of PPE through frequent safety campaigns can motivate them to put their own safety first at work.

Any plan to increase agricultural safety must include training initiatives. Employee education is necessary to inform them of potential risks and the best ways to reduce them. Numerous subjects should be covered in training, such as how to operate machinery safely, how to handle and apply chemicals correctly, and how to respond to emergencies in the event of an accident. To ensure that everyone gets pertinent and appropriate information, these programs should be customized to meet the needs of various worker types, from seasonal laborers to smallholder farmers. Additionally, training shouldn't be a one-time thing; workers can stay current on the newest safety procedures and technological developments with the support of refresher courses and continuous education. Even in rural and isolated locations, governments, non-governmental organizations, and agricultural businesses can work together to create thorough training programs and make them available.

Environmental factors and climate have a big impact on agricultural safety as well. Extreme heat, cold, and other severe weather conditions are frequently encountered by workers, which can result in health problems like heatstroke, dehydration, and hypothermia. Long-term sun exposure without appropriate protection can also raise the risk of developing skin cancer. It is crucial to put policies in place that shield employees from environmental risks to address these issues. Regular breaks, areas with shade, and access to potable water are all things that employers can implement. The dangers of sun exposure can be further decreased by promoting the use of hats, sunglasses, and sunscreen. Work schedules may need to be modified during extreme weather events to avoid the hottest or coldest times of the day to protect employees from harm.

Developing a culture of safety in the agriculture industry is crucial in addition to these tactics. The goal of preserving a safe workplace should be shared by employers, managers, and employees. This entails promoting the use of safety equipment, rewarding adherence to safety procedures, and fostering open communication about safety issues and incidents. Employees should be able to report risks without worrying about reprisals, and there should be defined protocols in place for handling such reports. Long-term gains can result from fostering a culture that values safety, which will lower accident rates and improve agricultural workers' general well-being.

To achieve these safety improvements, cooperation between different stakeholders is essential. Governments are required to uphold laws and supply funding for safety and training initiatives. Given that a healthy workforce is critical to productivity, agricultural businesses and cooperatives should make investments in worker safety. International organizations and NGOs can help with these initiatives by providing funds, knowledge, and support for improved working conditions. Together, these organizations can make the agriculture industry safer and more sustainable (FAO, 2017; FAO and ILO, 2018; Kerr and Camfield, 2018).

1.1.4. Uses and Limitations of Indicators

The degree to which a given goal or result has been accomplished is determined by indicators. As a result, they can be used to evaluate performance and track advancement toward achieving goals over time. They can also be useful for cross-national comparisons. Indicators are also widely used to test alternative theories regarding the connections between the various elements of decent work. Indicators should ideally offer a precise measurement of the stated goal. For example, the indicator should provide information on the number or proportion of people who are ill if the goal is a healthy population. Giving a general objective a precise meaning, however, is frequently challenging. For example, remunerative employment is one of the characteristics of decent work; however, before a suitable indicator to measure it can be developed, the term "remunerative" must be defined. An indirect measure may need to be used because it is frequently more expensive or difficult to obtain a direct measure. Children's nutritional status, for example, can be assessed directly by looking at how many different nutrients they consume through food, or indirectly (and more affordably) by looking at their weight or height for their age.

The intended outcome is rarely measured by a single indicator; instead, a combination of multiple indicators may provide a more precise measurement of a given goal. Therefore, wage disparities, training opportunities, promotion opportunities, and the distribution of work responsibilities may all reflect how much gender discrimination in the workplace has decreased. Furthermore, the indicators can be either quantitative or qualitative. For example, social security's quantitative indicators may relate to the percentage of recipients of various benefits, while its qualitative indicators focus on the efficacy and quality of services. Therefore, it might be necessary to combine multiple indicators into an overall indicator or index to get an accurate picture. The process of creating an index brings up issues regarding the relative importance of various indicators as well as the best way to combine qualitative and quantitative data. When indicators from various domains are combined to create synthetic measures, similar but more serious issues occur. One example of a synthetic indicator that combines income, education, and health metrics is the Human Development Index created by the United Nations Development Programme (UNDP). The Gross Domestic Product (GDP), which is frequently used to gauge economic growth and production levels, is also a composite indicator that is determined by adding up the monetary value added from a wide range of activities.

When choosing and employing indicators, accuracy and comparability are additional factors to take into account. It is commonly known that even some of the most widely used social statistics, like life expectancy, literacy, and infant mortality, have significant variations in accuracy. For data pertaining to work, this might be even truer. The definitions used in data collection differ not only between nations but also over time within a single nation, which could lead to measurement errors when comparing data over time and between nations. It is therefore unrealistic to expect that metrics for decent work will provide a reliable picture of performance within a nation, let alone across nations. Accordingly, it is best to view the indicators used here for the various aspects of decent work as offering merely an approximation of performance (FAO, 2017; FAO and ILO, 2018; Kerr and Camfield, 2018).

1.1.5. Decent Work, Decent Life for Women Campaign

The International Trade Union Confederation (ITUC) and the global union federations (GUF) started the two-year Decent Work, Decent Life for Women campaign on March 8, and International Women's Day. The campaign seeks to promote gender equality in trade union structures, policies, and activities as well as decent work for women in labor agreements and policies. The second goal is to increase the proportion of women in elected office and trade union membership. The need for the campaign is a result of the various forms of discrimination that women encounter daily in both policy and practice, including the gender pay gap, the absence of maternity protection, and the higher unemployment rates among women. Compared to their male counterparts, women are frequently paid less and do not have the same opportunities for career advancement because of gender bias. The majority of Asian women work in domestic labor, which is one of the least organized, least valued, and lowest paid occupations. Between 2006 and 2011, the wage growth rate for women in Asia, excluding China, was 0.9%. Currently, 56 countries and 81 national centers take part in this campaign's various events.

1.1.5.1. Background and Need for the Campaign

Women make up a significant portion of the global workforce, yet they often face inequalities in terms of pay, working conditions, and access to opportunities. In many sectors, especially informal and low-wage industries like agriculture, domestic work, and garment manufacturing, women are frequently subjected to long hours, poor pay, and unsafe working conditions. Gender-based discrimination, harassment, and the lack of social protection further exacerbate their vulnerability. The "Decent Work, Decent Life for Women" campaign recognizes these challenges and advocates for systemic changes to ensure that women have equal rights and opportunities in the workplace.

1.1.5.2. Key Objectives of the Campaign

- **Promoting Gender Equality in the Workplace:** One of the primary goals of the campaign is to eliminate gender-based discrimination. This includes advocating for equal pay for equal work, fair treatment, and ensuring that women have the same opportunities as men for promotions, training, and career advancement. It also involves tackling issues like the gender pay gap, which persists in many industries worldwide.
- **Ensuring Safe and Fair Working Conditions:** The campaign emphasizes the need for safe and healthy working environments. Women in sectors like agriculture, construction, and manufacturing often face hazardous conditions, and the lack of proper safety measures puts their health and lives at risk. The campaign pushes for stricter enforcement of safety regulations, better protective measures, and policies that address specific health concerns faced by women, such as reproductive health.
- **Strengthening Legal Protections:** Effective legal frameworks are crucial for safeguarding the rights of women workers. The campaign

advocates for the implementation and enforcement of international labor standards, such as those established by the ILO, which call for the elimination of forced labor, child labor, and discrimination. By pushing for stronger legal protections, the campaign aims to ensure that women have access to decent work without the fear of exploitation or abuse.

- **Promoting Access to Social Protection and Benefits:** Many women, especially those working in informal sectors, lack access to essential social protections, including maternity leave, healthcare, and pensions. The campaign calls for the extension of these benefits to all workers, regardless of their employment status, to support women's economic security and well-being. Ensuring access to childcare services and flexible work arrangements is also crucial for enabling women to balance work and family responsibilities.
- **Empowering Women through Education and Training:** Education and skills training are vital for empowering women to access better job opportunities. The campaign advocates for investment in education and vocational training programs that are inclusive of women, especially in male-dominated fields such as technology, engineering, and management. By equipping women with the necessary skills and knowledge, the campaign aims to break the cycle of low-wage, low-skill employment that many women find themselves trapped in.
- **Combating Gender-Based Violence and Harassment:** Gender-based violence (GBV) and harassment remain significant barriers to women's participation in the workforce. Many women face harassment, abuse, and violence at their workplaces, which not only affects their physical and mental health but also limits their career growth. The campaign works to raise awareness about GBV and advocates for strict policies to prevent and address harassment in all its forms, ensuring that women can work in safe and supportive environments.

1.1.5.3. Strategies and Initiatives

The "Decent Work, Decent Life for Women" campaign employs various strategies to achieve its objectives:

- **Advocacy and Policy Engagement:** The campaign works closely with governments, trade unions, and employers to advocate for policies that support gender equality in the workplace. This includes lobbying for the adoption of international labor standards, equal pay legislation, and better social protection frameworks.

- **Awareness Campaigns:** Raising awareness is a crucial part of the campaign. By organizing events, workshops, and media campaigns, the initiative seeks to educate the public about the importance of decent work for women. These efforts aim to challenge stereotypes, change perceptions, and highlight the contributions of women to the economy.
- **Capacity Building:** To ensure sustainable change, the campaign focuses on building the capacity of women's organizations, trade unions, and other stakeholders. This involves providing training on leadership, negotiation skills, and labor rights so that women can advocate for themselves and their peers.
- **Collaboration with International Organizations:** The campaign partners with various international organizations, including the ILO, UN Women, and other NGOs, to leverage global support for its cause. By aligning its objectives with international development goals, such as the Sustainable Development Goals (SDGs), particularly Goal 5 (Gender Equality) and Goal 8 (Decent Work and Economic Growth), the campaign seeks to create a global movement for change.

1.1.5.4. Impact and Success Stories

Since its inception, the "Decent Work, Decent Life for Women" campaign has made significant strides in various parts of the world. In countries where the campaign has been active, there have been improvements in labor laws, increased awareness of women's rights, and better access to social protections. For example, in several African and Asian nations, the campaign has successfully lobbied for stronger legal protections against workplace harassment and has worked with employers to implement gender-sensitive workplace policies. The campaign has also played a role in raising global awareness about issues such as the gender pay gap and the importance of women's economic empowerment. Through collaborations with companies, it has encouraged businesses to adopt inclusive practices, such as gender audits and equal pay reviews, to ensure fairness and transparency (FAO, 2017; FAO and ILO, 2018; Kerr and Camfield, 2018).

1.2. UNDERSTANDING DECENT WORK IN THE CONTEXT OF AGRICULTURE

The general framework for the current discussion of decent work in agriculture is provided by globalization. All around the world, agricultural workers' living and working conditions have been impacted by the intricate processes brought about by the globalization of agriculture. These effects have differed between and within regions, and a variety of factors, such as trade and macroeconomic policies, have influenced the unique experiences of individual nations. However, the effects have always been substantial for waged agricultural workers, who make up a sizable portion of the total rural labor force involved in agriculture. Self-employed farmers, waged agricultural workers, and wage-dependent small farmers make up the agricultural labor force. Recent data suggests that conditions are generally getting worse, which points to significant and concerning shortages in decent work for a large portion of this labor force.

For the representatives of the millions of men, women, and children who depend on waged labor in agriculture and who are actually finding it more and more difficult to rely on agricultural employment for their livelihoods, the magnitude and variety of deficits thus present new difficulties. Around the world, trade unions and organizations representing rural workers are tackling these issues and looking for solutions. Employers and businesses struggling with the concept and ramifications of corporate responsibility and socially responsible investment, as well as governments worldwide that are being held more and more accountable for responsible governance, face difficulties due to the magnitude and widespread nature of these deficits. Due to its unique mandate and tripartite structure, the ILO is also called upon to address the issue of decent work in agriculture.

1.2.1. Waged Agricultural Workers

If the concept of sustainable development is to have any tangible manifestation in the real world, the ILO Decent Work Agenda needs to be implemented in the agricultural sector. It is obvious that the current state of affairs cannot continue. The debate over decent work in agriculture is based on many important relationships, both directly and indirectly involving waged agricultural workers. First, since waged agricultural workers make up a large percentage of the populations affected by poverty, there is an evident link between efforts to reduce poverty and sustainable agriculture. Second, and connected to the first, is the connection between food security, livelihoods, and rural incomes. Thirdly, there are the connections that affect the status of waged agricultural workers among all of these important policy areas: sustainable agriculture, incomes, poverty alleviation, and food security. The overall policy context is defined in large part by the conditions of waged agricultural workers. Which groups are we talking about? This is the fundamental question that underlies all of these relationships. It is crucial to define and separate groups like the rural poor, wage-dependent small farmers, landless agricultural wage workers, and self-employed farmers to identify the groups most impacted by deficits and to create effective responses and strategies for resolving them. Self-employed workers, self-employed farmers, and waged agricultural workers make up the entire agricultural labor force. In contrast to small-holder farming, which involves a large amount of unpaid family labor, the definition of waged agricultural workers acknowledges that women and men in this group—and children in many situations—are part of the larger rural agricultural labor force and are primarily involved in cultivation and harvesting in plantations, commercial agriculture, horticulture, and primary agricultural processing. It is acknowledged that self-employed small farmers frequently overlap with waged agricultural workers, despite the fact that the latter are typically wage-dependent on a seasonal or casual basis and augment their meager earnings with paid labor on other farms or plantations. For instance, it has been estimated that up to 40.8 million small farmers in Mexico, or roughly 75% of the country's rural labor force, worked for pay for at least a portion of the year.

It goes without saying that small farmers who are frequently wage-dependent—which is undoubtedly the case for a large percentage of African women farmers—may share many needs with other waged agricultural workers who are employed on temporary, seasonal, or casual basis. Furthermore, these differences become less noticeable as waged employment in agriculture becomes more casualized. However, the nature of the employment relationship is crucial, and in this case, the differences in terms of rights, conditions, benefits, and protection are still very obvious. It is helpful to briefly outline some of the processes that have shaped the agricultural sector in every region over the past 10 years and continue to do so before examining recent data regarding these kinds of conditions.

1.2.2. Global Processes and Agriculture

Many interconnected processes and developments have led to globalization. The expansion and significance of foreign direct investment (FDI), which gives multinational corporations a larger role, the concentration of corporate power, the globalization of financial markets, the advancement and spread of transportation and communication technologies, deregulation, market liberalization, and market reform, and the privatization of the public sector are some of the important factors. Due to these processes as well as modifications to the international policy framework, which includes WTO agreements and other international instruments and arrangements, the international environment is always changing. With major effects on their agricultural sectors, particularly the restructuring carried out in subsectors like sugar, these processes have been and still are reflected in the continuous structural changes observed in national economies across all regions. The ILO and other organizations have been thoroughly documenting these changes since the middle of the 1980s.

The way that these processes have subsequently influenced, changed, and harnessed labor markets, labor relations, and employment practices is related to the aforementioned. Even though a great deal more context-specific research is required before we can make firm policy recommendations, there are already convincing signs that the type and scope of these impacts are causing many deficiencies, which are becoming more noticeable in the widening disparities on a local and global scale. The character of these deficits greatly influences the parameters of trade union activity in the agricultural sector worldwide, both in terms of its scope and its limitations.

1.2.2.1. Foreign Direct Investment (FDI) and The Role of Multinational Enterprises

Global inflows of FDI have been increasing quickly. Nonetheless, the FDI inflow pattern is extremely skewed, with less than 16% of the US$1,491,934 million in global FDI inflows in 2000 going to developing nations and territories. This broad category includes: 40 countries in South and Central America and the Caribbean received roughly 6% of FDI inflows; 48 countries in Asia, including China, received less than 9%; nine countries in Oceania received very little; and five countries in the former Yugoslavia received 1.67% of total inflows. African economies receive a minuscule portion of foreign direct investment inflows on a global scale. 52 countries in North Africa and the rest of Africa received only 0.58% of global inflows in 2000, down from 1.2% in 1999. The pattern of inflows also shows significant variation for many developing nations, even though data for inflows in 1980, 1990, 1995, and 2000 show a generally positive trend for the majority of developing nations. The percentage of FDI going to the agricultural sector is undoubtedly the main concern. The majority of FDI in Africa goes to mining, minerals, and petroleum, but in some nations, plantation agriculture—especially non-traditional export crops—seems to be drawing in foreign capital. This is the situation in Kenya, for instance, where one of the three primary industries drawing foreign direct investment is agriculture.

Multinational firms, also known as transnational corporations, provide a crucial conduit for national economies to access international markets and investment capital. Despite the comparatively low FDI inflows, this is especially true for African economies, where some national governments work hard to draw in foreign capital and support the operations of multinational corporations, joint ventures, and transnational corporations. Therefore, economic reforms aimed at increasing investor confidence—such as lowering corporate tax rates—usually come before foreign investment. To entice investment in Ugandan floriculture, many incentives are provided, such as exemptions from import duties on plant and equipment. From the standpoint of encouraging decent work in agriculture, national governments' attempts to draw in foreign investment have also usually involved limitations on trade union activity.

Even though FDI inflows to African nations are relatively small globally, they can have a big local impact in certain situations. Many East African nations, including Uganda, Kenya, and the United Republic of Tanzania, seem to be interested in agriculture, especially the horticulture subsector that produces non-traditional export crops. In fact, from less than \$500 million in 1987–1992 to over \$2.26 billion in 1993–1998 FDI inflows to 11 East African subregion countries increased fivefold. FDI tripled in Kenya, the biggest economy in East Africa, from \$42 million in 1999 to \$127 million in 2000. At least 96 foreign affiliates were active in the nation by 1999, and the rise in FDI in 2000 seemed to reflect investor confidence following recent economic reforms like lower corporate tax rates. Notably, one of the three primary industries in Kenya that drew the most foreign investment was plantation agriculture. However, FDI fell to \$60 million in 2001, which is another sign of the large swings in FDI that many developing nations face.

1.2.2.2. The Consolidation of Corporate Control: Aggravating Global Inequality

The IUF and many other analysts believe that the true cause of global inequality is not a straightforward North versus South divide, but rather the influence of transnational corporations (TNCs) with headquarters in the North and the backing they receive from domestic and international political elites. This is the result of analyses that show the degree of corporate monopolization and control over the global food system and the formulation of agricultural and food policies worldwide. Some striking examples include the following:

- The top 10 agro-chemical companies control about 80% of the \$32 billion global market;
- 80% of world grain is distributed by two companies;
- Approximately 75% of the banana trade is controlled by five corporations;
- Three corporations control 83% of the cocoa trade; and
- Three other corporations control 85% of the tea trade.

Each of the five large coffee roasters has annual sales of at least \$1 billion and enormous profit margins. Together, they purchase nearly half of the coffee beans produced worldwide. In the last three years, the price of coffee has dropped by nearly 50%, and exports from producer countries now account for less than 10% of the market value, down from a third a decade ago. To increase the markets for their produce in developing nations, agri-food TNCs aim to crush local competition, seize control of expanding markets, and make these nations more reliant on food imports. A paradoxical situation is created when land use is changed to non-traditional agricultural exports, increasing reliance on TNCs for distribution, markets, and inputs, including seed while importing heavily subsidized agricultural products to replace the traditional crops that were initially displaced. For instance, heavily subsidized American corn undercuts local corn prices in the Philippines by 20%. As a result of the Philippine government's push to export cash crops in place of rice and corn, 5 million hectares of land that were previously used to grow these two crops have been converted to livestock production, turning the country into a frequent importer of corn. In turn, the livestock industry is currently being opened up to provide US exporters of poultry and pork with more access. Thus, the cycle goes on.

There are examples of this process in practically every South American country, and each has experienced growing transnational corporate control over domestic food production and trade to varied degrees. Increased domestic food production has historically helped both developed and developing nations achieve food security. Nevertheless, the net cereal deficit for developing nations has increased over the last three decades to 103 million tons, or 9% of consumption from 1997–1999. It is anticipated that by 2030, that deficit will rise to 14% of consumption. Traditional grain exporters' increased surpluses and new exports from transition nations are two ways to close the gap. The debate surrounding biotechnology advancements, especially those of genetically modified seeds and crops (GMOs), which are regarded with skepticism in some places and alarm in others—including developing nations that are afraid of incorporating GMOs into their agricultural systems and practices—overshadows this potentially unstable trend.

Since food staples typically make up 45–60% of the diets of the rural poor, their output has a significant impact on poverty trends. While the reasons for declining food yields and outputs in developing countries are multifaceted, IFAD's recent report on rural poverty highlights that food staple output is a major factor.

This proportion is often even higher in less developed nations and areas. The yield growth for food staples and the rate of poverty reduction both significantly slowed in the 1990s. Conversely, food staple yields increased and poverty rates decreased during the 1970s and the majority of the 1980s. Examples of the growing corporate power in the world economy abound. One such example is the restructuring that has been occurring in the global sugar industry for the past 20 years. Alongside this restructuring, there has been a persistent concentration of industry ownership and control due to TNC consolidation within specific regions. A lot of market reform and liberalization programs that swept through the global economy in the early 1990s contributed to the overall policy environment that made this consolidation and control possible during the 1990s.

1.2.2.3. Market Reforms and Liberalization: Opening the Way for Market Forces

The agricultural sectors in numerous nations and regions have been significantly impacted by the market reforms and liberalization initiatives that were embraced and extensively executed in the early 1990s. A few industries were impacted more than others. The reform initiatives brought deregulation and privatization, which had an impact on the sugar industry, which was used to significant government involvement in both centrally planned and market economies. These impacts, along with significant political events like the fall of the Soviet Union, are represented in a few significant regional trends.

1.2.2.3.1. The Example of Sugar

Under the Sugar Protocol to the agreement with African, Caribbean, and Pacific countries (ACP), eight countries export sugar to the European Union, demonstrating the importance of sugar industries to many African domestic economies. Over 14% of global sugar imports in 2001–02 came from Africa, and many other African nations also import sugar. Asia accounted for nearly 40% of sugar exports. However, regional developments like the reform of the Common Agricultural Policy (CAP), which includes the domestic sugar regime, the expansion of the European Union to include Central and Eastern European Countries (CEECs), some of which produce sugar, and the free trade agreement with the western Balkans are causing changes in the sugar markets and pressure to reform the ACP Sugar Protocol.

Few crops have as much political content as sugar. Because it is regarded as a staple food in many nations, its consumption is susceptible to price fluctuations, and governments frequently heed organized pressure and lobbying from growers. In both market and centrally planned economies, sugar has also been characterized by a high degree of state intervention, either through direct ownership of production facilities (land and factories) or through industry regulation through both domestic and international trade, setting prices for sugar and cane or beet crops. For instance, when the government passed the new Farm Act in early 2002, lobby protection for the domestic sugar industry—which is very strong in the United States—was renewed and strengthened. In the future, there may be pressure on the tariff-rate sugar quota system, which currently gives the US preferential treatment to about 40 nations.

1.2.2.3.2. Free Market Prices – Some Local Impacts

Reforms to economic policies that impact agricultural production focused on exports have been widely adopted in Latin America, the Caribbean, Asia, and Africa. Policy reforms in Uganda included the dismantling of export marketing boards, the removal of export duties on export crops, the liberalization of producer prices on export crops linked to the world-based exchange rate, the reduction of tariffs on imported goods, the adoption of an Investment Code in 1991, and other measures that reduced barriers to foreign ownership and investment. The overall package was described as massive and extensive, with specific policy measures aimed at various aspects of the economy. Additionally, these programs were put into place during the period from the early 1990s to 2000–01 when the market prices of the majority of food commodities, beverages (coffee, cocoa, and tea), vegetable oils, and agricultural raw materials were steadily declining. Although there might be advantages for consumers, this led to cheaper imports than locally made goods, which had a significant effect on local markets and domestic economies.

A wide range of incidents, from effects at the microenterprise level to actual dangers to establish behemoths, like the Indian tea industry, have been documented. On the one hand, there is the common situation of a tomato processing company in Senegal being forced to shut down due to pressure from Mediterranean European tomato concentrate imports. Of course, there are additional variables at play, like labor productivity concerns, technological advancements, and management effectiveness. However, Senegal, where an estimated 47% of the population is estimated to be living in poverty, is no longer self-sufficient in rice and currently imports the majority of its rice from Thailand, Argentina, and Pakistan, which is indicative of a wider decline. India's tea industry, on the other hand, is the world's largest producer of tea (with a stagnant yearly production of 750 million kilograms), the largest employer in the nation (with over a million permanent employees, the majority of whom are women), and the world's largest consumer of tea. The industry's relative stability has been aided by the steady growth of the domestic market. However, the new threat to this industry is the lower prices of tea imports compared to domestically produced tea.

Workers in the Indian tea industry are bearing the brunt of the industry's declining yields and falling prices. Some 10,000 workers have reportedly not received their wages for over a year, and even more concerningly, trade union leaders were forced to negotiate a collective bargaining agreement that lowers wages to address the current crisis. In actuality, the tea global price index (with 1990 = 100) fell from 130.6 in 1998–103.5 in 2001. Given the sharp drops in the prices of coffee and cocoa in recent years, the price index for the three beverages combined—coffee, tea, and cocoa—declined even more, from 138 points in 1998–74 points in 2001. Coffee producers, who are primarily smallholders, have been selling their coffee beans for significantly less than what it costs to produce them, as the price of coffee has dropped by nearly 50% over the last three years. For example, beans in one Vietnamese province have been sold for 60% of their cost of production. About 25 million coffee producers worldwide are losing their livelihoods as a result of the collapsing coffee market, according to Oxfam (FAO, 2017; FAO and ILO, 2018; Kerr and Camfield, 2018).

1.2.2.4. Labor-Related Aspects of Restructuring in Agriculture

Measures taken to restructure agricultural businesses in numerous nations are a significant component of investment policies falling under the liberalization and deregulation umbrella. Usually, a significant effect of this reorganization is the reduction of the permanent workforce combined with a rise in casualization. The growing percentage of women in this group of temporary employees is a second, related factor. However, the increasing number of women in the category of casual wage workers seems to hold true even in circumstances where there is no ongoing reduction in the permanent workforce, such as in developing agricultural sectors that generate non-traditional agricultural exports. Since casual workers are rarely organized and represented and strategies to increase organization and trade union membership must specifically target the full potential of women wage workers, these two significant trends have important implications for how trade unions address the lack of decent work in agriculture. These topics are covered in more detail in the sections that follow.

Because agriculture is a seasonal industry, casual labor has always been a part of the job. The employment of casual labor is frequently allowed by collective bargaining agreements, which acknowledge the employer's right to do so for a predetermined amount of time before offering the employee a permanent contract. In many nations that produce a range of commercial and export crops, women have historically made up a sizable portion of this worker category. Nonetheless, a common and well-researched grievance has been that many women experience their employment being interrupted shortly before the end of this period and then resumed following a variable length of time off. The outcome has been nearly constant work for many years, albeit on a part-time basis. Since maternity and other benefits are not payable in these circumstances, there is undoubtedly another significant benefit to the practice in the short term: lower costs for employers. The counterargument, on the other hand, is predicated on agricultural productivity and long-term sustainable development (FAO, 2017; FAO and ILO, 2018; Kerr and Camfield, 2018).

1.2.2.4.1. An Example of Non-Traditional Agricultural Exports

Some developing industries exhibit a different variation of this equation, where expansion is predicated on the extensive use of temporary workers from the outset. The promotion of non-traditional agricultural exports, mainly fruit and vegetables and cut flowers, to the European and North American markets is a significant trend in the agricultural sectors of Africa, Latin America, and increasingly Asia. This is an intriguing example of this. Recent estimates indicate that these exports are currently a part of the growing global trade in floricultural and horticultural goods, which has a total value higher than the trade in cereals. This increase in non-traditional agricultural exports is remarkable because it seems to have relied heavily on the use of temporary laborers, the majority of whom are women, from the beginning.

For instance, in the floriculture industry, 85% of the workers on Tanzanian flower farms are casual workers who perform the primary duties of planting, harvesting, and grading. Only a small percentage of skilled and managerial positions are held by men; the majority of these workers are women. Commercial flower cultivation for export only began in Uganda in 1993, but plans are already in place to grow, boost exports, and take at least 13% of the European rose market. Women make up about 85% of the workforce, and nearly all of them work as casual employees. Of the 12,000 workers on flower farms in Kenya, which is currently the world's second-largest exporter of cut flowers after the Netherlands, nearly 70% are women, and 85% of the workforce is casual. In general, horticulture in Latin America seems to follow the same pattern. For instance, more than half of the 350,000 temporary employees in Chile's such businesses are women, and more than 85% of them are temporary employees. Of the 50,000 permanent employees, 95% are men.

It is evident from developments in the cut flower industry that the trade in cut flowers can significantly boost a nation's export revenue and employment. Commercial flower cultivation is practiced in many nations, such as the US, Japan, India, and China, primarily to meet domestic demand. In other nations, such as Kenya and Colombia, there is a sizable export market for flowers but a tiny domestic market. Colombia is a prime example, as in recent years, 95% of production has come from exports. The flower trade is the fourth-largest export earner in the nation, with an estimated 80,000 people directly and another 50,000 indirectly employed in the sector. The best illustration of a sizable domestic market with a sizable export share is the Netherlands. In 1996, flowers accounted for 71.4% of Ecuador's non-traditional agricultural exports, a growing contribution. Due in large part to the creation of jobs, the flower industry, the most active export subsector within the agro-industrial sector, has been acknowledged for its role in slowing rural migration. More than 70% of the industry's workers are women, and regional development plans and policies view the flower industry as a critical component.The feminization of waged agricultural labor is not limited to non-traditional export crops. According to a recent thorough study on how globalization has affected Indian agriculture, among other factors, machinery displacement, the switch to specific cash crops that require less labor input, and the conversion of agricultural land to other uses are all contributing factors to the decline in wage employment in the sector. According to an annual survey cited in the study, there is a noticeable increase in capital investment along with a general decline in employment opportunities per unit of agro-industries, indicating a clear labor substitution. These changes have led to a significant out-migration of men from rural areas in many areas, forcing women to work as marginal workers (defined as those who work fewer than 183 days a year) in low-paying agricultural jobs. The number of female agricultural workers in their regions has increased, according to worker unions throughout India. While the number of male main workers—those who work more than 183 days a year—has decreased, census data actually demonstrates a feminization and casualization of the workforce in rural areas, with marginal women workers becoming increasingly important (FAO, 2017; FAO and ILO, 2018; Kerr and Camfield, 2018).

1.2.2.4.2. The Increasing Role of Labor Contractors and Migration

Most regions are experiencing increased flexibility and the casualization of an even larger portion of the waged agricultural workforce due to fierce competition for a place in global markets. Research is starting to show that this drive for flexibility is mirrored in the concurrent growth of the role of labor contractors or intermediaries, which is causing employment in agricultural businesses all over the world to rapidly change. Increased cross-border labor contracting is also taking place in many ways, some of which may be on the verge of being illegal or in the gray area between legal and illegal human mobility. Such changes frequently lead to labor practice abuses, which exacerbate the variety of deficiencies in decent work. Although much more research is needed to fully understand the scope and ramifications of these developments, the evidence that is currently available points to growing livelihood insecurity and the lack of social protection for a larger number of agricultural workers.

One important aspect of comprehending the current state of agricultural workers seems to be the connections between international labor migration and agricultural globalization. Higher levels of exploitation and deregulation go hand in hand with increased international labor mobility brought on by modern globalization. For many migrants, agriculture is a major source of employment in many nations. Since migrant workers are forced to work in environments where respectable working conditions are not upheld, their lack of legal protection makes them more appealing as tools for preserving competitiveness. Because the fear of deportation prevents them from joining a union and raises the possibility that they will be exposed to hazardous working conditions, irregular migrants are particularly at risk. Thus, the effects of globalization are seen in the growing labor flows and movement both within and across national borders, as well as across continents, as well as in the progressive degradation of labor rights and protection. The ILO and trade unions have long been concerned about protecting migrant labor, but the sheer volume of these movements around the world—particularly in the agricultural sector, where it has historically been challenging to guarantee protection of workers' rights—is causing great hardship for many workers across the globe and posing significant challenges for trade unions.

Migration is one of the many issues being discussed that are not new. What is novel is how widespread certain exploitative practices are, and how intricately they are linked to one another under the influence of globalization, making it very challenging to come up with suitable and workable reactions. It is crucial to highlight the mounting pressure on workers' organizations to advance the fundamental rights of waged agricultural workers, as the task facing trade unions and representatives of these workers is formidable. Governments in the North and the South face a challenge as well, though, since they are responsible for upholding fundamental rights (FAO, 2017; FAO and ILO, 2018; Kerr and Camfield, 2018).

1.3. FUNDAMENTAL RIGHTS FOR WORKERS IN AGRICULTURE

The ILO Declaration on Fundamental Principles and Rights at Work recalls that specific rights and obligations developed and expressed in Conventions are recognized as fundamental both inside and outside the Organization. All Members, even if they have not ratified the Conventions in question, should respect, promote, and realize the principles concerning the fundamental rights that are the subject of those Conventions, namely:

- Freedom of association and the effective recognition of the right to collective bargaining;
- The elimination of all forms of forced or compulsory labor;
- The effective abolition of child labor; and
- The elimination of discrimination in respect of employment and occupation.

The main ILO standards that articulate these ideas are covered in this section, along with their applicability as essential rights for decent labor in agriculture.

The first pillar of the ILO Decent Work Agenda is the implementation of these values and fundamental rights.

Even if the statistical picture is not complete, it will be evident from some of the data regarding their use in the agricultural sectors of many nations that there are significant deficiencies with regard to this pillar of decent work. The other three pillars of decent work—access to employment and incomes, social protection, and social dialogue—will be covered in the sections that follow. In addition to these fundamental ILO standards, there are other international labor standards that are particularly pertinent to agricultural workers.

1.3.1. Freedom of Association and the Right to Collective Bargaining

The 1948 Freedom of Association and Protection of the Right to Organize Convention (No.) embodies this idea and fundamental right. 87), as well as the 1949 Convention on the Right to Organize and Collective Bargaining (No. 98). The Right of Association (Agriculture) Convention, 1921 (No. 1), one of the first international agreements ratified by the ILO, had previously expanded the concept to include agricultural workers. 11. Rural agricultural workers in several nations face both de facto and de jure barriers to organizing, despite the Conventions' guarantee that all workers have the right to form representative organizations to protect their rights.

The cases reviewed by the Committee on Freedom of Association demonstrate how common it is for agricultural workers to have their freedom of association violated. Regulatory limitations, anti-union actions, physical assault, repression, and even the murder of trade union leaders are examples of such transgressions.

For indigenous workers, who are frequently denied the opportunity to form or join such organizations, the situation is even more dire. The Americas account for more than 55% of all complaints about violations of these fundamental Conventions that were filed between 1990 and 2002. Africa accounts for about 15% of the cases, while Europe accounts for 15%. There have been the most reported violations (120) in the Andean subregion, which includes Bolivia, Colombia, Ecuador, Peru, and Venezuela. In addition, there are 111 cases in Central America, 97 cases in the MERCOSUR nations (of which Argentina accounts for nearly half), and 52 cases in North America, primarily in Canada. The fact that more than 62% of complaints in the Central and Latin American region are aimed at governments is noteworthy. Thus, the State is seen as a major infringer of fundamental liberties.

1.3.1.1. Issues Of De Jure and De Facto Access to the Fundamental Rights

The Committee of Experts noted claims regarding the precarious status of indigenous day laborers in Mexico in 1999 on the Application of Conventions and Recommendations. For instance, the Committee noted that independent organizations were routinely denied the opportunity to register after they began organizing agricultural workers. National legislation frequently contains requirements and restrictions that, in practice, restrict the activities of agricultural workers who are allowed to form and join organizations of their choosing and engage in collective bargaining. As in Uganda, where 51% of the workforce must be union members, these can include requirements that a particular union enlist a specific proportion of the workforce for the employer to recognize it. 6 In South Africa, the 1995 Labour Relations Act upholds the right to join a union and promotes employee involvement through workplace forums, following the 1993 introduction of basic employment conditions for agricultural workers. Nevertheless, the Act restricts the creation of workplace forums for large employers and does not ensure union access to farms.

Large groups of agricultural workers are frequently excluded by the labor laws currently in place. Agricultural and livestock businesses in Costa Rica that have fewer than five permanent employees are not covered by the Labor Code and, as a result, are not entitled to any kind of representation. Only a small percentage of the agricultural workforce is affected by the results of negotiations in some nations due to labor laws that restrict the ability to engage in collective bargaining, thereby excluding seasonal and temporary workers. Collective bargaining agreements are typically reached in industries and businesses with substantial permanent employment, and they typically represent the interests of permanent employees. Some workers, who frequently make up the majority, may not be covered by such agreements or may only be partially covered. These workers may be seasonal, temporary, weekly, or daily. Generally speaking, collective bargaining does not seem to be a significant feature in the agriculture sector of many countries, primarily because governments do not encourage such negotiations, the necessary institutions are lacking, and rural and agricultural workers' organizations tend to be weak, according to the 1996 ILO report on wage workers in agriculture, which asked to what extent collective bargaining is an important element in improving working conditions in agriculture. The practical implementation of the provisions seems to be restricted, even in Brazil, where collective bargaining in agriculture has a long history and negotiated agreements can be extremely comprehensive. This is caused by many factors, including a weak labor inspection system, a lack of severe penalties for violations, and intricate legal processes that undoubtedly exist in many nations.

1.3.1.2. Some Factors Affecting the Organization of Agricultural Workers

The distance of agricultural endeavors, communication barriers, and other factors have historically restricted the degree of organization among agricultural workers. These are now made worse by the seemingly common practice of firing permanent employees through reorganization and other methods, the growing use of labor contractors by employers, and the growing dependence on outside growers for the production of some crops, like tea and sugar. It is reasonable to assume that globalization has significantly impacted the degree of trade union representation among waged agricultural workers, despite the difficulty in obtaining systematic data. Except for sizable plantation industries in Asia and certain African nations with more formalized labor relations, representation has typically been low in most nations; however, there have been reports of membership declines even in these settings. Thankfully, some organizations, like the National Union of Plantation and Agricultural Workers in Uganda (NUPAWU), have shown a consistent rise in membership throughout the 1990s, defying the overall trend of declining levels of unionization. Its membership increased from 22,000 in 1990 to 32,838 in 1995 and to about 47,500 at present, which is significantly more than it was in 1973 before the political, economic, and social unrest that plagues the nation. This is the outcome of NUPAWU's efforts to address the long-term issues of relevance and membership, and it shows the application of a cogent plan to hire female employees that dates back to the early 1990s. Currently, there are about 15,000 female members or about 32% of the total membership.

Therefore, it is evident that the general political climate can have a significant impact on worker protection levels and unionization rates. For example, in 1973, Chile's agricultural unionization rate was 42.4%, which was higher than the country's average. Following the reversals under a military regime, the national unionization rate dropped to 5.2% by 1985. Some employment rights have been restored since democracy was restored in 1990, and labor laws have started to specifically include temporary workers. It is evident that the general institutional and legislative frameworks controlling labor rights and working conditions in each nation reflect the right to freedom of association and collective bargaining, albeit with different limitations. However, it is crucial to emphasize the function of trade unions and other workers' organizations within these frameworks, the degree to which governments support and acknowledge them, and the reality that in far too many nations, unions struggle to exercise their rights and play their full potential. In many cases, it is the state's responsibility to increase the opportunities for this kind of negotiating and to guarantee the circumstances necessary for the growth of productive social discourse. The continuation of forms of forced and compulsory labor, some of which are especially pertinent to the problem of decent work deficits in agriculture, is a reflection of some of the detrimental effects of restrictions on freedom of association and the fundamental right of trade union organizations to defend workers' rights (FAO, 2017; FAO and ILO, 2018; Kerr and Camfield, 2018).

1.3.2. Elimination of All Forms of Forced or Compulsory Labor

This idea is stated in the 1930 Forced Labor Convention (No. 29), as well as the 1957 Convention to Abolish Forced Labor (No. 105). The practice of forced labor continues in many new and old forms despite widespread ratifications among member states. While forced labor in the form of coercive recruitment is prevalent in many Latin American countries (a practice that is obviously linked to the violations of trade union rights mentioned above), as well as in parts of the Caribbean and elsewhere, slavery is still practiced in a small number of African countries, according to the ILO Global Report on Stopping Forced Labor. This type of mistreatment disproportionately affects Indigenous communities.

1.3.2.1. The Cycle of Debt and Bondage: A Feature of Forced Labor in Agriculture

A type of forced labor, bonded labor is used in a variety of industries, but primarily in agriculture. Despite initiatives in some nations, most notably India, to quantify and end the issue, bonded labor is still prevalent in South Asia. Due to various forms of debt, millions of men, women, and children throughout the subcontinent are dependent on their agricultural labor. These include the increasingly common practice of debt-based attachment of migrant and casual workers in commercial agriculture, both large and small, as well as personally bonded labor secured by debt (which can last for generations). Recent assessments of initiatives to help male laborers in India escape bonded labor conditions have brought to light instances where more and more women are becoming enslaved as they shoulder greater debt repayment responsibilities.

Once more, indigenous peoples frequently find themselves ensnared in the debt and bondage cycle. An illustration of this can be found in Peru, where some indigenous groups have been forced into debt bondage on sizable forestry and agricultural properties, serving as an unpaid or partially compensated labor force. The situation of a particular ethnic group in southwestern Nepal, where the government has recently outlawed the practice through legislation, is another example of semi-feudal agrarian bonded labor. In Pakistan, where the government passed legislation to end the system in 1992, there are also instances of bonded labor in particular communities. One of the primary characteristics of forced labor in many contexts is the use of the debt mechanism to bind workers to production on estates until they settle debts. These debts, which frequently include expenses for transportation, food, housing, tools for the job, etc., are imposed fraudulently. Without a formal agreement. When employees possess identification documents or working papers, the contractor may often keep them. Because labor inspection capabilities are noticeably lacking in the majority of the nations where forced labor is prevalent, this covert and unlawful violation of workers' rights is challenging to stop. As previously stated, it is likely that fewer than 5% of the 11.3 billion agricultural workers worldwide have access to systems for legal and health protection or labor inspection.

1.3.2.2. Coercive Labor Contracting

The current misuse of labor contractor recruitment systems encourages the practice of debt bondage, which in turn leads to forced labor. Even though labor contractors are sometimes legally forbidden, they are still important middlemen, especially in commercial agriculture where there is a high seasonal demand for labor. These contractors are widely used in several Latin American nations, including Brazil. Independent labor contractors frequently make advance payments to peasant communities during periods of scarcity, forcing them into debt before the harvest season. Advance payments and hiring from impoverished areas impacted by drought or seasonal unemployment are common practices in Guatemala, Bolivia, and particularly Brazil. Up to 200,000 people in Brazil are reportedly engaged in forced labor, according to various sources. A massive commercial enterprise that deals in forced labor uses recruitment practices and middlemen that completely violate people's rights. Landowners and farmers are rarely punished and seem to act without consequence.

In many nations, the production and harvesting of sugar cane is notorious for being a site of coercive labor contracting, where seasonal workers are primarily hired under verbal agreements.

Cane-cutters in Bolivia, for example, may borrow the monetary equivalent of 40 tons of sugar at the start of the harvest, which is difficult to repay by the end of the four-month harvest, according to ILO research on indigenous (internal) migrant labor during the sugar harvest. After a season, indigenous workers typically ask for another loan, promising to return the following season. Before privatization efforts, state enterprises controlled the production of sugar cane in many nations. This was the case, for example, in the Dominican Republic, where state-owned enterprises have been involved in systematic documentation of forced labor practices since the early 1980s and Haitian migrant workers are hired each year for the harvest. Actually, both governments were held accountable for these abuses by the ILO's supervisory bodies. The increase in human trafficking, which has taken on global dimensions, is also highlighted in the ILO Global Report of 2001. The coercive aspects of trafficking, such as forced labor, debt bondage, and slavery-like practices, are mentioned in the globally accepted definition of the term. In West Africa, trafficked children—boys in particular—are employed in agriculture.

1.3.2.3. Forced Labor Outcomes of Irregular Migration

We should draw attention to one more dimension. This relates to the effects of irregular migration on forced labor, such as labor trafficking for exploitation. Since human trafficking for labor exploitation constitutes a serious violation of labor rights, a failure of the labor market, a failure of employment systems and services, and a failure of migration management, the ILO is particularly concerned about this phenomenon. Since 2000, irregular migration has become a major political concern for many European nations, and awareness of the coercive conditions and forced labor that trafficked individuals may encounter in their destination nations is growing. Once more, the horticultural industry—this time in Europe—offers some insights into these labor exploitation circumstances.

The Ethical Trading Initiative took up the issue in early 2002, convening key players for seminars on seasonal and foreign labor in the UK food industry. Now, a working group has been established to investigate with the government the licensing and registration of contract labor suppliers and to create a framework for identifying best practices. Growers are obviously under pressure to reduce expenses due to price pressure from large supermarkets, so they look for deals with gang-masters who can supply inexpensive labor. The most vulnerable workers are forced to put up with the terrible conditions that are starting to be revealed in European horticulture, whether they are undocumented workers with illegal status or irregular migrants. The weakest links in the chain are thus squeezed by the commodity supply chain.

Where does legal responsibility lie? The ILO Global Report highlights the apparent impunity enjoyed by those responsible for enabling and extorting forced labor from millions of men, women, and children working in agriculture in many parts of the world. To avoid being perceived as complicit in the violation of fundamental labor rights, governments must take more proactive measures in this situation within the borders of many nations, as well as when migrant workers from other nations are directly impacted by illegal migration and covert recruitment practices. Clarifying the roles and responsibilities of those who profit from forced labor and human trafficking is critically important. Even though workers' organizations and trade unions are crucial to any attempt to create plans to stop these abusive practices, they can no longer handle the task alone.

1.3.3. Elimination of Discrimination in Respect of mployment and Occupation

This idea is stated in the 1951 Equal Remuneration Convention (No. 100), as well as 1958's Discrimination (Employment and Occupation) Convention (No. 111. In Article 1, the latter refers to discrimination that renders equality of opportunity or treatment in employment or occupation null and void or impairs it on the basis of race, color, sex, religion, political opinion, national extraction, or social origin. Discrimination as it appears in waged employment in agriculture has at least three overlapping dimensions. The first is discrimination against groups of men and women based on gender criteria, such as differences in female and male wages. In particular, migrant workers, indigenous, and tribal populations, and women and men within these groups may experience the second, which is based on racial or ethnic classifications, differently. The third type of age-based discrimination involves specific policies and practices that discriminate against children, adolescents, and the elderly in particular ways. Again, older women and men, as well as girls and boys, may be impacted and exploited in different ways. A brief analysis of each of these dimensions is provided below.

1.3.3.1. Gender-Based Discrimination

Even after ratifying these conventions, many female employees are denied supervisory positions, sexually harassed, and paid less than their male counterparts. Wage disparities between men and women are the most obvious example of gender-based discrimination. Once more, what matters is what actually occurs, even in cases where equality is protected by law. For instance, women have long held important positions in Sri Lankan plantations, where they make up roughly half of the regular workforce. In 1984, the daily wage rates for men and women working on tea and rubber plantations were deemed to be equal by law. On tea plantations, women's average daily earnings were 83.5% of men's in 1980, while on rubber plantations, they were 87%. But there's also the issue of how many hours are spent on various tasks. For instance, women who pluck tea put in more hours than men who perform various tasks. As time went on, it became customary for husbands to pick up their wives' paychecks earlier in the day. Because women had no control over their income, this practice was causing conflict within households. Several studies detailed the prevalence of debt and alcohol misuse in these plantation communities. Eventually, trade union worker education initiatives and negotiations for a clear clause mandating that women collect their own wages addressed this basic gender issue of wage collection.

There have been reports of discrimination in the Dominican Republic, where many female employees receive lower pay than their male counterparts.

This practice is especially prevalent on sugar plantations, where Haitian women workers, whether legally or illegally in the nation, face discrimination in their contracts and work in virtual slavery, with their employers keeping their wages and possessions to keep them from leaving. Additionally, more and more Haitian children—many of whom are under the legal minimum working age of 14 in the Dominican Republic—are employed on sugar plantations.

1.3.3.2. Discrimination Against Indigenous and Tribal Populations

For many years, the ILO has been concerned about the general state of affairs among indigenous and tribal peoples. The 1957 Convention on Indigenous and Tribal Populations (No. The 1989 Indigenous and Tribal Peoples Convention amended 107) (No. 169). Compared to other agricultural and rural workers, this group of workers—especially the women—frequently face even more difficulties. Indigenous peoples generally have less secure health protection than the general population. The extensive and careless use of toxic pesticides on tobacco plantations in Mexico poses health risks to indigenous workers, according to the ILO Committee of Experts on the Application of Conventions and Recommendations. On certain estates in Paraguay, where the minimum wage paid to indigenous workers is less than the statutory minimum and non-indigenous workers receive higher wages for the same work, the Committee of Experts has also found wage discrimination. According to a human rights organization's publication that was referenced in the Committee's 1999 observation on Mexico, the wives of day laborers—who are primarily indigenous women—are the ones who carry out the most taxing tasks. They may work 18–20-hour days, doing farm work and household chores, and their pay is always considered an addition to their husband's salary. They are required to work until the very end of their pregnancy and are not granted maternity leave or access to healthcare.

1.3.3.3. Discrimination Based on Age

The practice of child labor and children working in bonded labor have been discussed above, along with age discrimination. It is crucial to stress that child laborers are frequently subjected to discrimination and exploitation due to their age; their youth makes them vulnerable to abuse. Child laborers are among the most vulnerable groups of waged agricultural workers, and there is evidence that they are severely abused about compensation. According to studies, children's employment agreements are typically informal, unwritten, and benefit-free. Due to deductions for things like food, lodging, other necessities, medical costs, fines, etc., the actual amount received is frequently far less than expected. In certain instances, children who work as child laborers perform the same tasks as adults, but they are compensated significantly less and make less than the minimum wage (where applicable). This severe discrimination is demonstrated in a few instances from various geographical areas.

Gender inequality, discrimination against minorities, and age-based discrimination are the three aspects of discrimination in employment conditions and compensation that are obviously connected and overlap. Given these various levels of discrimination, the exploitation that an indigenous woman, adolescent, or child may endure may be particularly severe. Large groups of women and men, girls and boys, in many countries suffer from a lack of decent work in agriculture as a result of this. According to the preceding paragraphs, there may be detrimental synergies between the four areas of fundamental principles and rights as they relate to agriculture and agricultural workers. These include the lack of rights regarding freedom of association, the continuation of forced labor, the presence of some of the worst types of child labor in agriculture, and the continuation of discriminatory practices based on racial, ethnic, and tribal classifications as well as de facto and de jure discrimination. When combined, these fields represent the opposite of decent work (FAO, 2017; FAO and ILO, 2018; Kerr and Camfield, 2018).

1.4. EMPLOYMENT AND INCOMES FOR WOMEN AND MEN IN AGRICULTURE

Around the world, agriculture continues to be an essential industry for economic growth and livelihoods, especially in rural areas of developing nations. Additionally, it plays a significant role in advancing the idea of Decent Work, which seeks to offer opportunities for productive employment that guarantee equitable pay, safety, and rights at work. Because it tackles gender inequality, income inequality, and sustainable economic growth, the second pillar of decent work—employment and income—is especially important in the agricultural sector. Creating high-quality jobs and guaranteeing fair compensation for both men and women working in agriculture are the main goals of this pillar. A wide variety of activities, including crop production, animal husbandry, agro-processing, and agribusiness, define the agricultural sector. It employs a sizable percentage of the world's workforce, particularly in developing and low-income nations where it frequently serves as the main employer. Nonetheless, there are noticeable differences in the employment and income levels of men and women in this sector. A significant portion of the agricultural workforce is made up of women, who help with planting, weeding, harvesting, and processing, among other duties. Despite playing a crucial role, women frequently experience structural disadvantages when compared to men, which can result in notable disparities in pay, working conditions, and overall financial security. The prevalence of informal and precarious employment is one of the main problems with the second pillar of decent work in agriculture. A large number of agricultural jobs are part-time, seasonal, or temporary, which results in inconsistent pay and little job security. Particularly women are more likely to work in vulnerable, informal jobs where they are not entitled to social protections, health care, and other benefits that come with formal employment. Their income is impacted, but it also feeds into larger trends of poverty and unstable economies. Formalizing agricultural employment, enhancing contract terms, and providing social protections to all workers—male or female—are all necessary to address this issue. Another major issue is the difference in income between men and women in agriculture. Generally speaking, women are paid less than men for doing comparable work, which is indicative of larger problems with gender inequality. Limited access to resources like land, credit, training, and technology is one of the many causes of this wage disparity. Women frequently have less opportunity to gain the knowledge and resources necessary to increase productivity and bargain for higher pay. Furthermore, social norms and traditional gender roles may prevent women from pursuing higher-paying jobs in management or agriculture. Policies that support equal pay for equal work are necessary to address these disparities, as are programs that increase women's access to agricultural resources and opportunities.

The significance of entrepreneurship and the growth of small-scale agribusiness is also emphasized by initiatives to support the second pillar of decent work in agriculture. One of the most effective ways to increase employment and income levels is to assist men and women in starting and expanding their own agricultural businesses. Particularly for women, entrepreneurship provides a route to increased financial autonomy and empowerment. But when it comes to launching and growing their businesses, women encounter many obstacles, such as restricted access to capital, markets, and networks. The prospects for women in the agricultural industry can be greatly enhanced by specially designed support programs that address these obstacles, such as microcredit programs, business management training, and access to market data. Improving working conditions is another way to support fair employment and wages in agriculture. Workers in agriculture are frequently exposed to physical risks, long hours, and occasionally dangerous conditions as a result of their labor-intensive and hazardous jobs. Additional difficulties that women agricultural workers may encounter include juggling work and household duties, which can result in longer workdays and more physical strain. Enforcing labor laws, ensuring safe working conditions, and promoting health and safety training are all necessary to address these problems. Furthermore, gender-sensitive policies that acknowledge and take into account women's dual roles in managing the home and producing agricultural products are required.Enhancing employment opportunities and income levels in agriculture also requires investments in education and skill development. Training in sustainable farming methods, modern farming techniques, and technology use can help both men and women become more productive and get better-paying jobs. Women can also break free from traditional roles and gain access to more varied and lucrative agricultural pursuits through education and skill development. Initiatives from the private sector, NGOs, and the government all contribute to agricultural workers' access to education and skill development. Lastly, advancing gender equality in agricultural employment and income necessitates a multi-stakeholder, integrated approach. To create and execute policies that address the particular difficulties faced by women in agriculture, governments, employers, civil society organizations, and international agencies must work together. This entails promoting equal rights to resources and land, guaranteeing financial services and credit availability, and endorsing programs that encourage women to take on leadership roles in rural communities. The agriculture industry can become a catalyst for inclusive and sustainable development by fostering an environment that benefits both men and women (FAO, 2017; FAO and ILO, 2018; Kerr and Camfield, 2018).

1.4.1. Gender Inequality in Agricultural Wages and Job Opportunities

Millions of people, especially in rural areas, depend on the agricultural sector for their livelihoods, making it a vital part of economies around the world. Women continue to experience large wage gaps and unequal employment opportunities in comparison to men, despite the crucial role they play in agriculture. Deeply ingrained structural barriers that prevent women from fully participating and advancing in the industry are revealed by these disparities, which go beyond simply unequal pay to include restricted access to resources, training, and high-quality jobs. Developing programs and policies that can empower women, increase productivity, and advance gender equality in agriculture requires an understanding of these disparities such as:

- **Wage Disparities between Women and Men:** The gender pay gap is one of the most obvious problems in the agriculture industry. For doing the same or comparable work, women are frequently paid much less than men. This disparity in pay is noticeable in a variety of positions, including fieldwork, processing, and other agricultural pursuits. According to studies, women in agriculture can typically make between 20% and 30% less than men. Even when women do the same physically taxing and time-consuming tasks as men, like planting, harvesting, and caring for crops or livestock, the pay gap still exists. The causes of this disparity in pay are multifaceted. Stereotypes and traditional gender norms are important, as women's labor is frequently underappreciated in comparison to men's. Despite evidence to the contrary, there is also a widespread perception that women's labor is less skilled or productive. Women are underpaid for their contributions in many rural areas because their jobs are perceived as merely an extension of their household duties. Furthermore, in agriculture, formal contracts or employment agreements are frequently lacking, which makes it simpler for employers to underpay female workers.

- **Unequal Access to Quality Jobs and Training:** Women's unequal access to high-quality jobs in agriculture exacerbates wage disparities. Women are usually forced into the most labor-intensive, low-paying, and unstable jobs, usually as temporary or seasonal employees. On the other hand, men are more likely to hold managerial, permanent, and higher-paying positions. Women's prospects for professional advancement and financial security are restricted by this occupational segregation. Due to discriminatory hiring and promotion practices, women are less likely to be promoted to higher-paying positions, even when they possess the necessary training and experience. Women's ability to find good jobs in agriculture is also hampered by a lack of access to education and training. Many women lack access to the same educational opportunities as men, which hinders their capacity to acquire the technical skills required for higher-paying and more specialized positions. For example, women are frequently left out of training programs that teach farm management, modern agricultural techniques, or how to operate machinery. This exclusion keeps them in traditional, lower-paying jobs by limiting their ability to adopt new technologies that could boost productivity and earnings. Denying women access to training also deprives them of chances to increase their efficiency and productivity, which impacts not only their earnings but also the agricultural sector's overall output.
- **Barriers to Accessing Resources:** Unequal access to resources like land, credit, and inputs like seeds, fertilizer, and equipment is another major factor causing wage and employment disparities. Women find it challenging to own or inherit land in many nations due to cultural and legal constraints. Women have less negotiating power when they don't own land, which makes it more difficult for them to get loans for agricultural projects or negotiate higher wages. Their capacity to make long-term agricultural investments, like buying better machinery or diversifying their crop production,

which could raise output and revenue, is also impacted by their lack of land access. Credit availability is another important problem. Because of ingrained gender biases, perceived risks, and a lack of collateral, financial institutions are frequently hesitant to lend to women, particularly in rural areas. Women are less able to invest in their agricultural endeavors without access to credit, which restricts their capacity to grow their businesses, buy better inputs, or recruit more workers. Thus, the cycle of low wages and low productivity is maintained. Additionally, information and extension services that offer crucial knowledge on contemporary farming methods, pest control, and crop diversification are frequently less accessible to women. Their productivity and competitiveness in the agricultural industry are further hampered by this knowledge gap.

- **Structural Barriers and Societal Norms:** Broader structural and societal barriers serve as the foundation for the wage and employment disparities that women in agriculture experience. The gender roles and patriarchal norms that govern women's domestic duties frequently carry over into the workplace, where women's contributions are not valued. Women's mobility and access to employment opportunities that require them to travel or work outside of regular business hours may be restricted by these cultural norms. Furthermore, women frequently have to juggle unpaid household chores like cooking, child care, and fetching firewood or water with agricultural labor, which leaves them with less time for activities that generate income. Inequalities may also be sustained by legal frameworks. Certain nations have discriminatory laws or practices that limit women's access to formal employment, property ownership, and land inheritance. Women are more likely to work in informal or casual settings, making them susceptible to wage theft and exploitation even in cases where legal barriers are not overtly discriminatory due to the lack of enforcement of labor rights and protections.
- **Towards Gender Equality in Agriculture:** A multifaceted strategy is needed to address the wage and employment disparities that women in agriculture face. To advance gender-sensitive policies that guarantee equal compensation for equal labor, improve access to high-quality employment, and offer focused training programs that give women the tools they need to succeed, governments, non-governmental organizations, and international organizations must collaborate. To guarantee that women have the funds required to make agricultural investments, legal reforms are required to address

issues of land ownership and credit availability. Furthermore, questioning social norms and bringing attention to the importance of women's contributions to agriculture can help change attitudes and advance greater gender equality (FAO, 2017; FAO and ILO, 2018; Kerr and Camfield, 2018).

1.4.2. Access to Financial Resources and Land Ownership for Women in Agriculture

In the agricultural industry, long-term economic empowerment, income generation, and sustainable employment depend on having access to loans, financial services, and land ownership. However, access to these vital resources is frequently severely hampered for women, who comprise a sizable share of the agricultural workforce worldwide. In comparison to men, these obstacles hinder their economic empowerment, productivity, and capacity to grow their businesses. Promoting gender equality and encouraging inclusive growth in agriculture require an understanding of and commitment to addressing these obstacles are:

- **Limited Access to Financial Services and Loans:** The restricted availability of financial services, such as credit and loans, is one of the main issues facing women in agriculture. For a variety of reasons, financial institutions are frequently reluctant to lend to women, particularly those who live in rural areas. First, there are deep-rooted gender stereotypes that hold women to be less trustworthy or capable than men when it comes to borrowing. Despite data indicating that women typically have higher loan repayment rates, this perception still exists. Because of this, women are frequently shut out of formal financial systems, which forces them to turn to less secure and more costly informal lending networks. Furthermore, because property ownership is restricted by law and culture, many women lack the collateral—such as land or other assets—necessary to obtain loans. Because women frequently lack property ownership or control over family assets, they are unable to provide the tangible assets that financial institutions typically require as collateral. Due to their inability to obtain credit, women are unable to make investments in vital agricultural inputs that are necessary for raising incomes and productivity, such as seeds, fertilizer, machinery, and technology. Women are stuck in low-yield, subsistence farming because they lack the resources to expand their agricultural operations, diversify their businesses, or use modern farming methods.
- **Barriers to Land Ownership:** A key component of agricultural productivity is land ownership since it offers stability, a revenue

stream, and the capacity to make long-term investments. However, when it comes to land ownership, women encounter major obstacles in many areas. Land resources are largely controlled by men because legal, cultural, and social norms frequently prohibit women from inheriting or buying land. Certain nations have laws that specifically forbid women from inheriting land, but in other nations, patriarchal structures and customs guarantee that land is passed down through male lineage, thereby keeping women out of the process. Even when laws are in place to protect women's land ownership and inheritance rights, they are frequently not strictly enforced. When attempting to assert their rights, women may encounter bureaucratic, financial, or legal obstacles that discourage them from doing so. Furthermore, compared to men, women who do own land typically own smaller and generally lower-quality plots. Women find it challenging to access credit, invest in their agricultural endeavors, and realize economies of scale when their landholdings are smaller. Women are less likely to invest in irrigation, soil enhancement, and other productivity-boosting techniques when they lack secure land tenure because they are afraid of being displaced or losing access to the land.

- **The Impact on Economic Empowerment and Productivity:** Women's economic empowerment and agricultural productivity are significantly impacted by their inability to obtain loans, land ownership, and financial services. Crop yields are directly impacted when women lack the financial means to buy better seeds, fertilizer, or farming equipment. Women find it challenging to raise their standard of living or reinvest in their farms when productivity declines because lower incomes translate into lower productivity. Women's chances to diversify their agricultural pursuits, such as selling their goods to larger markets or participating in value-added processing, which could boost their revenue streams, are also restricted by their incapacity to scale up operations. Furthermore, other facets of women's economic empowerment are impacted by the absence of stable land tenure. Women who lack legal land rights are not allowed to participate in decisions that impact how land resources are used and managed. Their economic marginalization is strengthened by this exclusion, which also restricts their capacity to fully engage in agricultural value chains. Women also have trouble accessing other resources, like government subsidies, training programs, and extension services, which are frequently connected to land tenure, if they do not own their own land. Women thus continue to be trapped in a cycle of low productivity, restricted access to markets, and financial vulnerability.

- **Overcoming Barriers and Promoting Inclusion:** It takes a multipronged strategy that includes policy changes, capacity building, and cultural shifts to remove the obstacles that prevent women from obtaining loans, financial services, and land ownership. To develop gender-sensitive financial products that specifically address the needs of women in agriculture, governments, and financial institutions must act proactively. This can entail creating loan products with reduced interest rates, streamlined application processes, and no need for conventional collateral. Although community-based lending programs, savings groups, and microfinance initiatives have demonstrated promise in giving women the money they require to invest in their agricultural endeavors, these initiatives still need to be expanded and incorporated into official financial systems. To guarantee women's rights to land ownership and inheritance, legal reforms are also essential. Laws protecting women's land rights must be more strictly enforced, and women's legal rights must be made more widely known. Programs that offer legal aid to women who want to assert their land rights can assist them in overcoming some of the legal and administrative obstacles they encounter. Furthermore, modifying cultural norms that limit women's access to land requires community involvement. Women's rights to land ownership can be supported and attitudes can be changed by collaborating with men, community leaders, and other stakeholders to advance gender equality. Initiatives to build capacity, like training courses that improve women's business acumen, financial literacy, and understanding of contemporary farming methods, can also enable women to make better decisions, efficiently manage their resources, and produce more. Women are better able to negotiate advantageous terms with financial institutions, take part in decision-making, and increase their economic opportunities when they have the necessary resources and skills (FAO, 2017; FAO and ILO, 2018; Kerr and Camfield, 2018).

1.4.3. Impact of Seasonal Employment and Informal Work on Income Security

Millions of workers worldwide are impacted by the agricultural sector's prevalent seasonal and informal employment. Although both men and women are impacted by this kind of work, women are disproportionately impacted because of the socioeconomic and structural disparities that exist in many areas.

Instability, low pay, and a lack of official contracts are characteristics of seasonal and informal agricultural jobs that impede access to respectable working conditions and increase income insecurity. Addressing gender inequality and advancing more inclusive economic development in agriculture require an understanding of how these dynamics impact women more severely than men.

1.4.3.1. Seasonal and Informal Employment: Definitions and Challenges

Jobs that are only available during certain times of the year—typically associated with planting, harvesting, or other crop-related tasks—are referred to as seasonal employment. Conversely, the lack of official work agreements, contracts, or social safeguards like health insurance, maternity benefits, or pensions characterizes informal employment. These two types of employment frequently coexist in the agricultural industry since workers are employed for temporary, task-specific positions that lack legal recognition and job security. Informal and seasonal work is usually low-paying, physically taxing, and unstable, with limited prospects for skill or career advancement. This kind of work is often the only way for many employees to make a living, especially in rural areas where there are few other employment options. For many reasons, women are overrepresented in seasonal and informal agricultural occupations. Even when the reality of the work does not match this perception, traditional gender roles frequently require women to perform tasks that are seen as less demanding or skilled. Crop planting, weeding, and harvesting are examples of low-wage, temporary positions with few legal protections that are more likely to be filled by women. Furthermore, cultural expectations and caregiving duties like raising children and cleaning the house frequently compel women to work in informal jobs that provide flexibility but little security. As a result, the vulnerabilities linked to seasonal and informal employment disproportionately affect women which include:

- **Income Insecurity and Economic Vulnerability:** Women's income insecurity is exacerbated by the erratic nature of seasonal and informal agricultural jobs. Workers in these jobs are unable to depend on a consistent income because they are seasonal, which makes poverty and unstable finances worse, particularly for rural households. Due to their limited access to credit, limited financial resources to fall back on, and limited opportunities for economic diversification, women are particularly vulnerable to economic shocks. Women who don't have a steady source of income also find it hard to save money, make investments in their farms or businesses, or provide for their families during slow times when there is a shortage of agricultural labor. In addition, the low wages linked to seasonal and informal agricultural labor mean that even when women do find work, it is frequently insufficient to help them escape poverty. Even when they carry out identical agricultural tasks, women are paid less than men, according to studies. This discrepancy is exacerbated in unregulated environments where there is no legal guarantee that equal labor is compensated equally. To make ends meet, many women are forced to work multiple jobs or put up with exploitative conditions due to low wages and seasonal employment cycles. In addition to having an impact on women's immediate means of subsistence, this economic vulnerability also limits their capacity to spend money on healthcare, education, and other necessities for their families.

- **Lack of Decent Work Conditions:** The ILO defines decent work conditions as having a stable job, a good wage, a safe workplace, and access to social protections. Unfortunately, seasonal and informal agricultural jobs rarely meet these requirements, and the lack of decent work standards disproportionately affects women. Due to their frequent exclusion from labor laws and protections, informal workers—especially women—have no legal options in the event of harassment, wage theft, or hazardous working conditions. Their financial and physical well-being is further jeopardized by the fact that seasonal workers, who may only be employed for a few weeks or months, are unlikely to receive benefits like sick leave, health insurance, or pension contributions. Women who work in agriculture frequently face difficult circumstances, such as long hours in inclement weather, exposure to dangerous chemicals, and physically taxing jobs without the proper safety precautions or protective equipment. Notwithstanding these dangers, they usually do not have access to healthcare and are unable to take time off when ill or hurt because doing so would result in their loss of pay. Many women, especially those who are the primary caregivers for their families, must put up with subpar working conditions to support their dependents because they are unable to find decent employment. The fact that women are less likely to belong to labor unions or workers' associations, which restricts their ability to advocate for improved working conditions and equitable treatment, makes the situation worse.
- **Barriers to Economic Empowerment:** One of the biggest obstacles to women's economic empowerment in agriculture is the prevalence of seasonal and informal work. These low-paying, unstable jobs make it harder for women to save money, make investments in

their own growth, or start their own businesses, all of which can contribute to long-term financial security. Women are also deterred from pursuing education or training because of the lack of job security, which prioritizes short-term income generation over long-term skill development. Women are unable to escape the cycle of poverty if they are unable to obtain steady, well-paying employment. Furthermore, because their work is informal, women do not have the same access to social protections as those who work in formal jobs. Gender inequality is maintained by this exclusion because women without health insurance, maternity benefits, or social security are more susceptible to financial shocks like disease, crop failure, or other unanticipated costs. Targeted interventions are needed to remove these obstacles, including those that support formalization of employment, guarantee equal compensation for equal labor, and increase access to social protections for all workers, regardless of employment status.

- **Towards Equitable and Decent Employment in Agriculture:** To tackle the issues caused by seasonal and informal work in agriculture, comprehensive plans that prioritize giving women more secure, respectable, and inclusive employment opportunities are required. Working together, governments, businesspeople, and development groups can advance laws that formalize agricultural employment, raise wages, and protect workers' rights. More sustainable job opportunities for women in agriculture can be achieved through actions like aiding the establishment of women-led cooperatives, expanding access to training and skill development, and offering subsidies or incentives to employers who offer long-term contracts. Additionally, it is crucial to provide social protections to seasonal and informal workers to guarantee that women have a safety net in the event of illness or unemployment. Policies that provide health insurance, maternity benefits, and pension access irrespective of formal employment status may fall under this category. Encouraging women workers to organize through cooperatives or unions and spreading awareness of their rights can also enable them to demand better working conditions, equitable pay, and job security (FAO, 2017; FAO and ILO, 2018; Kerr and Camfield, 2018).

REFERENCES

1. Bramble, T., & Camfield, L. (2018). Decent work in the rural economy: Challenges and opportunities. *Labor Studies Journal, 43*(1), 3–19.
2. Cohen, M. J., & Garcia, J. (2016). Achieving sustainable development goals through decent work in agriculture. *Environmental Science & Policy, 66,* 32–41.
3. Duncan, J., & Laven, A. (2020). Decent work in global value chains: The case of agriculture. *Journal of Economic Geography, 20*(2), 293–311.
4. FAO & ILO. (2018). *Promoting Decent Work in the Agricultural Sector: A Toolkit for Policymakers.* Rome: FAO & ILO.
5. Graham, R., & Morsy, M. (2012). Women's empowerment and decent work in agriculture: The role of microfinance. *International Journal of Agricultural Management, 2*(2), 69–77.
6. Kerr, R. B., & McCulloch, N. (2021). The role of agriculture in promoting decent work. *Agricultural Economics, 52*(2), 231–245.
7. Mehta, L., & Movik, S. (2016). *The Politics of Water: A Global Perspective on Agricultural Water Management.* Earthscan.
8. Mellor, J. W. (2017). Agriculture on the move: Decent work in global agricultural value chains. *The Journal of Development Studies, 53*(5), 684–698.
9. Oya, C. (2013). *Rural Work in Africa: Labor Markets and Employment Strategies.* Palgrave Macmillan.
10. Timmer, P. C. (2009). *A World Without Agriculture: The Structural Transformation in Historical Perspective.* University of Pennsylvania Press.
11. United Nations Food and Agriculture Organization (FAO). (2017). *Women in Agriculture: Closing the Gender Gap for Development.* Rome: FAO.
12. Wiggins, S., & Keats, S. (2013). Food prices and labor in agriculture: A review of the evidence. *Food Policy, 43,* 10–21.

CHAPTER 2

LABOR CONDITIONS IN AGRICULTURE

Contents

INTRODUCTION

A vital component of the larger agricultural system, labor conditions in agriculture have an effect on the economy, food security, and sustainability of farming practices in addition to the workers themselves. Millions of people worldwide are employed in agriculture, many of whom endure difficult and frequently exploitative working conditions. Low pay, long hours, and insufficient access to fundamental rights and protections are just a few of the problems that these workers—including seasonal and migrant workers—frequently deal with. The prevalence of low wages is one of the main issues with agricultural labor. Many farmworkers make significantly less money than the minimum wage, especially in areas where labor laws are not strictly enforced. The seasonal nature of agricultural labor, which restricts access to benefits like health insurance and retirement plans, further exacerbates this economic inequality. Workers frequently find themselves caught in a never-ending cycle of poverty, striving to make ends meet while enduring long hours and unfavorable working conditions. Agricultural laborers usually face hazardous and physically taxing working conditions in addition to unstable finances. Their health is at risk when they are exposed to dangerous chemicals, severe weather, and large machinery without proper safety precautions. Workplace illnesses and injuries are more common because many workers lack the necessary tools and training. These risks are exacerbated when safety regulations are not followed, leaving employees at risk.

Another important component of the working conditions in agriculture is migrant labor. To cover seasonal labor shortages, many farmers turn to migrant laborers, but these workers frequently encounter additional difficulties. They might experience feelings of vulnerability and loneliness as a result of being cut off from their families and support networks. They may find it more difficult to advocate for their rights or obtain necessary services as a result of language barriers. Agricultural workers often have few legal protections. A lack of fundamental rights, like the ability to form a union or obtain overtime compensation, results from the exclusion of agricultural workers from labor laws that apply to other industries in many nations.

This legal loophole feeds the cycle of abuse and facilitates worker exploitation. Labor groups and activists have demanded changes to provide agricultural workers with additional protections, highlighting the importance of equitable pay, secure working conditions, and the ability to form unions. Food justice and immigration policy are two more general social issues that are linked to the problem of agricultural labor conditions. Movements for fair food systems, which aim to address the disparities experienced by those who produce our food, frequently cross paths with calls for improved working conditions. Customers are calling for reforms that put farmworkers' rights and welfare first and are demanding more ethical and transparent agricultural practices. Agricultural labor conditions continue to be a complicated and urgent problem that needs to be addressed right away. Resolving these issues is crucial for the sustainability and moral integrity of the agriculture industry overall, as well as for the welfare of agricultural laborers. To ensure that people who work to put food on our tables are treated fairly, with dignity, and with respect, policymakers, consumers, and advocacy groups must cooperate to improve working conditions. We can only hope to establish a more fair and just agricultural system by working together.

2.1. OVERVIEW OF WORKING CONDITIONS IN GLOBAL AGRICULTURE

Geographical location, economic development, labor laws, and the type of agricultural production all have an impact on the wide variations in working conditions in global agriculture. Agricultural workers encounter numerous obstacles that affect their general well-being, safety, and health in many areas, particularly in developing nations. A significant issue is the amount of unofficial work. Workers are at risk of exploitation because a significant amount of agricultural labor is performed without official contracts. Their financial instability is made worse by the fact that these informal workers frequently do not have access to benefits like social security, health insurance, and paid time off. Workers are frequently paid piece rates and receive low wages, which can cause them to overwork and earn too little to sustain their families.

In agriculture, physical labor conditions can also be difficult. Extreme weather, such as intense heat or heavy rain, frequently forces workers to perform taxing tasks. The manual handling of large loads, exposure to pesticides, and the use of dangerous machinery all increase the risk of mishaps and long-term health problems. Due to inadequate safety precautions and equipment, many agricultural workers report respiratory issues, chronic pain, and other health issues. Agricultural labor is fraught with psychological stressors in addition to physical risks. Many workers, especially seasonal workers, deal with pressure to meet production quotas, fluctuating demand, and job insecurity. Isolation can exacerbate this stress, particularly for migrant workers who might be separated from their communities and families. It can be challenging for employees to speak up for their rights when they are feeling disconnected and vulnerable due to language and cultural barriers.

Another crucial issue is labor rights in agriculture. Widespread infractions are caused by lax or poorly enforced labor laws in many nations. When workers try to band together for better pay or working conditions, they frequently encounter intimidation. This atmosphere of fear has the potential to silence workers' concerns and hinder efforts at collective bargaining. Anti-union sentiments are ingrained in some areas, which makes attempts to enhance working conditions even more difficult.

Furthermore, women play an important but frequently disregarded role in agriculture. Although women make up a sizable share of the agricultural workforce, especially in developing nations, they usually face additional obstacles like restricted access to resources, training, and decision-making authority. Discrimination based on gender can result in lower incomes and fewer opportunities for promotion, which feeds the cycles of inequality and poverty.

Global supply chains make matters more difficult. The gap between producers and consumers widens as more and more agricultural products are coming from other nations. This may result in a lack of responsibility for the circumstances surrounding food production. Companies may put profit ahead of the welfare of employees, and consumers may not be aware of the exploitative practices that can occur in the supply chain. Initiatives like fair trade, corporate social responsibility, and sustainable farming methods have all contributed to the growing movement in recent years to improve working conditions in the agricultural industry. These initiatives seek to advance increased respect for workers' rights, safer working conditions, and higher wages. There are still many obstacles to overcome, though, and governments, organizations, and consumers must work together to guarantee that agricultural workers receive just treatment and dignity to support a more sustainable and equitable food system on a global scale (Moore, 2015; Akram-Lodhi et al., 2021).

2.1.1. Why Global Agricultural Workers?

Historically, the largest economic sector in terms of both employment and output volume has been agriculture. Although they are incredibly unevenly distributed throughout the world, approximately one billion people of working age—or 26.7% of the global labor force—are still employed in agriculture today. Nevertheless, both in the "old" and "new" labor histories, agricultural workers have long been underrepresented. This academic disdain for agriculture may stem from what some have seen as the industrialization-related effects of agriculture's increasing productivity and the gradual demise of agricultural labor, or from a process of de-parasitization. In actuality, though, the declining significance of the agricultural sector has long been restricted to a small number of highly industrialized countries. Agriculture continues to be a significant economic sector and a source of income for a sizable portion of the populace everywhere else. While the percentage of workers in the agricultural sector has been falling globally and is currently (far) less than 5% of the working population in the majority of high-income countries, it is still a sizable portion in the majority of mid- and low-income countries, where the majority of the world's population currently resides. The study of agricultural laborers is motivated by many qualitative factors in addition to their quantitative significance. Three topics that require more thorough historical contextualization but are highly pressing in the modern era have been chosen for this volume.

First of all, the topic of food security has emerged as a top political priority in recent decades due to factors such as global population growth, rising food prices, and environmental change. It was one of the primary topics of the United Nations' "Millennium Development Goals" and "Sustainable Development Goals," and it was thoroughly discussed at the first World Food Summit (Rome, 1996). The idea of food sovereignty, which is "the right of peoples to healthy and culturally appropriate food produced through ecologically sound and sustainable methods, and their right to define their own food and agriculture systems," is inextricably linked to food security, but it is also ambivalent. Instead of focusing on market and corporate demands, it places the people who grow, distribute, and eat food at the center of food systems and policies.

Small-scale farming has given way to large agro-industrial companies that control food production in the name of ensuring "food security," especially in the Global South, over the past few decades, raising serious concerns about food sovereignty. Large corporations' monopolization of the food production industry raises urgent concerns about the quantity and quality of food produced, the conditions under which it is produced, and the system's effects on the environment. A prominent idea in academic and political discussions, food sovereignty "now constitutes a powerful anti-capitalist vision for food system transformation." In addition, peasants are essential to the political movement supporting food sovereignty. They are also viewed as a potential remedy for the numerous political, economic, and ecological issues that the "corporate food regime" has already brought about. Family-run, small-scale farming is a distinct type of agricultural activity that is very different from entrepreneurial farming. In addition to ensuring food sovereignty, bolstering small-scale peasant farming would improve the ecological sustainability of agricultural production. The field of agroecology, sometimes known as eco-agriculture, has advanced this viewpoint. Smallholders in the Global South make up more than 95% of those still working in agriculture today, despite the industry's expansion and the emergence of "agribusiness." To understand how and why smallholder and peasant producers have survived despite the ongoing upscaling and increasing capitalist relations of production, it is worthwhile to examine their resilience from a long-term historical perspective.

Coerced labor in the primary sector is a second urgent theme. According to recent estimates, two million people worldwide were engaged in private agricultural labor under bonded or slave labor conditions in 2016; this number does not include state-coerced programs. In addition to direct coercion, millions of other agricultural workers are employed under dubious contracts that give them advances that give labor contractors enormous control over when and how they employ them, all while living and working in appalling conditions. Three estimates, for instance, that vary from a few hundred thousand to 10 million migrant plantation workers in the Malaysia-Indonesian region are cited by Oliver Pye. Their numbers are hidden because many of them work as undocumented laborers, and because of their frequently illegal status, they are susceptible to exploitation. Similarly, migrant contract workers play a major role in the fruit and vegetable industries in Southern Europe. A research group focused on the agri-food supply chains, CRA, estimates that there are approximately 400,000 migrant workers in Italy's agricultural sector, accounting for nearly half of the country's labor force in the primary sector. Most of the laborers are undocumented immigrants who are primarily from South Asia, Eastern Europe, the Maghreb, and sub-Saharan Africa. Living conditions are appalling, and migrant workers are often housed in makeshift slums in remote rural areas, especially in Southern Italy where their employment is temporary (for example, during tomato harvesting). Contract labor was frequently presented as a voluntary substitute for slave labor following abolition. However, many of the contributions in this volume demonstrate that slavery and other forms of labor under (semi-)coerced conditions were ubiquitous in the global countryside during the period under investigation, as is the case with other recent historical studies. There appear to be distinct historical parallels in this regard. This begs the question of why forced labor in agriculture has persisted for so long and how significant changes in the economy, politics, and society have influenced its appearance over time.

Third, and closely linked to the first, is the relative lack of protection that states and governments provide for agricultural workers. Many states ignore the exploitative practices mentioned above, despite the fact that many of them occur in the private sector of small-scale commercial agriculture, agribusiness, or even households. In certain instances, like the Dominican Republic, the government has openly supported forced labor on plantations since the 1980s, according to the Ilo. Furthermore, because current laws frequently bar (seasonal) migrants from social security programs, millions of waged agricultural workers—men and women alike—do not receive essential social benefits in their host nations. More generally, many states' migration policies keep a sizable portion of the workforce in a highly precarious position, allowing employers to take advantage of workers excessively and fostering what some academics call "modern slavery" in the workplace. Therefore, in addition to peasants, who make up a disproportionately large portion of the world's agricultural labor force, a sizable landless agro-proletariat has emerged in recent decades. In addition to being ignored and even taken advantage of by governments, this proletariat—which includes some undocumented migrants—has also been infamously underrepresented in recent academic research. To what degree we can identify patterns and shifts in the way states interfere with labor conditions in the agricultural sector, and how we might account for them, is a significant historical question that emerges in the context of this volume. These three trends in agricultural labor—the peasantry's tenacity, forced labor in agriculture, and the specific state (non-)intervention from a historical perspective—are the focus of this volume. By combining research on agricultural labor relations from around the world, it seeks to deepen our comprehension of the causes of these changes (Moore, 2015; Akram-Lodhi et al., 2021).

2.1.2. Labor Relations in Agriculture

Workers on plantations and other agro-industrial complexes, small-scale peasant producers, and family farm employees are all included in the fairly broad category of "agricultural worker." They also have very different labor relations. They include wage labor and self-employment as well as different types of forced labor arrangements like slavery and indentured labor. This volume covers the majority of these various labor relations, including wage laborers in the industrializing United States of the 19th century, contract workers on colonial plantations in British India, and peasants under Eastern European serfdom. Who or with whom one works, as well as the regulations, are defined by labor relations. The type of work, the kind and amount of compensation, the working hours, the levels of physical and mental strain, and the degree of freedom and autonomy related to the work are all determined by these rules, which can be explicit or implicit, written or unwritten.

In the framework of the project "Global Collaboratory on the History of Labour Relations, 1500–2000," the International Institute of Social History (IISH) has convened scholars for the past 15 years, and they have collaboratively created a taxonomy of labor relations based on the aforementioned definition. A significant milestone in the history of labor relations worldwide was reaching such a common classification. It enables researchers to analyze the labor relations of a particular population at a given point in time or to ascertain the effects of specific social, economic, or political changes on labor using a common framework and language. Furthermore, it became simpler to compare different labor relations over time and space, allowing researchers to spot regional variations as well as changes over time.

It is difficult to classify agricultural workers in terms of labor relations because they usually combine multiple income streams. For example, a worker may grow millet for domestic use, sell her labor to a large landowner daily, and grow wheat for cash. Each situation has its own set of rules governing the kind of work, the quantity of work, the compensation, and the degree of freedom. It could be helpful to consider labor regimes in light of the intricate nature of labor relations in agriculture. A labor regime can include everything from forced labor services to so-called free wage labor, as well as subsistence and commodity production.

When examining the temporal and spatial changes in labor regimes, the following ideas and analytical instruments have proven essential. The World-Systems Analysis (WSA) developed by Immanuel Wallerstein is still a very effective method for comprehending the connections between changing agricultural labor relations, growing demand for cash crops, and emerging capitalism. WSA emphasizes how the rise of merchant capitalism and global trade are intertwined, leading to the extraction of agricultural and mineral products from the "periphery" to supply expanding producer and consumer markets in the "core." African slave labor was used to produce sugar and other cash crops on plantations in the Americas during the 17th and 18th centuries, thanks to thc support of their merchant capitalist states in Holland and Britain. The 18th and early 19th centuries saw the height of the slave trade, as demonstrated by the cooperative project SlaveVoyages. 95 million of the 10.64 million enslaved individuals listed in the transatlantic slave trade database set sail for the plantations in the Caribbean, South America, and North America between 1700 and 1866. A new source of inexpensive and exploitable labor was discovered following the gradual abolition of the slave trade during the 19th century, particularly in South and Southeast Asian agricultural laborers. Indentured laborers who were transported to plantations across the world and self-exploiting peasant families who grew produce on their own fields using their own means of production or through sharecropping were the two main labor relations that dominated the growing cash crop markets.

Understanding the global division of labor required knowledge of WSA. It caused us to focus on times when the world economy, which was dominated by Europe, expanded to include new regions. The labor force in the peripheries was increasingly coerced during these incorporation processes. This covers both the well-known instance of plantation slavery in the Caribbean and the forced cultivation of cash crops by peasant producers. India's colonial opium industry serves as a prime illustration of the mechanisms of coercion that many peasant producers in colonies had to deal with. Approximately 1.5 million peasant households grew poppies, the plant from which opium is extracted, for the Opium Department of the British Indian government at the height of the industry. The Opium Department set prices that were too low to even cover the costs of cultivation, which left the peasantry in poverty even though opium pumped the government's coffers. The mechanics of a triangle of debt, power, and dependency relations clearly forced the peasants of North India to produce opium at a loss. WSA makes it possible for us to see the coercion that peasants, indentured labor, and enslaved labor endure as aspects of the same process—the addition of new zones to the global division of labor. It's interesting to note that all of this happened as labor relations in northwest Europe, the capitalist "core," grew more open and focused on the market. A group of academics has emphasized the significance of researching "commodity frontiers" in the wake of WSA. Sven Beckert, Ulbe Bosma, Mindi Schneider, and Eric Vanhaute contend that the history of capitalism is a history of the spread of commodity frontiers on a global scale. These frontier areas are distinguished by the labor and land being commodities. This approach's emphasis on the ecological changes brought about by the commodification of labor and land enhances the WSA concept of incorporation. The commodity frontier approach begins the analysis in the countryside rather than in the industrial "cores" of the global economy, despite being strongly influenced by the idea of the commodity chain. Thus, by reorienting their attention from the urban industrial centers to the global countryside and, consequently, to labor relations in agriculture, it has inspired economic and social historians to investigate the history of capitalism.

Harriet Friedmann and Philipp McMichael's concept of the "food regime" is another crucial analytical tool for examining changes in agricultural labor relations, particularly in the more recent era. This idea enables us to identify significant hegemonic shifts and analyze food and agricultural systems from a world-systems perspective. McMichael distinguishes "three phases of geo-political ordering of international food production and circulation": the corporate food regime, which is centered around the World Trade Organization (WTO), which has controlled global agriculture and food since the 1970s; the US-centered food regime, which lasted from the 1940s to the 1970s; and the British-centered imperial food regime, which lasted from the 1870s to the 1930s. The state's evolving role is the primary distinction between the corporate food regime and the two previous food regimes. During the British and US-centered regimes, states controlled the food and agricultural markets; however, under the most recent regime, they were reduced to the status of mere servants of global markets. When it comes to the liberalization of international trade, the WTO has defended the interests of multinational corporations. Countries in the Global South were pressured to open up their agricultural markets under the pretext of "food security," which resulted in an increase in agrarian exports and the encroachment of large-scale businesses at the expense of small-scale farming. According to McMichael, "Southern lands in particular have been increasingly converted from local food provisioning to contract farming and agro-industrial estates." Similarly, to analyze "the global and oligopolistic networks that control increasingly large parts of the production, processing, distribution, and consumption of food," Jan Douwe van der Ploeg introduces the concept of "food empires." According to van der Ploeg, food empires aim to dominate the connections between agricultural and food systems, allowing agribusiness to centralize and appropriate value (Moore, 2015; Akram-Lodhi et al., 2021).

2.1.3. Coerced Labor Relations in the Global Countryside

As previously stated, one of the major mysteries in the history of agricultural labor is the persistence of coercion, with slavery as its most extreme manifestation, in this industry. The relatively new idea of the "commodity frontier" has been a helpful analytical tool in researching this issue. This idea is used to examine how labor and property relations, as well as the environment, have been impacted by capitalism in general and the exploitation of land and labor at the periphery of the growing global economy in particular. Our knowledge of the crucial role that forced labor relations played in the formation of capitalism has been enhanced by commodity-centered histories. Coffee, tea, cacao, tobacco, opium, and sugar were among the most significant psychoactive substances that began to dominate global trade in the early modern era. They were perfect commodities for the profitable transoceanic trade that was crucial to the establishment of empires in the early modern era due to their value-to-weight ratio and addictive qualities. Large sums of money were made by traders and states through the manufacture, sale, and use of psychoactive substances. "Cheap nature and cheap labor" were the primary drivers of these profits and revenues. With consumers in the center and low-cost production in the periphery due to the violent appropriation of labor and land, the political and economic sphere of empires provided plenty of opportunity for the growth of this type of maritime trade.

The persistence of coercive labor relations in agriculture can be explained by a few factors. The two most crucial elements in preindustrial agricultural production were labor and land. Since land was obtained through violent means, it was usually available in large quantities during colonial periods. By eradicating or driving out the native population, the Caribbean islands became accessible for the production of sugar. Similarly, the loss of native lands to aggressive frontier men was the only factor that allowed cotton production in 19th -century North America to rapidly expand westward and southward. Thus, there was never a land shortage in either situation.

The issue was labor. Because labor was the most expensive component in these agricultural sectors, entrepreneurs were highly motivated to keep labor costs as low as possible, which led to maximum exploitation through coercion. Although there was an abundance of land available due to the frontier zones' expansion, the local labor force was frequently displaced. For instance, after the native population had perished, there was just no local labor available on Caribbean sugar plantations. Workers had to be enlisted from far-off places and then confined for extended periods to labor on plantations in isolated regions of the world. Without a great deal of coercion, whether through slavery or indentured labor, this combination of mobility and immobility would not have been possible.

However, extreme forms of coercion are not appropriate for all agricultural production. For instance, throughout the year, vineyards require a high degree of skill and attention from their growers. High monitoring costs would arise from using forced laborers who lack the intrinsic motivation to carefully tend the vines. For this reason, compared to industries like sugar, coffee, or cotton, the production of wine was less likely to involve slavery. The latter crops require a lot of labor to grow, but the actual labor of picking cotton, coffee, or cutting cane is simple and easy to watch. This is especially true for plantations, where managing sizable workforces at minimal expense was made possible by long rows of a single crop. In labor relations, there is therefore a certain crop-dependency. This indicates that some crops are better suited for forced labor than others. In the early modern era, the majority of the crops that dominated global trade fit the description of "very suitable for coerced labor."

Agricultural labor in a colonial setting from the 17th century to the beginning of the twentieth century, encompassing the emergence and decline of the transatlantic slave trade and other forms of forced labor." They demonstrate two things: first, that the transition from slavery to less forced labor in colonial agriculture was not straightforward and linear; and second, that labor relations were greatly impacted by local ecological, economic, and political conditions, and vice versa.

For instance, in the Amazon, colonialists mainly depended on the traditional production techniques and knowledge of native people to extract and grow so-called "drogas do sertão" (forest products), like cacao. Slavery or indentured labor was not an option in this instance because cacao production required skilled labor from the local community. That does not imply, however, that these native laborers were truly free. To supply the metropole with drugs do sertão, the Jesuits established a network of farms or production centers that enabled them to take advantage of the labor and land of this enormous hinterland. The distribution of native workers was governed by an imperial law, and the Jesuit missions were one of the main recipients. Despite being formally recognized as "free" individuals, the workers were coerced into extremely cruel and exploitative working conditions. For example, in the seasonal gathering of forest products, the missionaries utilized the laborers' expertise and abilities. Native production techniques were integrated into the larger imperial economy and became a crucial component of the missionary network in the Amazon.

In Leida Fernandez-Prieto's case study of slavery on sugar plantations, the coexistence of wildly disparate types of labor relations within a single global economy is especially evident. Plantation agriculture based on slave labor emerged during the era of industrial modernization in Europe, which spanned the middle of the 18th century to the end of the 19th century. Fernandez-Prieto is particularly interested in the connections between traditional local knowledge and customs in tropical plantation areas and Western science and industrial modernity. Knowledge and practices were exchanged between tropical Cuba and industrializing Europe, from breeding and fertilization techniques to providing food for enslaved laborers. This demonstrated the connections between "traditional" and "modern" modes of production and labor relations (Moore, 2015; Akram-Lodhi et al., 2021).

2.1.4. State Intervention and Agricultural Labor Mobility

The hiring of agricultural laborers has been largely the responsibility of states. They establish networks, enact legislation, and supply resources to help their agricultural sectors find inexpensive labor. States have employed migration policies in particular as a potent tool to protect a cheap and mobile agricultural labor force. Since the early modern era, worker mobility has been a crucial aspect of agricultural production. This was due to many factors, including the fact that the growth of frontier zones forced production facilities to farther reaches of the globe, frequently reducing the supply of locally available labor. Additionally, seasonality—a characteristic that distinguishes agricultural labor to a far greater extent than industrial production—creates a labor demand that fluctuates greatly. When and where workers are needed depends on the season. In certain instances, these variations in labor demand have been leveled out by technology. For instance, fruit and vegetables can be grown all year round in Southern Spain's greenhouses. Nonetheless, the intensity of agricultural labor demand is still primarily determined by seasonality in the majority of cases. By providing temporary visas, for instance, migration policies can assist in meeting the seasonal labor demands of agriculture while also minimizing labor costs. Workers are kept in a precarious position and are forced to put up with low pay, long hours, and poor living conditions as a result of the creation of "illegality" through migration policies.

The relationship among citizenship, mobility, and labor, as well as the impact of state intervention on these are needed. Employers can establish coercive and exploitative labor relations by separating migrant workers' rights based on citizenship, which puts them in a highly vulnerable position. Whether in the French Empire's sugar cane industry or on modern-day oil palm plantations in Malaysia, the need for certain types of labor appears to be met by using migrants and other workers with lower citizenship status, whose exploitation is not stopped but rather made worse by either the absence of protection for migrant workers or even by policies specifically designed to target them. Furthermore, it appears that migrant workers, who are less able to resist such obligations than locals due to their citizenship status, are the target of recent colonial practices of communal labor in Ghanaian rural communities (Moore, 2015; Akram-Lodhi et al., 2021).

2.1.5. The Working Conditions in Agricultural Economies

The working conditions in agricultural economies are characterized by a complex interplay of economic, social, and environmental factors that significantly affect the lives of workers. These conditions can vary dramatically between developed and developing countries, but several common challenges persist globally.

2.1.5.1. Employment Structure and Labor Rights

A sizable section of the workforce in many agricultural economies works informally, frequently without contracts or job security. Workers in this unofficial status are more susceptible to exploitation and have less access to benefits like paid time off, social security, and health insurance. Numerous agricultural workers put in long hours for little pay in unfavorable conditions that frequently violate fundamental labor laws. Furthermore, widespread infractions such as inadequate safety precautions, wage theft, and unfavorable working conditions can result from lax enforcement of labor laws.

2.1.5.2. Health and Safety Risks

Work in agriculture is physically taxing and can be extremely harmful to one's health. Without the proper protective equipment, workers are frequently exposed to dangerous chemicals like fertilizers and pesticides. Farming-related manual labor, such as lifting large objects and spending a lot of time in the sun, can cause long-term health problems like respiratory and musculoskeletal disorders. Additional risks that seasonal and migrant workers must deal with include loneliness and limited access to healthcare, which makes it challenging for them to get the care they need.

2.1.5.3. Gender Inequality

The agricultural workforce is largely composed of women, especially in developing nations. They frequently encounter other obstacles, though, such as prejudice and restricted access to resources like land, credit, and training. In addition to generally earning less than men, women may not be allowed to participate in decision-making positions in cooperatives and households. In addition to impeding women's economic empowerment, this gender gap also restricts agricultural productivity and innovation.

2.1.5.4. Working Conditions and Migration

Many workers travel long distances for seasonal work, making migrant labor a prevalent feature in agricultural economies. Economic opportunities may arise from this, but unstable living conditions, such as subpar housing and restricted access to essential services, are frequently the result. Additionally, linguistic and cultural obstacles may prevent migrant workers from advocating for improved working conditions.

2.1.5.5. Global Supply Chains and Consumer Awareness

The working conditions in these economies are complicated by the global nature of agricultural supply chains. There is a disconnect between agricultural practices and consumer expectations in developed countries because consumers frequently don't know the conditions under which their food is produced. Some businesses are starting to adopt fair trade and corporate social responsibility programs in response to the growing demand for moral and sustainable business practices. To guarantee that all employees are treated fairly, these initiatives must be strengthened and implemented.

2.1.5.6. Moving Towards Improvement

Improving working conditions in agricultural economies has received more attention in recent years. Organizations and non-governmental organizations are promoting improved pay, safer working conditions, and stronger labor rights. Fair trade certification, community-driven projects, and sustainable farming methods are also becoming more popular as ways to improve the standard of living for agricultural laborers.

Even though working conditions in agricultural economies are extremely difficult, there is hope for improvement. To solve these problems, advance ethical labor standards, and guarantee that agricultural workers can prosper in their positions—all of which contribute to a more sustainable and just food system—governments, corporations, and consumers must work together.

Obstacles have a major effect on their well-being and means of subsistence. The prevalence of informal employment, which frequently results in low pay, job insecurity, and limited access to social security and healthcare benefits, is one of the main problems. Without proper safety precautions, many workers put themselves through long hours in hazardous conditions where they are exposed to physical risks like pesticide exposure and dangerous machinery. Furthermore, because agricultural labor is seasonal, workers may encounter periods of underemployment or unemployment, which leads to instability. The situation is made more difficult by gender disparities, which frequently result in women receiving lower pay, experiencing discrimination, and having less access to resources and decision-making authority. It can be challenging for migrant workers to advocate for their rights because they often struggle with loneliness, communication difficulties, and subpar living conditions. These issues are made worse by the fact that many areas have lax enforcement of labor laws, which results in widespread worker rights violations and exploitation. All things considered, these elements put agricultural workers in a risky situation, emphasizing the pressing need (Moore, 2015; Akram-Lodhi et al., 2021).

2.2. SEASONAL AND MIGRANT LABOR IN AGRICULTURE

Seasonal and migrant labor plays a crucial role in agricultural production, particularly in regions where crops are time-sensitive and require intensive labor during specific periods. This type of labor is characterized by its temporary nature, with workers often traveling considerable distances to meet the demands of harvest seasons.

2.2.1. Characteristics of Seasonal and Migrant Labor

Seasonal laborers are typically employed for specific agricultural tasks, such as planting, tending, and harvesting crops. These workers may be local residents or migrants who move to areas with agricultural demand. Migrant laborers often cross borders or travel within countries to find work, contributing significantly to the agricultural workforce in many regions.

2.2.1.1. Economic Impact

The reliance on seasonal and migrant labor is driven by economic factors, including the need for flexible and cost-effective labor solutions. Agricultural producers often prefer this model to manage fluctuating labor needs without the long-term commitments associated with permanent employees. For many migrant workers, these jobs offer vital income opportunities, particularly in low-income regions where employment options are limited.

2.2.1.2. Challenges Faced by Seasonal and Migrant Workers

Despite their significance, migrant and seasonal workers encounter many difficulties. Many endure difficult working conditions, including long hours, subpar housing, and restricted access to medical care. They frequently don't have official contracts, which leaves them vulnerable to exploitation and job insecurity.

It can be challenging for migrant workers to advocate for improved working conditions or pursue legal action in cases of abuse due to language and cultural barriers that isolate them. Seasonal and migrant workers are subject to many health and safety hazards due to the nature of agricultural work. They frequently have to handle dangerous chemicals and large machinery without the necessary protective gear or training. Due to limited access to healthcare services, this increases the risk of accidents and long-term health problems.

2.2.1.3. Legal and Social Protections

In many countries, labor laws designed to protect workers do not adequately address the unique circumstances of seasonal and migrant labor. Enforcement of existing laws can be weak, leaving workers vulnerable to exploitation. Additionally, social protections, such as access to healthcare and legal support, are often insufficient for this transient workforce.

2.2.1.4. Efforts for Improvement

To address these issues, many advocacy groups and organizations are trying to make things better for migrant and seasonal workers. Among the initiatives are those that advocate for stricter enforcement of labor laws, promote fair labor practices, and increase public awareness of workers' rights. To guarantee improved pay and working conditions for all supply chain employees, some farmers are also implementing fair trade and ethical sourcing policies. Global agriculture depends heavily on seasonal and migrant labor, which is crucial during periods of high production. But to protect these workers' rights and welfare, it is imperative that the difficulties they encounter are addressed. To encourage fair labor practices and enhance working conditions and eventually contribute to a more sustainable agricultural sector, governments, organizations, and consumers must keep up their efforts.

2.2.2. The Realities of Seasonal and Migrant Workers in Agri-Food Value Chain

Over the past 15 years, there has been a steady decline in agricultural employment. To meet the demands of planting, harvesting, and processing crops, the agri-food sector significantly depends on the contributions of seasonal and migrant workers. At a seminar hosted by the European Economic and Social Committee (EESC) and organized by Safe Habitus, the current state of European policies regarding seasonal and migrant workers was examined, along with best practices, opportunities, and the state of play. Depending on the demands of their domestic labor market, their relationships with third countries, and their overall immigration system, EU member states oversee their own seasonal worker programs. Seasonal agricultural labor is a harsh reality, with typically subpar living and working conditions. Both legal and undocumented migrants may become victims of contemporary forms of slavery or unlawful gang-master tactics. In certain places, women are exploited.

Nearly 2 million domestic and intra-EU seasonal workers and over 0.4 million non-EU seasonal workers are estimated to be present in the agriculture sector across EU Member States, according to a recent EESC study on which Ms. Merlo worked. The fact that an estimated 71% of these non-EU seasonal workers in agriculture were not employed under the auspices of the Seasonal Workers Directive in 2021 and were therefore exempt from its rights and social protection provisions is among the most significant pieces of evidence derived from this sizing exercise. Essential social protections like healthcare, unemployment insurance, and retirement pensions are frequently inaccessible to seasonal and migrant workers. They may not be eligible for these benefits due to their temporary or migrant status, which leaves them open to social exclusion and financial instability, especially when they are unwell or unemployed.

Member states are in charge of defining and enforcing national labor and social laws, and they have the authority to enact stronger provisions than those found in EU legislation. This national "flexibility" suggests that different strategies for ensuring social protection exist throughout the EU. Furthermore, information about how labor laws are implemented at the national level is sparse and dispersed for the European Commission (EC).

2.3. WAGES, HOURS, AND JOB SECURITY CHALLENGES

The agricultural sector is pivotal to the global economy, providing food and employment to millions of people. However, the working conditions within this sector often raise significant concerns, particularly regarding wages, hours, and job security. These challenges affect not only the livelihoods of agricultural workers but also the overall sustainability of agricultural practices.

2.3.1. Wages: A Struggle for Fair Compensation

The pay gap is one of the most important problems facing agricultural workers. Low pay is a defining feature of many agricultural jobs, frequently falling short of basic necessities of workers and their families. Workers are paid piece rates in many areas, which means that rather than an hourly wage, they are paid according to the amount of produce harvested. Although this system can encourage productivity, it can also result in erratic profits, particularly in seasons with low crop yields or when harvests are impacted by unfavorable weather.

The situation is frequently made worse in developing nations by the absence of minimum wage legislation or the lax enforcement of current rules. Many workers are vulnerable to exploitation because their employment is informal and lacks contracts that specify wage agreements. Agricultural labor is usually exempt from certain protections, allowing employers to pay less than the minimum wage, even in nations with well-established labor laws.

Furthermore, there is serious concern about the gender pay gap in agriculture. Despite making up a sizable share of the agricultural workforce, women are frequently paid less for doing the same work as men. Systemic gender discrimination is the root cause of this discrepancy, which prevents women from advancing into higher-paying positions.

2.3.2. Hours: Long and Unpredictable

Another issue is how many hours agricultural workers put in. Long hours are necessary for many agricultural jobs, especially during busy times of the year when crops must be picked promptly to prevent spoiling. Workers frequently start earlier in the morning and work well into the evening, going above and beyond the required number of hours without receiving extra pay.

This can exacerbate the difficulties faced by employees by causing physical exhaustion and health issues. Furthermore, agricultural labor is by its very nature unpredictable. Work schedules can be impacted by weather, pest infestations, and market fluctuations, leaving employees unsure of their weekly hours. Many seasonal workers might experience joblessness during off-peak times, which could lead to unstable finances and make future planning challenging.

2.3.3. Job Security: A Precarious Situation

Because agricultural work is seasonal and frequently temporary, job security in this industry is especially precarious. To meet certain harvest demands, many workers are hired on a temporary basis, which can result in unstable and stressful work cycles. Unemployment is a common occurrence for seasonal workers, which can negatively impact their general well-being and financial security.

Concerns about job security are made more difficult by the use of migrant labor. Although they frequently deal with unstable employment situations, many migrant workers are drawn to agricultural jobs in search of better opportunities. One important consideration may be their legal status; undocumented workers may be reluctant to report unfair treatment or hazardous working conditions out of fear of deportation. Because these workers believe they have no options if they are mistreated or paid too little, this fear can result in a vicious cycle of exploitation.

2.3.4. The Intersection of Wages, Hours, and Job Security

For agricultural workers, the relationship between pay, hours, and job security creates a difficult work environment. Workers who receive low pay are frequently forced to work multiple jobs to make ends meet, which results in longer hours and more physical strain.

This can lead to a vicious cycle in which employees' health and well-being are jeopardized to maintain financial stability.

Furthermore, because they are afraid of losing their current job, workers may be discouraged from fighting for their rights or looking for better-paying opportunities. It is challenging for employees to demand fair pay or reasonable hours because of this dynamic, which upholds a culture of silence and tolerance for subpar working conditions (Moore, 2015; Akram-Lodhi et al., 2021).

2.3.5. Efforts Toward Improvement

The conditions faced by agricultural workers are being improved by many organizations and advocacy groups in recognition of these difficulties. In this endeavor, labor unions are essential because they fight for improved pay, hours worked, and job security. Although employers may oppose organizing efforts, their goal is to enable employees to collectively bargain for just compensation and better working conditions.

Improving agricultural conditions also requires local, national, and international legislative changes. Their livelihoods can be greatly impacted by policies that uphold minimum wage laws, control working hours, and provide protections for agricultural workers.

Initiatives that support fair trade also incentivize businesses to adopt ethical sourcing, guaranteeing that workers are fairly compensated for their labor. Programs for education and training can also assist employees in developing their skills, which will increase their marketability and raise their chances of landing steady, well-paying employment. Agricultural economies can develop a more skilled workforce that is better able to handle the difficulties of the sector by investing in worker education.

2.3.6. The Role of Consumers

Consumers have a significant impact on the working conditions in agriculture. Customers can help generate demand for policies that put worker rights first by endorsing fair trade and ethical sourcing products. Businesses may be encouraged to implement fair labor practices if consumers are more aware of the conditions faced by agricultural workers and make more thoughtful purchases as a result. Additionally, organizations and governments can be influenced by consumer advocacy to uphold labor laws and defend the rights of agricultural workers. A growing number of consumers are prepared to pay a higher price for goods that guarantee fair wages and humane working conditions, which will eventually lead to changes in the agriculture industry. For workers in the agricultural industry, the issues of pay, hours, and job security pose serious challenges.

A multifaceted strategy including legislative reforms, consumer education, and ardent labor organization advocacy is needed to address these problems. We can promote a more sustainable and equitable agricultural system by establishing conditions where agricultural workers receive fair compensation, have stable employment, and work reasonable hours. In addition to being morally required, ensuring that workers receive the respect and dignity they are entitled to is crucial for the long-term stability and prosperity of the agricultural sector as a whole. For everyone who depends on agriculture for a living, a more equitable and sustainable future can be achieved through teamwork (Moore, 2015; Akram-Lodhi et al., 2021).

2.3.7. Analyze Labor Conditions in Agriculture Worldwide

Globally, labor conditions in agriculture are characterized by notable obstacles that differ greatly between areas and farming methods. A significant amount of agricultural labor in many developing nations is informal, meaning that workers lack contracts, job security, and access to benefits like healthcare. Particularly for seasonal and migrant workers who deal with precarious employment and uncertain income, wages are frequently low and sometimes fall below minimum requirements. Long workdays are common, particularly during busy times of the year, which raises the risk of physical fatigue and other health issues.

Without proper safety precautions or training, many agricultural workers are exposed to dangerous situations, such as dangerous machinery and dangerous chemicals. Despite making up a sizable share of the workforce, women frequently earn less than men for equivalent roles, resulting in stark gender disparities. They encounter further obstacles that sustain systemic injustices, like restricted access to resources and decision-making authority. The agricultural industry depends heavily on migrant laborers, who frequently face social isolation, exploitation, and a lack of legal protections. These issues are made worse by language barriers and the fear of being deported.

Agricultural workers' legal protections differ greatly from nation to nation, and enforcement is frequently lax, allowing for widespread labor rights violations. The situation is made more difficult by the complexity of global supply chains, which can mask the realities of working conditions due to consumer demand. Nonetheless, a growing trend toward progress is being propelled by workers' rights advocacy, corporate social responsibility, and fair trade initiatives. Governments, organizations, and consumers must work together to address these problems to build a more sustainable and equitable agricultural system that upholds the rights and dignity of all workers.

2.3.8. The Impact of Low Wages and Job Insecurity on Agricultural Workers

Agricultural workers are heavily impacted by low wages and job insecurity, which leads to a cycle of poverty and vulnerability that compromises their well-being and means of subsistence. Many agricultural laborers, particularly those in developing nations, make wages that are insufficient to cover their essential expenses for housing, food, and medical care. Many people are forced to work multiple jobs or put up with difficult living conditions to make ends meet because of this financial strain, which frequently leads to mental and physical exhaustion. This problem may be made worse by the use of piece-rate pay, in which employees are paid according to the amount of produce they harvest. Although it might encourage productivity, it also results in erratic income, which makes it challenging for employees to save for the future or make financial plans.

The difficulties faced by agricultural workers are exacerbated by job insecurity. There are times of unemployment in between harvests because many are hired on a seasonal or temporary basis. Because of this instability, workers lead precarious lives where they constantly worry that they won't be able to provide for their families during slow times. The situation is even more precarious for migrant workers. They frequently leave their communities and families behind in search of employment, only to return the next season with no guarantee of employment. Their mental health is impacted by this instability as well as their financial status because anxiety and depression can be brought on by the stress of job insecurity.

The ability of workers to advocate for better conditions is further limited by low wages and a lack of job security. Many are afraid that by speaking out against unfair practices, they could lose their jobs, which would reinforce the culture of silence and tolerance for subpar working conditions. This vulnerability prolongs cycles of exploitation by making it challenging to advocate for changes in pay, working conditions, and hours. Furthermore, the effects of poor pay and unstable employment go beyond a single employee.

It has an impact on communities and families, resulting in limited social mobility, difficult schooling for kids, and poor health outcomes. The sustainability of the agricultural sector as a whole is threatened when workers are forced into low-paying, unstable positions. In addition to being vital for workers' well-being, ensuring fair pay and job security is also necessary to build a more resilient and equitable agricultural system that can better support economic stability and food security. Governments, organizations, and consumers must work together to address these problems to support ethical labor practices and defend the rights of agricultural workers (Moore, 2015; Akram-Lodhi et al., 2021).

REFERENCES

1. Akram-Lodhi, A. H., Dietz, K., Engels, B., & McKay, B. M. (2021). *Handbook of Critical Agrarian Studies.* Edward Elgar Publishing.
2. Avena, N. M., Rada, P., & Hoebel, B. G. (2008). Evidence for sugar addiction: Behavioral and neurochemical effects of intermittent, excessive sugar intake. *Neuroscience and Biobehavioral Reviews, 32*(1), 20–39.
3. Banaji, J. (1980). Summary of selected parts of Kautsky's *the Agrarian Question.* In H. Wolpe (Ed.), *The Articulation of Modes of Production* (pp. 45–93). Routledge.
4. Bauer, R. (2019). *The Peasant Production of Opium in Nineteenth-Century India.* Brill.
5. Beckert, S., Bosma, U., Schneider, M., & Vanhaute, E. (2021). Commodity frontiers and the transformation of the global countryside: A research agenda. *Journal of Global History, 16*(3), 435–450.
6. Bosma, U., & Hofmeester, K. (2018). *The Life Work of a Labor Historian: Essays in Honor of Marcel van der Linden.* Brill.
7. Chayanov, A. V. (1966). On the theory of non-capitalist economic systems. In D. Thorner, B. H. Kerblay, & R. E. F. Smith (Eds.), *A. Chayanov: The Theory of Peasant Economy* (pp. 1–28). AEA.
8. Chesney, T., Evans, K., Gold, S., & Trautrims, A. (2019). Understanding labor exploitation in the Spanish agricultural sector using an agent-based approach. *Journal of Cleaner Production, 214,* 696–704.
9. Claeys, P., Desmarais, A. A., & Singh, J. (2021). Food sovereignty, food security and the right to food. In A. H. Akram-Lodhi, K. Dietz, B. Engels, & B. M. McKay (Eds.), *Handbook of Critical Agrarian Studies* (pp. 238–249). Edward Elgar Publishing.
10. Corrado, A. (2017). Migrant crop pickers in Italy and Spain. *E-Paper Heinrich Böll Foundation.*
11. Courtwright, D. T. (2001). *Forces of Habit: Drugs and the Making of the Modern World.* Harvard University Press.
12. Domar, E. D. (1970). The causes of slavery or serfdom: A hypothesis. *The Journal of Economic History, 30*(1), 18–32.
13. Friedmann, H. (1980). Household production and the national economy: Concepts for the analysis of agrarian formations. *The Journal of Peasant Studies, 7*(2), 158–184.
14. Friedmann, H. (2021). Origins of peasant studies. In A. H. Akram-Lodhi, K. Dietz, B. Engels, & B. M. McKay (Eds.), *Handbook of Critical Agrarian Studies* (pp. 15–24). Edward Elgar Publishing.

15. Hofmeester, K., & van der Linden, M. (2018). *Handbook Global History of Work.* De Gruyter Oldenbourg.
16. Hopkins, T. K., & Wallerstein, I. (2000). Commodity chains in the world-economy before 1800. In I. Wallerstein (Ed.), *The Essential Wallerstein* (pp. 221–233).
17. Lucassen, J. (2021). *The Story of Work: A New History of Humankind.* Yale University Press.
18. McMichael, P. (2021). Food regimes. In A. H. Akram-Lodhi, K. Dietz, B. Engels, & B. M. McKay (Eds.), *Handbook of Critical Agrarian Studies* (pp. 218–231). Edward Elgar Publishing.
19. McMichael, P. (2005). Global development and the corporate food regime. *New Directions in the Sociology of Global Development, 11,* 265–299.
20. McMichael, P. (2014). Historicizing food sovereignty. *The Journal of Peasant Studies, 41*(6), 933–957.
21. Moore, J. W. (2015). *Capitalism in the Web of Life: Ecology and the Accumulation of Capital.* Verso.
22. Nieboer, H. J. (1900). *Slavery as an Industrial System: Ethnological Researches.* M. Nijhoff.
23. Patnaik, U. (2021). An alternative perspective on the agrarian question in Europe and in the developing countries. In A. H. Akram-Lodhi, K. Dietz, B. Engels, & B. M. McKay (Eds.), *Handbook of Critical Agrarian Studies* (pp. 45–51). Edward Elgar Publishing.
24. Pye, O. (2021). Agrarian Marxism and the proletariat: A palm oil manifesto. *The Journal of Peasant Studies, 48*(4), 807–826.
25. Saunders, K. (1984). Introduction. In K. Saunders (Ed.), *Indentured Labor in the British Empire, 1834–1920* (pp. i–vii). Croom Helm.
26. Stanziani, A. (2013). Introduction: Labor, coercion, and economic growth in Eurasia, seventeenth–early twentieth centuries. In A. Stanziani (Ed.), *Labor, Coercion, and Economic Growth in Eurasia, 17th–20th centuries* (pp. 1–26). Brill.
27. Topik, S. (2009). Historicizing commodity chains: Five hundred years of the global coffee commodity chain. In J. Bair (Ed.), *Frontiers of Commodity Chain Research* (pp. 37–62). Stanford University Press.
28. Van Schendel, W. (2018). Beyond labor history's comfort zone? In U. Bosma, K. Hofmeester, & M. van der Linden (Eds.), *The Life Work of a Labor Historian: Essays in Honor of Marcel van der Linden* (pp. 174–208). Brill.

29. Vanhaute, E. (2018). Agriculture. In K. Hofmeester & M. van der Linden (Eds.), *Handbook Global History of Work* (pp. 229–247). De Gruyter Oldenburg.
30. Wallerstein, I. M., Aguirre Rojas, C. A., & Lemert, C. C. (2012). *Uncertain Worlds: World-Systems Analysis in Changing Times.* Paradigm Publishers.
31. Wallerstein, I. M. (1989). *The Second Era of Great Expansion of the Capitalist World-Economy, 1730–1840s.* Academic Press.

CHAPTER 3

CHILD LABOR IN AGRICULTURE

Contents

INTRODUCTION

Around the world, the agricultural sector accounts for 70% of child labor. In particular, 112 million boys and girls work as children today in forestry, fisheries, aquaculture, livestock, and crop production—many of which entail long hours and dangerous working conditions. To address child labor in agriculture, governments, civil society, and the private sector must use all of their resources, even though many of the problems can be linked to underlying causes like poverty, sociocultural factors, and a lack of awareness about the problem.

3.1. CHILD LABOR DEFINITION

A common definition of child labor is work that is unsuitable for a child's age, interferes with their education, or has the potential to negatively impact their safety, health, or morals. This type of work negatively impacts children's health or impedes their education, growth, and future employment opportunities (Figure 3.1).

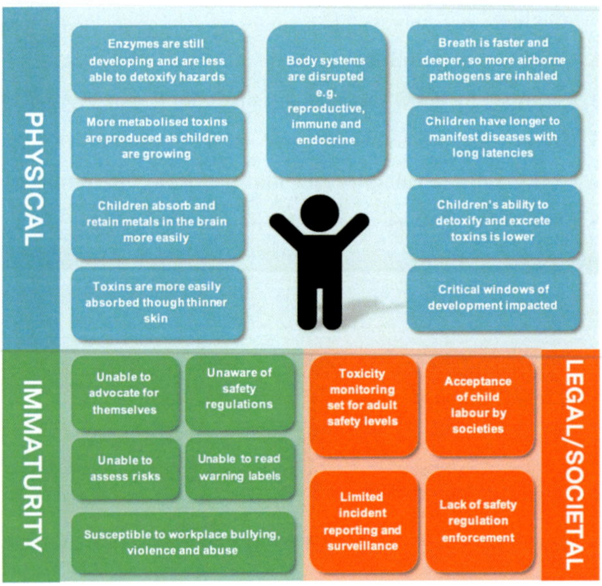

Figure 3.1. *Definition of child labor.*

Source: Scott NB, Pocock NS. The Health Impacts of Hazardous Chemical Exposures among Child Labourers in Low- and Middle-Income Countries. International Journal of Environmental Research and Public Health. *2021; 18(10):5496. https://doi.org/10.3390/ijerph18105496.*

In addition to dangerous work, the worst types of child labor include forced or compulsory labor, including the recruitment of children for use in armed conflict, the sale and trafficking of children, debt bondage, serfdom, and the use, procurement, or offering of a child for prostitution or pornographic performances or materials. These most severe types endanger a child's physical, mental, or moral health due to their inherent characteristics or the circumstances surrounding their execution. (Basu, K., 2000).

A child's health, safety, or morals could be harmed by hazardous work. This work is hazardous or takes place in unhygienic conditions that could cause a child to be killed, hurt, or become ill due to inadequate health and safety regulations and working conditions. Permanent disability may be the result of certain illnesses or injuries. Frequently, health issues brought on by child labor may not manifest or develop until the child reaches adulthood. It is important to identify hazardous work at the national level.

The 14–17-year-old overlap group falls into both the youth (15–24) and child (0–17) age groups. Children in the overlap group are considered to be of legal working age, which is set at 14 in certain developing nations and 15 or 16 in the majority of them. They may be younger workers engaged in youth employment or child labor. The risk of the jobs those younger employees perform is the decisive factor. When a child in this age range sprays dangerous pesticides, they are engaging in child labor. A younger worker would be a child in this age range using biopesticides safely (Figure 3.2).

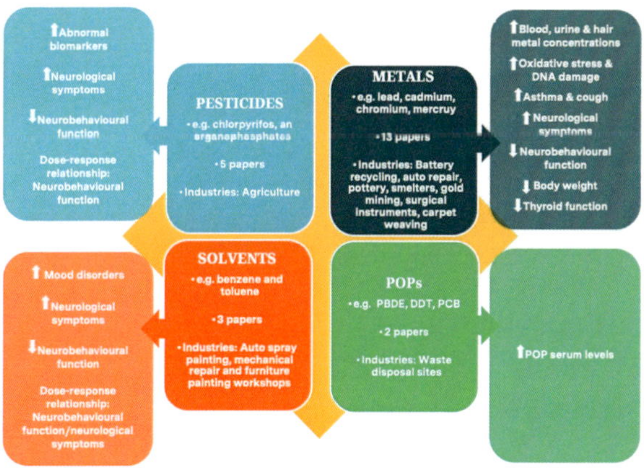

Figure 3.2. *The effects of child labor.*

Source: Scott NB, Pocock NS. The Health Impacts of Hazardous Chemical Exposures among Child Labourers in Low- and Middle-Income Countries. International Journal of Environmental Research and Public Health. *2021; 18(10):5496. https://doi.org/10.3390/ijerph18105496.*

Following its mandate, FAO's work in this area focuses on child labor in the agricultural sector, including family-based agriculture, and pays close attention to the relationship between poverty and child labor. Not every task performed by kids in the agricultural industry qualifies as child labor. Certain activities can guarantee knowledge transfer between generations, help children develop critical life and agricultural skills for the future, and support their family's livelihood.

However, knowing what constitutes child labor and what does not is essential to effectively tackle the issue, as agriculture is one of the three most hazardous industries to work in at any age in terms of work-related illnesses, accidents, and fatalities (Figure 3.3).

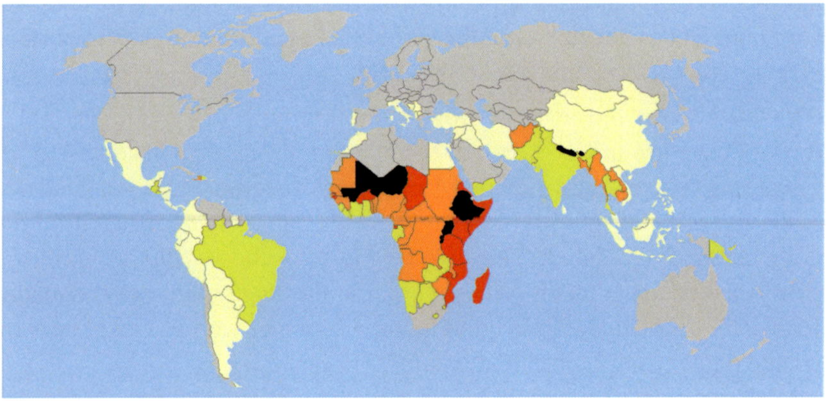

Figure 3.3. *Child labor statistics.*

Source: By historicair 19:58, 5 December 2006 (UTC) - World Bank World Development Indicators 2005 data series: Labour force, children 5–14 (% of age group, SL.TLF.CHLD.ZS), CC BY-SA 3.0, https://commons.wikimedia.org/w/index.php?curid=1435048.

The line between permissible activities and child labor to be eliminated therefore depends on the child's age, the activities performed (and related hazards) and (non-) interference with compulsory education:

- As long as it doesn't endanger their health and safety or interfere with their education or training, children over the age of 13 (or 12, in developing nations) are allowed to perform light labor. Although few nations have done so, national governments are supposed to decide what constitutes appropriate light work at the local level. Two significant universal provisions for light work are the requirement to work during daylight hours and a weekly threshold of 14 hours.

- As long as it doesn't endanger their health and safety or interfere with their education or training, children over the age of 15 (in developing nations, this can be 14) or younger workers are permitted to work full-time.
- It is forbidden for children under the age of 18 to perform hazardous jobs or work in unsanitary or dangerous environments where there is a risk of disease, injury, or death. This is referred to as dangerous labor.

Developing nations that have ratified the 1973 Minimum Age Convention can choose to set a higher age or, in extreme circumstances, one year below the norm. For instance, considering the unique difficulties of the fishing industry, the ILO suggests that the minimum age for employment in fishing vessels be 16.

3.1.1. Characteristics of Child Labor in Agriculture and Rural Areas

The characteristics of agriculture as an economic sector and occupation greatly influence child labor in this field, which is driven by both push (supply side) and pull (demand side) factors.

Some features of agricultural work may present additional challenges in controlling the way it is carried out, especially among children. It may be:

- **Seasonal**: Agricultural labor demand depends on production cycles and seasonal movements of animals;
- **Informal**: Many economic activities in agriculture are, by law or practice, not registered or insufficiently covered by formal regulations;
- **Hazardous:** Agricultural work often involves the use of hazardous tools or equipment, and toxic substances such as pesticides, and agricultural workers can also be exposed to extreme weather conditions, rough seas, biological hazards, long/irregular working hours, or carrying heavy loads;
- **Under-Regulated:** Workplaces can be in remote locations or fragmented, or concealed for illegal purposes, making law enforcement difficult

Since the majority of children work as unpaid family laborers in scattered small-scale farms or rural businesses or are purposefully concealed by employers thanks to the limited reach of labor inspectors in rural areas, child labor in agriculture is frequently invisible (FAO, IFAD, and ILO). Children involved in child labor are frequently referred to as "helpers," helping their parents on the family farm, on fishing boats, in plantations, in mountainous regions, and when herding livestock—although their jobs may be similar to those of adults and just as demanding—data on girls' and boys' labor in agriculture are scarce and not included in traditional surveys. Additionally, the majority of national surveys continue to ignore household duties, which leaves out the children's "double burden," which is typically carried by girls. Another feature is that kids as young as 6 or 8 years old start working very early. The distribution of household chores and gender factors are important factors that influence why, how, and under what circumstances children work. Additionally, the effects of child labor are strongly gendered, which perpetuates patterns of gender discrimination in agriculture. Another aspect of girl child labor in agriculture is that girls who marry young are no longer regarded as children, regardless of their age.

Another factor that contributes to child labor is the seasonality of agricultural work. Due to the informal or irregular status of seasonal migrant workers, seasonal work sites may be located far from schools and other services, like daycare centers. Additionally, seasonal admission may be difficult or impossible. To repay debts incurred during the migration process, workers may be compensated based on their output or may require the labor of every family member. As a result, kids may be taken to the fields to help their parents work, where they are subjected to many of the same risks as their parents. The risk of child labor in their places of origin may be further increased by challenges they face when reentering the educational system. Children who stay at home while adult family members migrate run the risk of being hired to do farm work in their place, especially if the family incurred debt to finance the migration or isn't getting any money or goods from the migrant. Although migration can offer families and their kids new opportunities (such as higher incomes and easier access to education), it can also significantly increase the likelihood of child labor, especially for kids whose parents are seasonal agricultural migrants.

Economic vulnerability and household poverty are significant motivators for child labor. To meet basic needs and cope with uncertainty, impoverished households without access to credit may find it more difficult to delay their children's employment and educational investments. Instead, they may turn to child labor. Furthermore, rural families might not have as much access to insurance and financial markets, which would limit their ability to maintain or raise productivity and, consequently, affect their reliance on child labor. In this situation, kids can also actively look for employment opportunities to become independent, get a taste of adulthood, make money, and get away from parental supervision or academic obligations. In rural areas, households may have little incentive to send their children to school due to limited employment opportunities or a lack of access to basic, meaningful, high-quality education and skills training.

Uncertainties, risks, and adverse shocks that impact households can lower incomes and lead to children quitting school and working to support their families, which further pushes children into child labor. Disasters related to the economy, agriculture, environment, and climate, such as drought, crop failure, or flooding, can have an impact on agriculture and cause significant fluctuations in income, such as when an adult family member loses their job, or health-related shocks, such as the rise in diseases brought on by climate change, e.g., like a serious illness or an injury sustained at work, diseases like dengue or malaria can force children to work to provide for their families. Another push factor is conflicts.

Furthermore, child labor in agriculture is significantly influenced by cultural, social, and demographic factors in rural areas. Children's participation in agriculture, for example, may be seen by rural families as a cultural practice and a means of "helping out" and education. In addition, there are instances where children raised by wealthier members of the extended family are expected to perform household chores and work in the fields in return for an education or to provide younger siblings with an education. Parents or other caregivers might not be aware of the negative effects of particular situations and activities, such as missing school. To comprehend and value local knowledge, attitudes, and practices, as well as to raise awareness of detrimental situations and encourage workable solutions, it is crucial to create a space for local dialogue and discussion (Edmonds and Schady, 2012).

3.2. RATIONALE FOR FAO ENGAGEMENT ON CHILD LABOR IN AGRICULTURE

Although there are many factors that contribute to child labor in agriculture, household poverty is one of the main causes. On the other hand, child labor prolongs the cycle of future poverty by depriving children of education and an opportunity for healthy development. Since improving household economic conditions and agricultural productivity can increase families' income security and lessen reliance on child labor or additional labor, reducing rural poverty—one of FAO's core mandates—is the first step towards effective child labor prevention strategies.

A sustainable food system provides nutrition and food security for everyone without compromising the social, economic, or environmental foundations needed to provide nutrition and food security for future generations. Sustainable food systems cannot be realized through the use of child labor. More specifically, FAO's comparative advantage resides in the following points:

- Mandate, which covers programs that are essential to ending child labor in agriculture, especially those that lower poverty in rural areas, promote more inclusive and effective food systems, and make livelihoods more resilient to threats and emergencies. As the UN specialized agency with intergovernmental status in food and agriculture, FAO has the full scope of its mandate, authority, and ability to work on any aspect of agriculture on a global scale.

- Technical expertise in agriculture is essential for helping nations create and carry out effective and sustainable policies, programs, and measures. For instance, this includes promoting agricultural technologies and practices that lessen or eliminate labor demands and work-related risks. FAO employs interdisciplinary staff members with a wide range of expertise in its mandate areas. Supporting partners, Member States, and other international, national, and local development actors to gain access to knowledge, information, training resources, best practices, and services in capacity building within its mandate areas is something FAO has done for a long time.
- Privileged connections with national governments and ministries involved in agriculture, among other agricultural stakeholders who are crucial to the fight against child labor. In accordance with development priorities and the needs and demands of agricultural stakeholders, FAO has supported the updating and revision process of relevant legislation, strengthened the effectiveness of public institutions, and facilitated stakeholder participation and processes to develop proposals for designing national agricultural strategy and policy.
- A long history of uniting and promoting communication and cooperation between various actors, including governments, producer groups, extension agents, civil society organizations, and research institutions. To promote policy dialogue, encourage agreement negotiation, and bring together a diverse range of stakeholders to discuss policy and reach consensus, FAO possesses networking capacity at the national, regional, and international levels.

3.2.1. Addressing Child Labor in Agriculture to Reduce Poverty

Child labor is a multifaceted problem that affects all of FAO's strategic initiatives to eradicate hunger and poverty, especially in rural areas where child labor is common.

3.2.1.1. How Poverty and Child Labor Are Interconnected

It is challenging for many low-income rural households to achieve sustainable livelihoods and to meet their immediate basic needs. The effects of climate change exacerbate this condition because they only make rural communities that depend on healthy ecosystems for their subsistence even more vulnerable. They believe that the only viable option is to use children to augment or replace adult labor. Children are frequently pulled from school to work in agriculture, or their academic performance deteriorates as a result. If they stay in agriculture, they have a lower chance of finding respectable employment as adults, are less likely to innovate and embrace new methods and technologies, and are more likely to become impoverished and experience the long-term consequences of the dangerous circumstances they endured as children. Communities suffer as a result, and the poverty cycle is maintained.

3.2.1.2. Breaking the Vicious Cycle

Prevention and reduction are key to ending child labor in agriculture. Enabling rural children to receive training, education, and healthy development can be one way to prevent this. When they become young and adult producers or workers, this gives them more economic potential. If they stay in agriculture, they are also more likely to embrace new, sustainable practices and technologies. Young people with skills are better able to negotiate better working conditions for waged employees or to sell their goods as young business owners. Consequently, this will boost agricultural productivity and the availability, caliber, and ability of a trained labor force, resulting in higher wages and improved food and nutrition security.

Child labor can be decreased by agricultural, food security, and nutrition programs that seek to improve livelihoods, lower poverty, and boost productivity. This means that from the planning stage to the monitoring and assessment stage, those programs must incorporate concerns about child labor.

3.2.2. Ending Child Labor: A Prerequisite for Sustainable Food Systems

As stated in SDG Target 2.4, ensuring resilient agricultural practices and sustainable, productive food systems are essential components of the 2030 Agenda. The growth of small-scale food producers' output and incomes is also connected to SDG 2.

A sustainable food system is one that provides nutrition and food security for all while maintaining the social, economic, and environmental foundations necessary to provide nutrition and food security for future generations (Fors, 2012). This means that:

- It is profitable throughout (economic sustainability);
- It has broad-based benefits for society (social sustainability); and

- It has a positive or neutral impact on the natural environment (environmental sustainability).

Considering children as a unique group, how can food systems be beneficial to them as part of social sustainability? Children's food security and wholesome diets should be taken into account by sustainable food systems. Simultaneously, sustainable food systems should address the struggles and stresses that food producers and other system members face, which affects economic sustainability, to eradicate child labor in agriculture and food production.

Child labor and productivity in family and small-scale farming have a complicated relationship. In the absence of mechanization and an accessible or reasonably priced labor force, practitioners might contend that children's temporary participation in food production is necessary to achieve a minimum level of production. Another presumption is that the coping strategy of using child labor would automatically stop once a certain level of productivity is attained, with brief and limited detrimental effects on the child's ability to find respectable employment and earn enough money as a young person and adult. Evidence, however, suggests otherwise. The implementation of strategies to reach productivity levels by investing in knowledge, technologies, and innovative practices, as well as by diversifying income sources and avoiding the use of child labor, are the focus of sustainable food systems and efforts to increase productivity to feed the world and ensure a respectable income for food producers.

FAO has led efforts to promote sustainable agriculture and food production for many years. Five guidelines for sustainable food and agriculture were established in a 2014 FAO report. "Agriculture that fails to protect and improve rural livelihoods, equity, and social well-being is unsustainable," states Principle 3, which upholds human systems. "Farmers, pastoralists, fishermen, and foresters have decent employment conditions and work," according to the report's vision. Only if child labor in agriculture is addressed will this be feasible. Some FAO documents that serve as guidelines for the creation and execution of FAO programs already incorporate the prevention of child labor.

For the millions of impoverished households in developing nations, the creation of sustainable food value chains may provide a crucial means of escaping poverty. The principles that support the creation of sustainable food value chains (SFVCs) are examined in the FAO Guiding Principles for Developing SFVCs. Evaluating value chain performance from the standpoint of the triple bottom line—economic, social, and environmental sustainability—relates to the first three principles. Regarding social sustainability, the improved value chain model should produce more value and not have any effects that are deemed unacceptable by society, like child labor or unhygienic working conditions. The Principles for Responsible Investment in Agriculture and Food Systems (CFS-RAI), created by the Committee on World Food Security (CFS), are applicable to agricultural investments of all shapes and sizes. To guarantee that agricultural investments support sustainable development, including food security and nutrition as well as the eradication of poverty, the principles offer a framework of guidelines for public and private stakeholders involved in investment promotion, regulation, planning, and implementation. Increased responsible investment in agriculture and food systems is necessary to address the four aspects of food security and nutrition: availability, access, stability, and utilization. When applicable, principle 2 encourages the efficient application of additional international labor standards, with a special focus on those that pertain to the agri-food industry and the eradication of the most severe types of child labor.

A comprehensive worldwide framework for evaluating sustainability along food and agriculture value chains is offered by the Sustainability Assessment of Food and Agriculture Systems (SAFA) guidelines. SAFA's overarching goal is for food and agriculture systems around the world to be distinguished by four aspects of sustainability: social well-being, economic resilience, environmental integrity, and good governance. Here, the absence of child labor is a component of well-being.

The Rotterdam Convention, the World Health Organization's (WHO) and FAO's International Code of Conduct on Pesticide Management, the FAO Voluntary Guidelines for Securing Sustainable Small-Scale Fisheries in the Context of Food Security and Poverty Eradication, the Organization for Economic Co-operation and Development (OECD), and the FAO Guidance for Responsible Agricultural Supply Chains all incorporate the significance of preventing child labor through particular means. Additionally, the CFS is creating voluntary guidelines for food systems and nutrition with the goal of encouraging sustainable practices.

3.3. DRIVERS OF CHILD LABOR IN AGRICULTURAL COMMUNITIES

Throughout the world, child labor in agricultural communities is still a major problem, especially in developing countries. The agricultural industry remains one of the most difficult sectors to eradicate child labor, despite international efforts to do so through legislative frameworks, educational programs, and social interventions. Poverty, educational disparities, family dynamics, social norms, and lax regulatory frameworks are some of the many complex factors contributing to the continued prevalence of child labor in agricultural communities. These factors will be thoroughly examined in this essay to comprehend the interrelated reasons that continue to support child labor in agriculture as well as possible solutions to these problems.

3.3.1. Poverty as the Primary Driver

Poverty stands as the most significant driver of child labor in agricultural communities. In many rural areas, families rely on agriculture as their primary source of income, and the economic contributions of children are seen as necessary for the family's survival.

The cyclical nature of poverty, wherein generations are trapped in low-income jobs with little access to education or resources, perpetuates child labor. The wages earned by children working in the fields can make a crucial difference in family income, helping to meet basic needs like food, shelter, and clothing.

In many cases, children work alongside their parents to maximize household earnings, particularly during harvest periods, when labor demands peak. This dynamic creates a situation where the value of a child's labor outweighs the potential long-term benefits of their education, as the immediate need for income is more pressing than future opportunities. Children often face the choice of working or seeing their families go without essential resources. As a result, despite legal restrictions, many families prioritize short-term economic survival over compliance with child labor laws.

3.3.2. Economic Pressures in the Agricultural Sector

In addition to poverty, broader economic pressures in the agricultural sector contribute to the reliance on child labor. The agricultural industry, especially in developing countries, is marked by low wages, high labor demands, and fluctuating prices for crops.

Farmers and laborers often face financial uncertainty, as their incomes depend on factors beyond their control, such as weather conditions, market prices, and crop yields. This economic volatility can exacerbate the need for additional labor, leading families to involve their children in farm work as a cost-saving measure.

Moreover, smallholder farmers, who make up a significant portion of agricultural producers, frequently operate on narrow profit margins. The inability to hire adult workers due to limited financial resources further encourages the use of child labor. The relatively low cost of employing children compared to adults, combined with the informal nature of labor arrangements in agriculture, makes it an attractive option for struggling farmers trying to maintain their livelihoods. This dynamic reinforces the cycle of poverty and child labor in rural communities.

3.3.3. Family Dynamics and the Role of Children in Agriculture

In many agricultural communities, child labor is normalized as part of family life. Farming is often a family-based activity, where children are expected to contribute to the household's productivity. This expectation is rooted in cultural traditions where children learn farming techniques from a young age by working alongside their parents. For some families, this practice is seen as essential for transmitting agricultural knowledge and skills to the next generation.

Family farms, particularly in rural and isolated regions, blur the lines between labor and family duties. Children may not be seen as workers in the formal sense but rather as helpers contributing to the family business. This mindset complicates efforts to enforce child labor laws, as parents may argue that their children are learning valuable life skills or helping to ensure the family's survival. In these cases, the distinction between educational opportunities and labor is less clear, as children might not be formally employed but are still engaged in labor-intensive activities.

While some forms of child participation in family farming can be relatively harmless and even beneficial in terms of skill acquisition, the line is crossed when this labor becomes exploitative, hazardous, or interferes with the child's education and overall development. For example, children working long hours in physically demanding tasks, operating dangerous machinery, or being exposed to harmful pesticides face significant health risks and are deprived of educational opportunities that could improve their future prospects (Kielland and Tovo, 2006).

3.3.4. Gaps in Education Access and Quality.

Access to quality education is a critical factor in determining whether a child is likely to be involved in labor activities. In many rural agricultural communities, schools are either absent, under-resourced, or too far away for children to attend regularly. The lack of infrastructure, trained teachers, and teaching materials exacerbates the problem, leading to low enrollment and high dropout rates. Without access to quality education, children, and their families often see little value in prioritizing schooling over work, as the immediate economic benefits of child labor seem more tangible than the long-term, uncertain benefits of education.

Furthermore, seasonal agricultural work creates an additional challenge for education. Children may attend school during the off-season but are required to work during planting or harvesting times. This irregular attendance disrupts their learning, making it difficult for them to keep up with their peers and leading many to eventually drop out altogether. The agricultural calendar often takes precedence over the school calendar in rural communities, with children being pulled out of school to meet the labor demands of the farm.

In cases where schooling is available, the quality of education may be so poor that it offers little incentive for children to continue attending. Overcrowded classrooms, outdated curricula, and a lack of career prospects for graduates contribute to the perception that education is not a viable path out of poverty. As a result, families may choose to engage their children in labor activities instead of sending them to schools that seem unlikely to provide a meaningful future.

3.3.5. Cultural Norms and Social Acceptance of Child Labor

Cultural norms and traditions in agricultural communities often reinforce the acceptance of child labor. In many societies, it is common for children to contribute to the family's work from a young age, and this participation is seen as an integral part of their upbringing. In these contexts, child labor is not viewed as exploitation but rather as a way for children to learn responsibility, contribute to the household, and prepare for adulthood. This cultural acceptance makes it challenging to implement laws and policies aimed at reducing child labor, as families and communities may resist outside interventions that conflict with long-standing traditions.

In some cases, community members may believe that children are better off learning practical skills through work than pursuing formal education. This belief is particularly strong in areas where the quality of education is poor or where there are limited employment opportunities for educated individuals. For many families, the economic realities of agricultural life mean that labor is a more immediate and practical solution than investing in an uncertain future through education.

Moreover, traditional gender roles can also influence the types of labor children are expected to perform. In many agricultural communities, boys may be more likely to engage in physically demanding tasks such as planting or harvesting, while girls may be tasked with caring for younger siblings or managing household duties. These gendered expectations further entrench the division of labor within families and contribute to the perpetuation of child labor across generations.

3.3.6. Weak Regulatory Frameworks and Enforcement

Weak regulatory frameworks and poor enforcement of existing laws are significant factors that contribute to the persistence of child labor in agriculture. While many countries have ratified international conventions aimed at eliminating child labor, the reality on the ground often falls short of these commitments. Agricultural work, particularly in rural and remote areas, is difficult to regulate due to its decentralized nature, the informal economy, and the fact that many farms operate without formal labor contracts.

Labor inspectors are often limited in their capacity to monitor agricultural work, as farms are spread over vast areas, making it logistically challenging to enforce child labor laws. Even when laws exist to protect children, enforcement is often weak, and penalties for violations may be minimal or non-existent. This lack of oversight creates an environment where child labor can thrive, as there are few consequences for those who employ children.

Furthermore, in many countries, agricultural work is subject to different labor laws than other sectors. Children may be allowed to work at younger ages and for longer hours in agriculture than in other industries, reflecting the historical and cultural significance of farming in many societies. These legal exceptions undermine efforts to combat child labor and create a legal gray area where children can be exploited without recourse.

3.3.7. Global Supply Chains and Consumer Demand

The demand for cheap agricultural products on the global market also plays a role in driving child labor. Agricultural communities, particularly in developing countries, are often integrated into global supply chains where the pressure to produce crops at low costs is immense. Large agribusinesses and multinational corporations may indirectly benefit from child labor, as smallholder farmers seek to cut costs and meet production quotas.

Consumers in wealthier nations may be unaware of the conditions under which their food is produced, contributing to a system that perpetuates child labor. While some companies have made efforts to address labor issues in their supply chains, such as through certification programs like Fair Trade or initiatives that promote ethical sourcing, these measures are not always effective in reaching the most vulnerable workers, including children.

3.3.8. The Seasonal Nature of Agricultural Work

Agriculture is highly seasonal, with peaks in labor demand during planting and harvesting periods. This seasonal nature of agricultural work is a key driver of child labor, as families often require additional help during these busy times. Children may be called upon to work long hours during harvest season, often at the expense of their education and well-being.

Seasonal labor patterns also contribute to the irregularity of employment for adult workers, which can create an economic need for children to step in and fill gaps in the labor force. In some cases, entire families may migrate to different regions during harvest periods, and children are expected to work alongside their parents to make ends meet. This migratory lifestyle further disrupts children's education, as they may be unable to attend school regularly or at all.

3.4. FAO GUIDING PRINCIPLES ON CHILD LABOR IN AGRICULTURE

The four guiding principles of FAO's approach to child labor in agriculture—inclusion, sustainability, integrated approach, and collaboration—should direct planning and operations in this field.

3.4.1. Inclusiveness

3.4.1.1. Mainstreaming

Child labor prevention is a cross-cutting issue in the design and planning stages, as well as in monitoring and evaluation systems. This is implied by the inclusiveness principle, which requires that FAO's interventions and project implementation systematically include the issue of child labor from the beginning during assessments and field studies to identify potential child labor situations.

3.4.1.2. Directing Attention also to Overlooked Situations

While child labor in small-scale production, including in local or regional value chains, is still largely ignored despite its prevalence, the majority of financial resources currently allocated to combating child labor are focused on addressing child labor in global value chains (cocoa, coffee, cotton) and their upper tiers. It is crucial to expand resources for eradicating child labor in all contexts, including local and regional value chains and subsistence farming, rather than concentrating only on a few global supply chains. This entails keeping an eye on child labor in all agricultural subsectors, particularly in areas where little is known about its prevalence and features.

3.4.1.3. Age Sensitivity

Being inclusive also entails considering various age groups. Youth who are older than the legal working age but under the age of 18 and who are not employed, enrolled in school, or undergoing training, for instance, require special consideration. Their own social and economic rights depend on them being in good youth employment (as younger workers), continuing high-quality education, or a vocational training/apprenticeship program. Given the unique characteristics of children who have attained legal working age (typically 14 or

15 years), it is crucial to guarantee coherence between child labor reduction and youth employment interventions (Nielsen, 2002).

3.4.1.4. Gender Sensitivity

Most social systems treat boys and girls differently, which means that they perform different tasks and encounter different risks (FAO, IFAD, and ILO). For example, boys are more likely to engage in risky agricultural tasks like operating machinery, herding animals in isolated areas, catching fish on ships, and diving to untangle nets.

On the other hand, girls are frequently expected to assist with household chores. Their workload frequently combines both reproductive and productive activities, and it may expose them to longer workdays as well as other risks unique to their jobs, like carrying heavy loads when gathering fuel or water, breathing in smoke from cooking stoves or fish-smoking kilns, as well as sexual harassment and gender-based violence. The local farming system, poverty levels, demographics, and social norms are some of the variables that affect gender disparities in child labor. It is important to consider less visible forms of child labor, such as domestic work frequently assigned to girls. Tasks that are considered household chores (e.g., water and firewood collection) are very often linked with farm work, blurring the division between girls' productive and reproductive work.

3.4.1.5. Recognition of Indigenous Identity

Children's traditional labor is seen by indigenous communities as crucial to the preservation of indigenous groups' cultural identities and the transfer of knowledge and skills.

To guarantee that FAO makes every effort to respect, incorporate, and advance indigenous issues in its work, the FAO Policy on Indigenous and Tribal Peoples was developed. Regarding child labor, it's critical to distinguish between indigenous customs of assigning children to specific tasks and what exactly qualifies as child labor. Indigenous children are more likely to experience exploitation, sexual abuse, and other types of maltreatment in situations where their communities are marginalized.

3.4.1.6. Recognition of the Voice of Children

Children should be given the chance to participate in focus groups to help design programs and to be taken into account when creating questionnaires about household task distribution in agricultural settings.

3.4.2. Sustainability

FAO has created a shared vision and a comprehensive strategy for sustainability in all agricultural subsectors. To guarantee sustainability, this viewpoint—which forms the cornerstone of FAO's action on the ground—must fairly address social, economic, and environmental factors. The achievement of sustainability in the other areas is put at risk when one area is neglected.

The common vision is built around five principles which provide a basis for developing national policies, strategies, programs, regulations, and incentives that will guide the transition to an agriculture that is highly productive, economically viable, environmentally sound and based on the following principles of equity and social justice:

- Improving efficiency in the use of resources;
- Conserving, protecting, and enhancing natural ecosystems;
- Protecting and improving rural livelihoods and social well-being. Enhancing the resilience of people, communities, and ecosystems, addressing the changes induced by climate change; and
- Promoting good governance of both natural and human systems.

According to these tenets, ending child labor in agriculture is a crucial part of social sustainability and a requirement for the accomplishment of sustainable food and agricultural initiatives. A do-no-harm strategy is necessary but insufficient. It is crucial to adopt a deliberate strategy to help eradicate child labor in agriculture and include particular activities and indicators to address the issue, particularly for interventions carried out in areas or value chains where child labor is common.

As part of a responsible and sustainable approach, FAO has created many guidance documents that specifically acknowledge the necessity of adhering to international core labor standards and offer guidance on a variety of environmental, social, and economic issues, such as the abolition of child labor and the advancement of decent employment.

3.4.3. Integrated Approach

Child labor is a complex issue that calls for an all-encompassing, integrated strategy. This entails involving various stakeholders and putting various interventions into place to address the supply and demand sides of child labor. Enhancing the safety and health of families and children in farm and agricultural environments, facilitating access to high-quality education, easing the transition from school to the workplace, and helping families increase their income and access to social protection are all examples of integrated approaches. To incorporate child labor prevention into policies and programs, this strategy also calls for increasing agricultural stakeholders' awareness and strengthening their capacity. Child labor issues should be mainstreamed into national and local development strategies, particularly those aimed at reducing poverty, according to the 2010 Roadmap for Achieving the Elimination of the Worst Forms of Child Labor by 2016. Education, social protection, labor market policy, national legislation and enforcement, and sustainable and rural development are the areas that the Roadmap highlighted as supporting an integrated approach to preventing and reducing child labor in agriculture.

FAO has created and been implementing an Integrated Country Approach as part of the Decent Rural Employment work program to assist nations in embracing and putting into practice more employment-centered and youth-inclusive agri-food system development policies, strategies, and programs. Along with the chance to improve coherence and synergies between policies meant to address child labor and decent youth employment, the approach offers another avenue for countries to support their efforts to address child labor in agriculture. In policies and interventions aimed at the diverse group of youth and rural youth, it is especially crucial to address the needs of children between the ages of 14 and 17.

3.4.4. Collaboration

Fostering cooperation between divisions and between headquarters and decentralized offices within FAO. In addition to helping regional and national office staff deal with child labor issues in their work when communicating with traditional FAO stakeholders, including agricultural line ministries, labor-related stakeholders, other agencies, and resource partners, the goal is to improve coherence and synergies throughout the Organization. Taking into account the Cooperation Framework's priorities, the Country Programming Framework (CPF) elaboration can serve as the best procedure and means of communication between the country office and the host government. Through the use of this crucial planning and programming tool, FAO can offer technical assistance to integrate child labor concerns and improve the CPF's sustainability and accountability.

At the national level: promoting coordination and communication between ministers. The Ministry of Labor, which organizes and leads steering committees and other mechanisms, is primarily in charge of child labor policy and action in the majority of nations. However, to improve the coherence of policies and programs, FAO collaborates with the ILO to promote interministerial dialogue, particularly between the Ministry of Labor and Agriculture (and other subsectors), as well as between other pertinent ministries like Social Development and Education, at the national and decentralized levels.

The cooperation of ministries outside of agriculture and labor is necessary to effectively combat child labor (e.g., health, education, and social protection). This can be accomplished by incorporating child labor into the Cooperation Framework, as the goal of ending child labor is consistent with the framework's tenets (i.e., leaving no one behind, a development, sustainability, and accountability strategy grounded in human rights. Establishing a formal contract (letter of agreement or memorandum of understanding) or using national coordination arrangements (focal points, technical working groups, committees) may be necessary for cooperation between agencies and institutions. To guarantee coherence in policies and interventions, these formal procedures are crucial for facilitating discussions, exchanging pertinent information, and identifying connections.

Working together with producer organizations (POs) and extension and advisory services (EAS) providers, utilizing their capillary presence in communities and on the frontlines, where they regularly engage with farmers, fishermen, and livestock raisers. Crucially, EAS providers encompass not only public extension but also a broad spectrum of non-state and informal actors, including development projects, producer groups, agri-inputs dealers, non-governmental organizations (NGOs), and farmer leaders. Because of this pluralism, they offer a vast array of pertinent services about production as well as the advancement of safer and labor-saving technologies and the enhancement of living standards. Additionally, diverse EAS providers can contribute to data collection, raise awareness of the negative impacts of child labor, and give researchers and policymakers feedback on the situation of child labor on the ground if they are properly capacitated and sensitized. Because they can be crucial in uniting diverse stakeholders and promoting cross-sectoral cooperation at the national and local levels to eradicate and minimize child labor in agriculture, EAS providers are also referred to as "bridging institutions." Therefore, FAO strives to increase the ability of both public and private EAS providers, such as POs, to prevent child labor by exchanging information on risks and identifying substitute methods.

Together with small and medium-sized businesses engaged in the food and agricultural sectors, The FAO is working to end child labor, promote social dialogue with agricultural workers, and create working conditions that respect their fundamental rights. Child labor in agriculture can be effectively addressed by promoting inclusive and fair business practices, transparent, and equitable transactions, and supply chain tracking initiatives.

At the international level, involving various UN agencies and other players helps achieve SDG Target 8.7 and end child labor. Through the IPCCLA, FAO collaborates with IUF and the UN system's ILO and IFAD. FAO promotes the inclusion of child labor in agricultural issues in international frameworks and backs the Alliance 8.7. Furthermore, FAO and UNICEF have been working together for a long time to study the effects of cash transfer programs. More recently, FAO and IFAD have extended their partnership to include the prevention of child labor in agriculture in the development of their investment programs.

3.5. FAO AREAS OF WORK ON CHILD LABOR IN AGRICULTURE

Programs related to agriculture, rural development, food security, and nutrition are crucial in tackling the underlying causes of child labor. FAO assists nations in improving the standard of living for impoverished rural residents, smallholders, and family farmers, especially by boosting their productivity and diversifying their sources of income. This helps to lessen the reliance of these families on child labor.

FAO's standard outreach and rural education programs, including farmer field schools (FFS), nutrition field schools, farm business schools, and junior farmer field and life schools (JFFLS), can incorporate important ideas and messages regarding child labor.

3.5.1. Promoting the Adoption of Safe Practices in Agriculture

Children are particularly susceptible to dangerous jobs and other risks because of their unique physical, cognitive, behavioral, and emotional development. Approximately 48% of children globally engage in hazardous jobs that could compromise their morals, safety, or health. 42% of these children are girls, and nearly half are in the youngest age group (5–11 years). Furthermore, the agricultural industry employs nearly half of all children between the ages of 15 and 17 who work in dangerous jobs .

Long workdays, heavy lifting, repetitive motions, exposure to high temperatures, hazardous chemicals, hazardous tools and equipment, and mistreatment or harassment are common risks in all agricultural subsectors. There are additional risks unique to forestry, livestock, fishing, and farming. Family-based agricultural labor can be just as dangerous as work done outside the home because children often perform tasks that adults and caregivers do. Children and adults face similar risks depending on the type of agricultural production, but children are more at risk from those risks (Swaminathan, 1998).

3.5.1.1. Approaches

FAO aims to increase awareness of children's exposure to hazardous work, promote safer agricultural practices and sustainable agricultural production, and encourage risk management and alternative practices. Entry points and approaches include:

- Creating educational resources for kids about dangerous jobs and strategies to lessen risks (e.g., pesticides). Visual aids, films, or educational gaming applications are examples of materials, e.g., a mobile application created in Lebanon that examines the dangers and risks connected to various agricultural environments, including greenhouses, field crops, and horticulture. The creation of higher education and vocational training curricula that incorporate occupational safety and human health issues, with a focus on children, offers a rare chance to better prepare the next generation of professionals to handle child labor, hazardous work, and decent work in their future roles.
- Implementing occupational safety and health (OSH) measures that address working children's exposure to identified hazards and increasing the knowledge and ability of agricultural stakeholders at the national and local levels regarding health and safety risk assessments to identify work hazards for children and young workers.
- National capacity-building workshops, training programs for plant protection officers and agricultural extension officers (including private extension services offered by buyers or development and education services offered by NGOs), and the inclusion of child labor issues in extension guidelines, agribusiness curricula, or FFS curricula are some ways to achieve this. Raising awareness at the community level can be accomplished through a variety of means, including Dimitra Clubs, public, and village meetings, radio, and television programs, participatory assessments, and Theatre for Development.
- Encouraging the addition of agricultural risks to national hazardous work lists for children. All nations that have ratified the 1999 Worst Forms of Child Labor Convention are obliged to provide these lists. After consulting with the representatives of employers and employees, they include sectors and dangerous work activities that are off-limits to minors. FAO can assist in determining which hazards related to agriculture should be on the lists and in facilitating the participation of agricultural stakeholders in the consultation procedure. The different EAS providers who regularly collaborate with producers at the national level can help gather data on child activities and associated risks and disseminate it as part of this legislative endeavor.

- Promoting cooperation between labor and agricultural stakeholders to find complementary measures to address children's exposure to agricultural work hazards, particularly in rural areas and the informal economy where labor inspectors are typically less active. This covers programs like the Clear Cotton Project and FAO interventions within the framework of the IPCCLA.
- Encouraging alternative methods and providing a supportive legal framework for environmental and sustainable agriculture issues that lower risks to health and safety while working in agriculture. FAO supports an integrated pest management (IPM) program that integrates various management techniques to produce healthy crops and support natural pest control mechanisms to lessen occupational hazards associated with pesticide use. FAO serves as the Secretariat for the Rotterdam Convention, which helps its parties regulate the import and export of specific dangerous chemicals and pesticides, in collaboration with the UN Environment Programme. To strengthen national stakeholders' capacity for pesticide management, it gathers information on vulnerable groups, distributes it through comparable and diverse digital channels, and offers technical assistance and capacity-building training. Through its efforts, policymakers are made aware of the dangers farmers face when using pesticides, allowing for the implementation of regulations. It raises awareness of these dangers even further on a worldwide scale.

3.5.2. Fostering Sustainable Labor-Saving Technologies and Practices

Labor-saving techniques and technologies can alleviate labor shortages, cut down on the amount of time and effort required to complete particular tasks, and boost output. Labor-saving technologies and practices can increase labor productivity and help families become less dependent on their children's labor in situations where family labor still accounts for the majority of farm labor. Additionally, they can offer safer working conditions for young people and children who assist on the farm. The adoption of new technologies, particularly for post-harvest activities, can boost family incomes and livelihoods while relieving children of household chores. It should be noted that different technologies and practices may have different effects on the workloads of men and women, as well as boys and girls, depending on the division of labor and the particular tasks assigned to each, as well as social norms. Adoption of labor-saving technologies and practices, as well as the formulation of needs related to them, should be accompanied by strategies to overcome certain social norms and behaviors that are deeply embedded in local contexts.

3.5.2.1. Approaches

To decrease the need for child inputs, FAO seeks to increase agricultural productivity and encourage the adoption of suitable labor-saving technologies and practices. The gender aspect of agriculture must be taken into consideration when designing and implementing interventions in a socially responsible manner. Entry points and approaches include:

- Examining labor-saving technologies for tasks that children typically perform. For instance, children are frequently involved in weeding in many nations. Mechanical or manual weeders are examples of technologies that save time and effort when weeding. Mulching (e.g., composed of plastic or crop waste) or cover crops are also effective methods of weed control and soil water retention. Numerous labor-saving technologies are available for use with various cropping systems. These include portable cutters, reapers, rice transplanters, and drum seeders in rice production systems. Training on how to use and maintain labor-saving technologies must go hand in hand with their promotion. For these technologies to be sustainable over time, support services like repair shops and spare parts must be accessible where they are being promoted.
- Encouraging the delivery of mechanization services (e.g., by individual farmers) within rural areas. As an agribusiness opportunity to boost revenue and create jobs while tackling the problem of child labor, mechanization can assist in addressing significant labor supply bottlenecks and responding to labor demand peaks.
- Enhancing accessibility and cutting down on the amount of time needed for water collection. Given that girls and women typically fill this role, there may be several benefits associated with child labor, particularly for girls. Child labor in water fetching can be decreased by water harvesting at the household level. Women can devote their time to economically productive activities that boost household income and lessen the pressure on children to work (and/or perform other household chores, which again frees up girls' time for school). Girls also spend less time collecting water and have more time to attend school.
- Putting money into and spreading labor-saving postharvest technologies. For instance, lowering the amount of labor needed to store, process, and transform agricultural products can help reduce child labor in agriculture and the food chain. A methodical evaluation of this indirect effect would be required (e.g., through food loss analysis), to pinpoint important loss points that need to be fixed to improve livelihoods and revenue production.

- Lowering the workload for women. Most women in developing nations must balance economically productive employment with childcare responsibilities and child care. Children may choose to help out in the field or take care of their siblings when their mothers must work away from home rather than attend school. In general, labor-saving and mechanization technologies can increase worker productivity while lessening the monotony of field operations. The applications of these technologies are not limited to on-farm activities (e.g., planting, moving, and harvesting), as well as for post-harvest and processing tasks that are generally performed by women (e.g., utilizing mechanical shellers to shill maize). When paired with a gender-transformative strategy, such as the Gender Action Learning System and the Dimitra Clubs, such interventions can result in behavioral changes at the household level.
- Monitoring the disruptive impact (including unintended consequences) of labor-saving technologies. While labor-saving technologies or practices reduce the demand for child labor in one particular area of activity, the child may not necessarily attend school instead. The supply and demand for child labor might be transferred to other areas of agricultural activity, unless awareness raising about the benefits of education takes place and possibilities to easily access quality education are fostered.

3.5.3. Including Agriculture, Food, and Nutrition in School Curricula in Rural Areas

Child labor and the available measures of school quality are correlated. In addition to enrollment and indirect costs, poor quality and limited relevance of education, as well as a lack of amenities (such as safe restrooms), all contribute to children leaving school and engaging in child labor. Including or enhancing agricultural and nutrition-related subjects in school curricula can help kids learn more about these subjects, become more environmentally conscious, and alter their perspective on farming.

3.5.3.1. Approaches

FAO acknowledges the critical role that schools can play in helping member nations combat poverty, hunger, and illiteracy. Studies conducted globally demonstrate that incorporating food and nutrition into curricula as early as primary school increases students' interest in agriculture and the food industry. To increase school attendance by making education more applicable to local lifestyles, FAO collaborates with governments to incorporate agriculture, food security, and nutrition education into the curriculum. There is proof that providing meals for school-age children increases attendance rates, thereby upholding their right to an education. Entry points and approaches include:

- **Promoting Practical Learning:** Children can learn about growing crops and livestock by learning about gardens and farms. By incorporating knowledge and skills related to agriculture, livelihoods, and nutrition into the curriculum and encouraging active learning, they can make education more relevant for kids. Additionally, they give kids practical experience in food production and natural resource management, which inspires them to come up with ideas for their own farms and gardens when they get home. Learning gardens at school can give kids a more varied meal made with fresh ingredients when combined with nutrition education. Families and communities must be involved in the design of the school's learning garden. Monitoring their effects is necessary to reduce the possibility of child labor. Lastly, children's perceptions and knowledge of agriculture and job opportunities along value chains can be altered by incorporating or further emphasizing agricultural topics into school curricula.
- **Supporting Local and Inclusive Procurement:** To give local farmers a regular market (schools) for their produce, FAO created the HomeGrown School Feeding Resource Framework. Child labor is recognized by the framework as a hindrance to education. This strategy can support community socioeconomic development, strengthen local agriculture, and open up business opportunities for smallholder farmers and other vulnerable producers, such as women, young people, and members of traditional communities. Including child labor indicators (e.g., attendance and retention) in interventions incorporating this strategy is essential for comprehending its effects.

3.5.4. Empowering and Building the Skills of Youth Aged 15–17

It is extremely difficult for young people in rural areas of developing nations to get ready for and find respectable employment, including in agriculture. For young people under the age of 18, these difficulties are even more severe. When it comes to how young people will move from school to the workforce and their chances of escaping poverty, this stage of life is usually pivotal.

Frequently, there exists a disparity in the legislative framework between the mandatory education age and the legal working age, particularly in rural areas. Adolescents below 18 may fall into the trap of child labor if their work is perilous. These young individuals encounter additional challenges in securing decent employment due to insufficient skills, weak bargaining power, limited knowledge about labor laws, scanty access to information about labor regulations, and the presence of legal obstacles and discrepancies in labor policies and regulations. Consequently, they become more susceptible to exploitation. They often find themselves in the informal sector, working without a formal contract, earning low wages, or enduring unsafe working conditions.

One of the biggest legal obstacles to entering into contracts and using financial services is being younger than 18. Social and cultural norms, such as early marriage and pregnancy, present additional obstacles for girls. They often restrict their interests to traditional skills or give up on their education early. Additionally, the 15–17 age group is typically left out of government policy and program design. Because of their emphasis on entrepreneurship, the majority of youth-related programs and interventions currently in place target the 18–30 age group. Programs for youth employment and child labor prevention frequently overlook the 15–17 age group.

The numbers of children in the worst forms of child labor can be dramatically reduced by putting in place effective protection measures for 14–17-year-olds who work in agriculture. Strategies include both measures specifically targeting children and those more broadly improving health and safety for all workers:

- Promoting and advocating for the national hazardous job registers. As an illustration, in light of a study on child labor, the Government of Côte d'Ivoire revised its national hazardous work list and established a catalog of less hazardous tasks (suitable jobs) in various agricultural sectors for children between 13 and 16 years old, which was approved in 2017.
- Putting in place OSH measures that address exposure of working children to work hazards.
- Enhancing working conditions and arrangements, such as keeping children away from specific work hazards.

3.5.4.1. Approaches

FAO encourages young people to find decent work in rural areas, acknowledging that young people are a diverse group and that those between the ages of 15 and 17 require extra care to keep them from engaging in risky jobs and to make it easier for them to access activities that will help them develop their skills.

Entry points and approaches include:

1. **Implementing Junior Farmer Field and Life Schools (JFFLS):** The JFFLS methodology, which integrates business, life, and

agricultural skills, was created by FAO. JFFLS's goal is to improve the livelihoods and long-term food security of vulnerable children and youth so they can have a better future. The targeted audience and local needs are taken into consideration when determining the specific content of a given JFFLS training course. In cooperation with the ILO, a specialized training manual on the prevention of child labor in agriculture has been created. It includes exercises and information for incorporating child labor prevention into JFFLS curricula and as a component of the overall modular approach.

2. **Integrating Youth Aged 15–17 in Decent Rural Employment Initiatives Targeting Youth:** The goal of FAO's Integrated Country Approach for Boosting Decent Jobs for Youth in the Agri-Food System is to assist nations in embracing and putting into practice more youth-inclusive policies, strategies, and programs related to the development of the agri-food system.
3. **Supporting Youth Employment in Agriculture:** It including financial inclusion for youth. For initiatives to be successful and for no one to be left behind, assessments, research, and interventions must take into consideration the unique needs and limitations of the various age groups of youth. For young people in rural areas, financial inclusion can be seen as a means of fostering self-worth and confidence in addition to resilience and economic independence. However, the minimum age to open a bank account is 18 in many sub-Saharan nations, and access to financial services and financial literacy would be extremely beneficial for young people under 18 who have already begun working. In the context of youth employment in agriculture, interventions that seek to increase financial inclusion should take into account how the needs of various age groups could be satisfied.
4. **Adopting a Life Cycle Approach:** Strategies that address child labor and youth employment must be closely related in a life cycle approach because they are points on a continuum. Child labor lowers the likelihood of finding respectable work as a young person and as an adult, and it has a negative impact on long-term health, education, and the development of higher-level skills. The cycle of poverty, illiteracy, and child labor is perpetuated when former child laborers become adults and depend more on the labor of their own children to cover basic household expenses. To address the changing needs of children and young adults, programs that prevent child labor and encourage youth employment must be created with a life cycle approach in mind.

3.5.5. Addressing Child Labor in Crop Production

Children's pesticide exposure is a significant problem in farming. Since children's bodies, brains, and nervous systems are still developing, pesticide poisoning is particularly dangerous to them. Children are still developing, so they have fewer natural defenses and exposure to pesticides can cause major health issues. For instance, children's bodies can absorb more chemicals due to their larger pores, and because their organs are still developing, exposure to toxins may be more harmful to them. Compared to adults, children are usually more susceptible to the risks associated with pesticides for both biological and behavioral reasons. It is not appropriate for children under the age of 18 to handle dangerous pesticides directly. Children can also be exposed to other health and safety risks, such as injuries related to sharp tools and machinery, back pains from long working hours (e.g., weeding), heavy work during land preparation, falls in unguarded wells and asphyxiation from working in grain solos or pits.

3.5.5.1. Approaches

To increase productivity, FAO places a high priority on promoting safe and sustainable farming methods while also increasing awareness of the need to protect children from pesticides and other OSH concerns. Entry points and approaches include:

- **Advocating for the Need to Protect Children From Pesticides:** FAO collaborates with partners to reduce children's exposure to pesticides by considering their unique vulnerability in international conventions and codes. For instance, children are identified as one of the groups most at risk of pesticide exposure in the FAO-WHO International Code of Conduct on Pesticide Management. A minimum international standard for observing the standards of governments and the pesticide industry is established by the voluntary Code of Conduct. The only way to successfully lower children's exposure to pesticides is to implement targeted pesticide management initiatives that are concurrently implemented by governments, pesticide control boards, researchers, producer organizations, and all other stakeholders. Raising awareness of this risk within the framework of the Rotterdam Convention improved comprehension of the problem and the necessity of group effort.
- **Raising Awareness at the Community Level:** It is important to raise awareness of child labor among farmers and communities. This can be carried out, for example, through rural radio programs, farmer listeners' clubs (e.g., Dimitra Clubs) and farmer field schools.

Producer groups and extension agents can be trained to advise farmers on how to lower health and safety risks, and they have a significant role to play in this regard. Extension agents, local facilitators, and leaders of farmer groups and organizations, for instance, received training in Ghana on labor-saving technologies, child labor, OSH, and the dangers of hazardous pesticides. FAO has developed a series of knowledge materials on protecting children from pesticides:

- JFFLS visuals, posters, and facilitator's guide (FAO, 2010).
- e-learning course "Pesticide management and child labor prevention."
- Facilitators' visual guide, Protect children from pesticides! – with the technical and financial support of the Rotterdam Convention – used successfully in different settings to raise awareness and promote alternatives: Uganda and Guinee-Bissau.
- Short visual story about the dangers of pesticides for younger, potentially illiterate children. Developed in Lebanon, it targets especially Syrian refugee children.
- **Promoting Safe and Sustainable Farming Practices:** Farmers should be encouraged to use IPM to control pests to reduce the use of pesticides and minimize health and environmental risks. This will help to reduce worker exposure to pesticides and, in turn, contribute to the reduction of hazardous child labor. For instance, in the Bla region of Mali, where FAO established a field school program in 2003, a community of cotton growers has all but eliminated the use of toxic pesticides thanks to field schools that teach farmers alternative pest control techniques. By lowering labor needs, using fewer pesticides, and increasing farmer incomes, other farming methods supported by conservation agriculture and agro-ecology have the potential to reduce child labor. Without the use of extra agrochemicals, the FAO conservation agriculture program in the United Republic of Tanzania reported an increase in average maize yields from 1 tonne/ha to 6 tonnes/ha. Additionally, farmers have been able to expand into commercial crops, which has improved food security by increasing income generation and providing a more balanced diet for their families. During the busiest planting and plowing seasons, farmers in both nations reported a notable decrease in labor demand.
- **Combining Technologies and Approaches:** Using less input (seeds, water, chemical fertilizers, and pesticides), the System of Rice Intensification (SRI) can boost rice productivity. The outcome

is higher yields and more input savings. Although SRI cuts down on the amount of time needed for transplanting and pesticide application, it may lengthen the time needed for weeding. Weeding is frequently done by children, so this may have an impact on the demand for child labor. To minimize weeding time and guarantee the greatest possible benefits for the household, SRI can be combined with other methods or technologies, such as basic mechanical weeders. Other technologies that save labor (e.g., specific planting and weeding equipment) or methods (e.g., mulching) can reduce the quantity and kind of labor that children perform, saving farmers' time and energy. Technologies for processing on a small scale (e.g., peanut shellers) may also enable kids to spend more time in school and less time working. Combining strategies can boost small-scale agriculture productivity and open doors to advanced value chain positions or diversification of revenue-generating ventures, all while increasing community awareness of child labor.

3.5.6. Addressing Child Labor in Capture Fisheries and Aquaculture

Among the most hazardous jobs in the world is fishing. Because fishers, fish farmers, and fishing and aquaculture communities are frequently impoverished and vulnerable, child labor is pervasive in the industry globally. Children are sometimes hired as workers and a source of inexpensive labor because of the depletion of fisheries resources.

Fishing, pre-trip planning, post-harvest activities (processing and marketing), feeding, protecting, and harvesting fish in ponds and cages, boat building, net making, and mending are just a few of the many jobs that children perform in the fisheries and aquaculture industry. Boys are typically more involved in fishing and girls in post-harvest activities, which is consistent with the gender roles that adults play in fisheries and aquaculture. Both types of occupational hazards exist. Compared to adults, children are more vulnerable to safety and health risks due to their developmental stage and skill deficiencies.

3.5.6.1. Approaches

The FAO seeks to establish a sustainable aquaculture and fisheries industry that takes social, economic, and environmental factors into consideration. By increasing knowledge on the topic, educating people about the dangerous work in the industry, encouraging the inclusion of child labor considerations in international instruments and guidelines, and assisting in adherence to these standards and guidelines, we can prevent child and forced labor and contribute to decent employment in fisheries and aquaculture for all. Entry points and approaches include:

- **Ensuring Child Labor Issues are Taken into Account in International Fisheries Instruments:** Certain FAO guidelines about fisheries and aquaculture include considerations for decent work and child labor. For instance, "child labor should not be used in a manner inconsistent with ILO Conventions and international standards," according to the FAO 2011 Technical Guidelines on Aquaculture Certification, which were created to guide the creation, administration, and execution of reliable aquaculture certification programs. Child labor eradication is specifically mentioned in the chapter on social development, employment, and decent work of the Voluntary Guidelines for Securing Sustainable Small-Scale Fisheries in the Context of Food Security and Poverty Eradication, which were approved by the FAO Committee on Fisheries in 2014. To enhance living and working conditions at sea, the FAO and the ILO collaborated to create the 2007 Work in Fishing Convention. The Convention lays out the minimal requirements that must be met and sets standards for decent work on fishing vessels. The Convention's standards can also be used to stop the industry's unacceptable forms of labor, such as child labor and forced labor. To improve the ability and coordination of labor and fisheries authorities at the national level in the implementation of this Convention, cooperation between the FAO and the ILO is crucial. The Joint FAO/ILO/IMO Ad Hoc Working Group on Illegal, Unreported, and Unregulated Fishing and Related Matters was joined by the ILO in 2020. In 2015, the Subcommittee on Aquaculture of the FAO Committee on Fisheries examined issues related to decent work in aquaculture, specifically child labor. To address poor working conditions in the seafood sector, new international guidelines are being developed.
- **Integrating Child Labor Issues in Fisheries Policies:** FAO assists governments at the national level in incorporating concerns about child labor into their national fisheries policies.
- **Protecting Children From Hazardous Work Through Risk Assessments and Appropriate National Hazardous Work Lists:** Child labor situations can be converted into respectable employment opportunities by enhancing OSH in fisheries and aquaculture and lowering the risks faced by young workers. FAO carries out risk assessments with an emphasis on the vulnerabilities of children and promotes OSH in fisheries and aquaculture at the national level. Risk assessments can help governments create or update their

national hazardous work lists and can help prevent children from being exposed to dangerous jobs.

- **Promoting Technologies and Practices to Reduce the Sector's Demand for Child Labor:** Implementing methods and technologies that do away with the need for child labor can help lower the demand for child labor. This could involve, for instance, better community infrastructure in terms of roads, transportation, landing sites, and water supplies to prevent carrying heavy loads. In the informal economy, it is especially crucial for communities to interact with organizations that represent fishermen, fish farmers, employers, and other sectoral institutions.
- **Building Knowledge on Child Labor in Fisheries and Aquaculture:** To comprehend the causes and effects of child labor in fisheries and aquaculture, more quantitative and qualitative data are required. Cost-effective ways to improve basic information include:

a. Adapting and integrating aspects of child labor in fisheries and aquaculture into standard household and living standard measurement surveys through the introduction of sector modules or specific questions; and

b. Ensuring that pertinent surveys have adequate data disaggregation (all data should be disaggregated by profession/occupation, age, and sex).

To support policy coherence, cross-sectoral capacity development is crucial. Fisheries and aquaculture policies and programs should address child labor concerns, and child labor strategies should take these sectors' unique characteristics into account.

- **Strengthening Fisheries Management and Governance to Offer Long-Term Solutions:** The main causes of fishermen cutting labor costs and searching for free or inexpensive labor, including child labor, are low return on labor and declining income as a result of overfishing and overcapacity. Child labor, poverty, and the depletion of fisheries resources are all intertwined. Environmental and legal violations in fisheries are also caused by declining revenues, weak management, and the inability of fisheries administrations to enforce the law. An effective approach to addressing some of the underlying causes of child labor and its most severe manifestations, including dangerous labor and illicit, unreported, and unregulated fishing, is through the management and governance of fisheries. Strong fisheries management can support the sector's economic viability and further mitigate the underlying causes of child labor, especially when combined with strategies to increase value addition, address waste and loss in fisheries value chains, and facilitate market access.

3.5.7. Addressing Child Labor in Livestock

Children are frequently involved in livestock keeping, which has deep historical and cultural roots in developing nations. At a very young age, both boys and girls usually begin herding and taking care of livestock. In addition to herding, children assist with feeding and cleaning animals and sheds, gathering water and fodder, handling draught animals, and processing. Although there is a dearth of information on the scope and regions of occurrence, reports from around the world indicate that children work in slaughterhouses, dairy production, poultry, and other small livestock (on farms and in their homestead), and other meat processing operations. The working conditions for kids who herd animals vary widely depending on the situation. Youngsters may herd for their families or for an employer. It seems that child herders who work outside the home are most at risk of being taken advantage of and subjected to verbal or physical abuse by their employers. When children are compelled to make up for lost cattle and damaged crops, they may even incur debt. Situations in which children are trafficked to engage in herding both inside and outside of national borders or are ensnared in bonded labor are especially concerning.

Compared to adults, children are more susceptible to the occupational risks associated with handling livestock. Long workdays in harsh weather, inadequate sanitation and hygiene, handling manure can contaminate the environment, and using chemicals can all lead to health issues(i.e., veterinary medications and disinfectants), exposure to hazardous animal handling, physically, and biologically dangerous slaughterhouse operations, inhaling livestock dust, and the spread of animal-to-human diseases (zoonoses). Additionally, handling sharp objects and animals carries a significant risk of injury. While herding, children may be attacked by wild animals, get bitten, kicked, or stamped on, or suffer from musculoskeletal conditions. Those who lead oxen to plow the fields—mostly boys, but also girls—run the risk of getting gored. To lessen the need for children to walk in front of them, some communities in Mali have trained older oxen to do the plowing. Long-term seclusion, fear of cattle raiders and employer retribution, or an overwhelming sense of duty for the family's assets can all lead to psychological stress, which poses a serious risk to children's wellbeing. Since herding activities keep kids busy for a significant amount of the day, they are typically challenging to integrate with education. Without adapted services, migration may lead to new challenges. Some herders have never attended school, and many have dropped out. Children are less likely to return to school after they leave.

3.5.7.1. Approaches

Through its continued work on livestock, FAO hopes to increase knowledge and awareness about child labor in the livestock industry and assist governments and partners in tackling the problem. Entry points and approaches include:

- Supporting research projects on the subject and assisting national governments in creating important indicators linked to child labor that should be incorporated into national surveys and censuses to help close the knowledge gap on child labor in the livestock industry. It is difficult to determine the exact level of child labor in livestock because there is a dearth of data broken down by subsector. Designing effective policies and programs to address child labor in agriculture, including livestock, is made much more difficult by this informational gap. Information on child labor in the livestock industry could be gathered and relayed by EAS providers collaborating with herders.
- By connecting FAO's efforts to improve the health and welfare of working animals (transport and traction) with child labor, we can raise awareness, influence behavioral changes, and strengthen capacity on OSH at the national and local levels with pertinent stakeholders.
- Primary (and secondary) school curricula that are pertinent to the rural and agricultural context, and pastoralist societies, and can support decent rural employment should be developed and implemented in cooperation with government agencies and partners. For both poverty alleviation and sustainable livestock development, the connections between education and livestock are essential. If the efficiency of livestock production systems is improved (e.g., through improved livestock practices and management), the children of poor, livestock-dependent households can be freed from child labor and gain better access to education (FAO, 2020).

REFERENCES

1. Alston, P. (2005). *Child Labor in Developing Countries: Policy Implications for the Global South.* Oxford University Press.
2. Basu, K. (2000). The International Labor Organization's convention on the worst forms of child labor. *The World Economy, 23*(6), 827–843.
3. FAO. (2020). *FAO Framework on Ending Child Labor in Agriculture.* Rome.
4. Basu, K., & Van, P. H. (1998). The economics of child labor. *American Economic Review, 88*(3), 412–427.
5. Bourdillon, M., Levison, D., Myers, W. E., & White, B. (2010). *Rights and Wrongs of Children's Work.* Rutgers University Press.
6. Brown, D., Deardorff, A., & Stern, R. (2002). The determinants of child labor. *World Development, 30*(5), 849–864.
7. Edmonds, E. V., & Schady, N. (2012). Poverty alleviation and child labor. *American Economic Journal: Economic Policy, 4*(4), 100–124.
8. Edmonds, E. V. (2007). Understanding child labor in agricultural settings. In P. Pinstrup-Andersen & F. Cheng (Eds.), *Agricultural Development and Nutrition.*
9. Edmonds, E. V., & Pavcnik, N. (2005). Child labor in the global economy. *The Journal of Economic Perspectives, 19*(1), 199–220.
10. FAO & ILO. (2018). *Child Labor in Agriculture: A Review of the Evidence and Policy Response.* Rome: FAO; Geneva: ILO.
11. Fors, H. C. (2012). Child labor: A review of recent theory and evidence with policy implications. *Journal of Economic Surveys, 26*(4), 570–593.
12. Garcia, B., & Chamlian, M. (2010). Child labor and agricultural productivity: An empirical study from Brazil. *Journal of Agricultural Economics, 61*(3), 543–560.
13. Hazell, P., & Haggblade, S. (2005). Child labor and the transformation of the agricultural economy. *Development Policy Review, 23*(2), 157–172.
14. Kielland, A., & Tovo, M. (2006). *Children at Work: Child Labor Practices in Agriculture in West Africa.* Social Protection Discussion Paper No. 0615, World Bank.
15. Kis-Katos, K., & Sparrow, R. (2011). Child labor and trade liberalization in Indonesia. *Journal of Human Resources, 46*(4), 722–749.
16. Maher, S. (2005). Children's rights and child labor in agricultural settings. *Children's Rights Quarterly, 27*(2), 56–74.

17. Morrow, V., & Vennam, U. (2012). Children's work in agriculture in rural India: Views from children and adults. *Journal of Agricultural Studies, 12*(1), 15–29.

18. National Bureau of Economic Research (NBER). (2019). *Child Labor and Schooling in Agriculture.* NBER Working Paper No. 24617.

19. Ngajilo, D., & Jeebhay, M. F. (2019). Occupational injuries and diseases in aquaculture: A review of literature. *Aquaculture, 507*, 40–55.

20. Rohlman, D. S., Ismail, A. A., Rasoul, G. A., Bonner, M. R., Hendy, O., Mara, K., Wang, K., & Olson, J. R. (2015). A 10-month prospective study of organophosphorus pesticide exposure and neurobehavioral performance among adolescents in Egypt. *Cortex, 74*, 383–395.

21. Nielsen, M. (2002). Child labor in agriculture in Africa. In P. Leroy (Ed.), *Labor Markets and Agriculture in Africa.*

22. Patrinos, H. A., & Psacharopoulos, G. (1997). Family size, schooling, and child labor in Peru: An empirical analysis. *Journal of Population Economics, 10*, 387–405.

23. Patrinos, H. A., & Psacharopoulos, G. (1997). Family, school, and work: Evidence from child labor in agriculture in Latin America. *Journal of Development Economics, 27*, 35–62.

24. Smillie, I. (2000). The child labor trap: Lessons from Africa. *African Economic History, 28*, 165–177.

25. Swaminathan, M. (1998). Economic development and the persistence of child labor: Evidence from rural India. *World Development, 26*(9), 1513–1528.

CHAPTER 4

GENDER EQUALITY AND WOMEN'S RIGHTS IN AGRICULTURAL WORK

Contents

INTRODUCTION

Gender equality and women's rights in agricultural work are critical issues that intersect with food security, economic development, and social justice. Despite women playing a vital role in agriculture—producing a significant portion of the world's food—they often face systemic barriers that limit their contributions and potential. In many regions, women are responsible for planting, weeding, and harvesting, yet they frequently lack access to essential resources such as land, credit, technology, and training. This inequity not only hinders women's economic empowerment but also undermines agricultural productivity and sustainability.

Cultural norms and discriminatory practices further exacerbate these challenges, often relegating women to subordinate roles and limiting their decision-making power. In many societies, agricultural policies and programs fail to address the specific needs and interests of women, perpetuating cycles of inequality. Consequently, women's unpaid labor remains largely unrecognized, and their contributions to food systems are undervalued.

Advancing gender equality in agriculture requires a multifaceted approach that includes legal reforms, targeted investments, and capacity-building initiatives. Empowering women through access to land rights, financial resources, education, and technology can enhance their productivity and foster greater economic resilience. Moreover, promoting women's participation in agricultural decision-making processes is crucial for developing policies that reflect the needs of all stakeholders.

As the global community seeks to achieve the Sustainable Development Goals (SDGs), particularly Goal 5 on gender equality and Goal 2 on zero hunger, addressing the disparities faced by women in agriculture is paramount. By fostering an inclusive agricultural sector that values and leverages women's contributions, we can pave the way for a more equitable and sustainable future for all (World Bank, 2012, 2015).

4.1. GENDER EQUALITY IN AGRICULTURE

Achieving sustainable development and food security in agriculture requires gender equality. Women are essential to agricultural production, making a substantial contribution to global food systems and rural economies. Women frequently encounter structural obstacles that restrict their access to resources, decision-making authority, and fair treatment in agricultural contexts, despite their crucial contributions. Improving livelihoods and promoting inclusive economic growth require addressing these disparities. The significance of women's contributions to economic progress cannot be overstated. They are the primary force behind the agricultural sector in many developing nations, devoting a significant amount of time to planting, weeding, ridging, and harvesting while juggling their daily responsibilities. However, women are frequently found to be less productive than men in the agricultural sector, regardless of the sub-Saharan African (SSA) nation under study. According to empirical data, women's agricultural productivity deficits vary from 4%–50% worldwide, but they are between 20% and 30% in the SSA region (FAO, 2011; Kilic et al., 2013).

In general, the discriminating factors include: limited access to advisory and extension services; low stocks of human and physical capital; exclusion from credit and financial markets; low application of modern inputs (such as chemical fertilizer, improved seeds, and pesticides); and land constraints (small land size, unequal land tenure or property rights).(Akresh, 2005; Goldstein and Udry, 2008; Alene et al., 2008; Peterman et al., 2011; Vargas Hill and Vigneri, 2011; Kilic et al., 2013; Croppenstedt et al., 2013; Aguilar et al., 2014; Backiny-Yetna and McGee, 2015; Ali et al., 2015).

The majority of SSA nations now understand how important it is to combat gender bias in agriculture to maintain economic growth and guarantee food security. This is especially important in nations where agriculture accounts for the great majority of the population's income. According to recent data, agriculture employs more than 65% of the labor force, contributes roughly 45% of earning sources, and accounts for 30% of GDP on average in SSA countries (World Bank, 2015). There would be significant effects at the household and national levels if SSA nations could fully utilize their agricultural sector.

Although the magnitude of gender productivity gaps in SSA is widely known, studies examining gender bias in agriculture frequently omit an essential component: an evaluation of the possible advantages that would result from reducing or completely eliminating gender inequality in agricultural productivity. This is a significant flaw in terms of policy. Indeed, if policymakers are not persuaded of the true advantages of dedicating national resources to combating agricultural gender bias, it will be especially difficult to persuade them to commit to reducing gender yield gaps. Listing the direct and/or indirect benefits of gender equality in agriculture is insufficient. A qualitative assessment of these effects provides valuable insights, but it does not provide much information about the extent of potential benefits that nations could receive.

In light of this, the objective of this brief is to provide a preliminary evaluation of the anticipated advantages of progressively minimizing or eliminating gender productivity gaps in SSA. This brief builds on the authors' recent study (Mukasa and Salami, 2016) on the causes of gender productivity gaps in SSA nations, which revealed that agricultural lands under the management of female farmers in Nigeria, Tanzania, and Uganda were 18%, 27%, and 30% less productive, respectively, than those under the control of male farmers. The current section focuses on three categories of benefits from gender equality in agriculture: production, consumption, and poverty gains. It does so using the same nationally representative data from the Living Standards Measurement Study – Integrated Surveys on Agriculture (LSMS-ISA). The relative percentage increases in production and consumption levels that result from empowering female managers to be just as productive as male farmers are referred to as production and consumption gains. Closing or gradually eliminating gender productivity gaps in SSA leads to a relative decrease in the number of impoverished households, which is referred to as poverty gains.

4.1.1. Gender Productivity Differentials in SSA

There are two primary lines of research that link gender disparities and agricultural productivity. To explain the observed variations in productivity levels, the first strand focuses on the role that constraints faced by women at various stages of the production process play. The productivity gap is probably explained by women's limited access to productive resources, such as land, contemporary inputs, technology, or financial services, which are essential in determining the level of agricultural productivity.

However, the gender gap nearly always vanishes when authors model equal access to land and other productive inputs (World Bank, 2012; Kilic et al., 2013). For instance, Udry et al. (1995) compared about 4,700 agricultural plots in their study of agricultural productivity in Burkina Faso and discovered that women's yields for vegetables and sorghum were 20% and 40% lower than men's, respectively. However, these significant differences were primarily caused by women's lower use of productive inputs, which they attributed to gender-based social norms. The same patterns of results were found in Ethiopia by Tiruneh et al. (2001), who attributed the 35% productivity gap to limited access to extension services and lower levels of input use. Horrell and Krishnan (2007) also attributed the gender yield disparities in Zimbabwe to a lack of extension services, fertilizer use, and experience. Studies by Saito et al. (1994) for Burkina Faso, Kenya, Nigeria, and Zambia; Alene et al. (2008) for Kenya; Oladeebo and Fajuyigbe (2007) for the Osun State in Nigeria; and Timothy and Adeoti (2006) for the Ondo and Ogun States in Nigeria also yield similar findings and causal explanations.

According to a second body of research, market inefficiencies—specifically in the labor and credit sectors—have a more significant impact on female plot managers' productivity by deterring them from engaging in off-farm activities and limiting their access to inputs. A wedge between the marginal product of labor and the going market wage rate for the same type of work is created by labor market imperfections, and the size of this wedge varies between male and female farmers, according to research on the gender wage gap (Biswanger and Rosenzweig, 1986; Barrett, 1996; Barrett and al, 2008). Palacios-López and López (2015) used data from Malawi to demonstrate that labor market imperfections that affect agricultural productivity accounted for 34% of the estimated 44% gender productivity differences between plots headed by men and females.

The country, sample representativeness, crop selection, unit, and analysis methodology all affect how much gender productivity disparities exist and how significant their drivers are. The use of production/yield function estimates and decomposition techniques to quantify the magnitude of agricultural productivity differences and identify their possible causes has been a commonality among the studies that have already been conducted. In terms of the unit of analysis, the vast majority of empirical research has compared households headed by men and women to find gender differences in yields, a common indicator of agricultural productivity in the literature. The practical impossibility of clearly assigning ownership and responsibility to a single individual based on the majority of available data may help to explain this decision (Croppenstedt et al., 2013). However, the underlying assumptions of this approach—that all household members have the same productive capacity, have the same access to information, and have negligible differences in the quantity and quality of input uses—will ultimately determine whether the conclusions drawn from it are valid (Kilic et al., 2013; Oseni et al., 2013). The second option is to differentiate between plots owned or managed by men and women in the same household by using information at the plot level instead. The benefit of this method is that it can reveal intrahousehold dynamics and the degree of power sharing between household members.

The reality in the agricultural sector is more complex, even though studies employing data from all regions have consistently found a gender wage gap. Actually, while the vast majority of research revealed that female farmers are much less productive than their male counterparts, some studies found no discernible differences between the two groups (Gilbert et al., 2002) or even suggested that women are more technologically proficient than men (Akresh, 2005).

According to the aforementioned studies, if the barriers that women farmers encounter in obtaining resources, land, inputs, and agricultural financing are removed, they may be just as productive as their male counterparts. Therefore, there may be significant benefits to eliminating or at least lessening the gender productivity gap, especially in nations where women own and/or manage a larger proportion of the land. For example, according to FAO estimates, agricultural output in developing nations would increase by an average of 2.5–4% and the number of undernourished people would decrease by 12–17% if women-managed agricultural lands used the same amounts of inputs as plots managed by men (FAO, 2011).

The LSMS-ISA project, which was carried out by each nation's national statistics agency with technical assistance from the World Bank and funded by the Bill and Melinda Gates Foundation, provided the data used in this brief. Given the dearth of recent data on gender disparities from the six SSA nations with LSMS-ISA data currently available (Ethiopia, Malawi, Niger, Nigeria, Tanzania, and Uganda), we use data from Nigeria, Tanzania, and Uganda. A variety of demographic, health, economic, agricultural, and community topics are covered by the nationally representative datasets. The analysis is carried out at the level of the land manager in each nation. The final sample for this brief includes 1,160 parcels in Uganda, 2,530 plots in Tanzania, and 4,017 agricultural plots in Nigeria, totaling 2,029 agricultural plots.

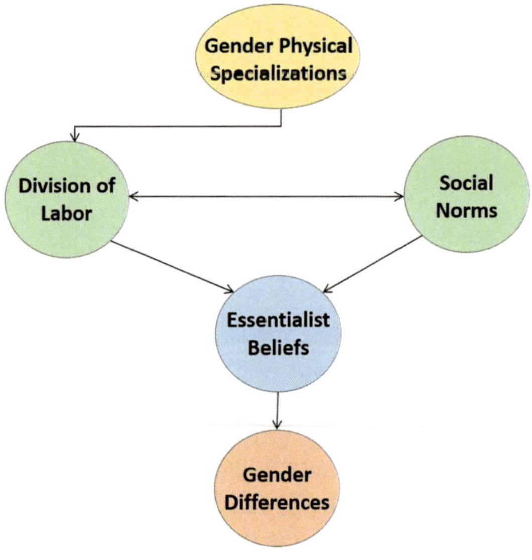

Figure 4.1. *Overview of social-role theories of gender differences.*

Source. Balducci M (2023) Linking gender differences with gender equality: A systematic-narrative literature review of basic skills and personality. Front. Psychol. 14:1105234. doi: 10.3389/fpsyg.2023.1105234.

Using kernel density estimates, Figure 4.1 is an overview of social-role theories of gender differences. Regarding the productivity distribution of lands under female management in Uganda is primarily on the left side of the male distribution, indicating that, on average, female managers perform worse than their male counterparts. After the middle of the productivity distribution, the disparities are especially noticeable. With the exception of the middle, where there are significant differences, Tanzania's male and female managers' productivity distributions almost exactly overlap. Last but not least, in Nigeria, the distribution of male managers is somewhat to the left of that of female managers, indicating lower productivity levels, and somewhat to the right, indicating higher productivity levels. The middle of the distribution shows notable disparities.

4.1.2. Models of Potential Benefits of Gender Productivity Equality

Mukasa and Salami (2016) discovered in a recent study that agricultural lands managed by women from Nigeria, Tanzania, and Uganda consistently yield less than those managed by men, regardless of whether the women were categorized as "average" farmers or moved along the entire productivity distribution. They also found many factors that contributed to these disparities, such as variations in the amount and caliber of land that farmers farmed, as well as variations in the inputs that farmers used and the characteristics of their households. "What if we could reduce or completely close all women's endowment and structural disadvantages in agriculture? How large would be the production and welfare gains if we could achieve gender equality in agriculture?" is the question that immediately jumps out. The greater these potential gains, the more urgent it is to address gender inequality in agriculture.

However, accurate estimates of these potential gains are typically impossible to obtain. First off, a large portion of agricultural production and yield levels may be explained by variables outside of farmers' control, such as weather, pest infestation, or plant diseases. Agricultural production is notoriously unpredictable. Second, the researcher may not be able to observe it, but the development of entirely different managerial and/or agricultural skills by male and female managers is likely to have an impact on levels of yield and production. Lastly, even if productive resources are distributed equally between male and female managers, there may still be a nonzero gender productivity gap because of the agronomic circumstances in which they work. Variations in physiological crop characteristics and local soil properties (such as soil nutrients, water retention rate, and soil nitrogen content) may be related to these agronomic conditions. We can only estimate the anticipated benefits of progressively lowering or doing away with the agricultural productivity deficits of female managers due to these and other factors.

The approximation exercise is predicated on the idea that the primary causes of gender productivity disparities in SSA are observable characteristics of the soil, farm, farmer, and household; unobservable heterogeneity undoubtedly contributes, but for simplicity's sake, its impact is deemed insignificant. We will use the percentage of male-managed agricultural lands, the areas under cultivation, and the estimated yield gaps between female- and male-managed plots/parcels to get reasonable approximations of these potential advantages of gender equality in agriculture (Mukasa and Salami, 2015).

4.1.3. Approximation of Production Gains

Here, "production gains" refers to the relative percentage increases in production levels that occur when all observable barriers to female managers' productivity are gradually removed, allowing them to reach the productivity levels of men.

Let Y_m, Y_f, and Y_T be respectively, current male, female, and total gross values of yields in the sample. Let also Q_m, Q_f, and Q_T denote current male, female, and total gross values of outputs in the sample, respectively. Let finally g be the estimated gender productivity gap in the country, as obtained in Mukasa and Salami (2016); p, the proportion of plots/parcels managed by men; A_T, the total cultivated land in the country; and i (with $0 \le i \le 1$), the percent of the estimated gap that the government would like to reduce.

This means that the following relationships hold:

$$Q_T \equiv Y_T * A_T \quad , \qquad (4.1)$$

$$Y_f = \frac{1}{[1+(1-i)g]} Y_m \quad , \qquad (4.2)$$

$$Y_T = p * Y_m + (1-p) * Y_f \quad . \qquad (4.3)$$

Using Equation 4.2 and 4.3, the relationship Equation 4.1 can also be written as:

$$Q_T = \left[p + \frac{(1-p)}{1+(1-i)g} \right] * (A_T * Y_m) \quad . \qquad (4.4)$$

If in a country there is a perfect equality in both endowments and returns between female and male managers and that the impacts of unobserved differences between the two groups are negligible, then g = 0 and $Y_f = Y_m$, so that:

$$Q_T^n \equiv Y_m * A_T \quad , \qquad (4.5)$$

where Q_T^n represents the new production level after closing the gender productivity gap. Hence, the difference between Q_T^n and Q_T gives the approximate production gains. Subtracting Equation 4.4 from Equation 4.5 and then dividing the result by Q_T, we get the relative expected production gains for various targets of i:

$$\underbrace{\left(Q_T^n - Q_T\right)/Q_T}_{production\ gains} = \frac{1}{Q_T}\left[\frac{(1-p)(1-i)g}{1+(1-i)g}\right]*\left(A_T * Y_m\right) \quad , \tag{4.6}$$

Equation 4.6 states that the percent increase in the production levels due to gender equality in agriculture would be greater the larger the proportion of female-managed plots, (1–p). Moreover, when comparing two countries, production gains will be more important in the country with larger gender productivity differentials, g, or a higher policy target, i, holding other things constant. Finally, the larger the level of production attained by male managers relative to their female counterparts,(A_T*Y_m), the greater the expected benefits from closing gender productivity gaps. Hence, under the model assumptions, Equation 4.6 shows that there might be considerable potential production benefits that could result from increasing women's access to agricultural lands and modern inputs, improving their managerial and agricultural skills, as well as their access to extension and advisory services.

4.1.4. Approximation of Consumption and Poverty Gains

Other possible benefits from minimizing or eliminating the gender productivity gaps can also be estimated. One of the most crucial of these is whether, in addition to increasing production, closing the gender gap could also improve the welfare of households with female-managed plots or parcels. To keep things simple, we will only look at two additional advantages of gender equality in agriculture: poverty alleviation and consumption. If monthly consumption per adult equivalent is higher than it was before reducing or closing gender gaps, there will be consumption gains. Similarly, after taking into consideration the consumption gains brought about by gender equality in agriculture, there will be poverty gains if the number of households below the poverty line in households with female-managed plots or parcels decreases.

Let τ, with $0 \le \tau \le 1$, be the share of production normally allocated to home consumption and that the additional production from reducing/closing the gender gaps obtained in Equation 4.6, i.e., Q_T^n-Q_T, be distributed equally across all households in the sample with female managers. Let also C_T^n and C_T represent the consumption values after and before reducing/closing the gender productivity gap. This means that:

$$\underbrace{\left(C_T^n - C_T\right)/C_T}_{consumption\ gains} = \frac{\tau}{n(1-p)C_T} * \left(Q_T^n - Q_T\right)$$

$$= \frac{\tau(1-i)g}{n[1+(1-i)g]C_T} * \left(A_T * Y_m\right) \tag{4.7}$$

where n is the sample size.

The intuitions underlying Equation 4.7 are simple. Production levels and consumption gains are positively correlated, according to the model's assumptions. The amount of food available for consumption at home increases when female-managed plots reach the productivity of male-managed plots, thereby improving household food security. Because consumption values were found to be consistently lower in households with female managers, this is especially significant in the context of the three countries under investigation (Nigeria, Tanzania, and Uganda) (Mukasa and Salami, 2016). Furthermore, the potential consumption effect per household decreases as the number of female managers in the economy increases.

Finally, there will be poverty gains from reducing/closing the gender gaps if:

$$\#(C_T^n + C_N < \overline{w}) < \#(C_T + C_N < \overline{w}), \qquad (4.8)$$

where C_N represents non-food consumption values. The left-hand size of Equation 4.8 stands for the number of households with female-managed lands in which the total value of household consumption of food and non-food, $C_T^n + C_N$, is smaller than the poverty line $\overline{w}$. The right-hand side represents the number of poor households with female managed lands before reducing/closing the gender productivity gap.

4.1.5. Potential Gains from Closing Gender Productivity Differentials in SSA

Figure 4.2 plots the approximate production, consumption, and poverty gains that would be attained if male and female managers had equal access to all productive factors and if the gender productivity differentials of 18.6% in Nigeria, 27.4% in Tanzania, and 30.6% in Uganda were totally eliminated. This is based on Equation 4.6 and findings from Mukasa and Salami (2016).

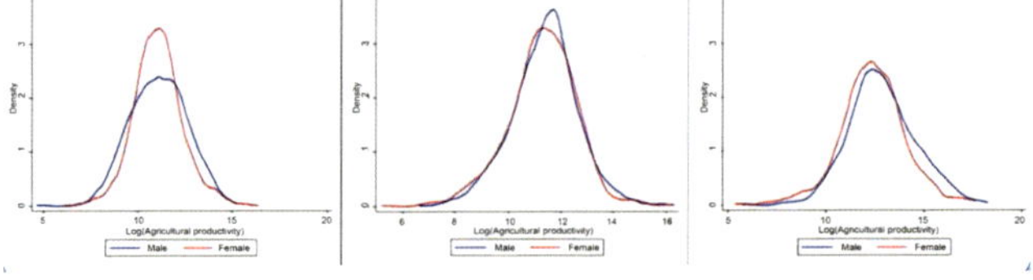

Figure 4.2. *The plots of the approximate production, consumption, and poverty gains.*

Source. Balducci M (2023) Linking gender differences with gender equality: A systematic-narrative literature review of basic skills and personality. Front. Psychol. 14:1105234. doi: 10.3389/fpsyg.2023.1105234.

As demonstrated earlier, the potential production gain from closing the gap increases with the size of the current gender productivity gap. The maximum production gains, as a percentage of the current production levels, are 2% in Nigeria, 8% in Tanzania, and 10% in Uganda. An average potential production increase of 6–6% is obtained by weighting these effects by the proportion of female managers in each nation. We might anticipate a roughly comparable increase in global production if these three nations are typical of the yield disparities, input usage, and land distribution patterns between men and women in other SSA nations. Given that SSA nations are faring worse than other regions in terms of agricultural productivity, these production gains are not insignificant. Therefore, removing obstacles that keep female managers from reaching male productivity levels could be a significant step in raising SSA countries' agricultural output and guaranteeing its structural transformation.

In terms of consumption gains, research shows that, at the highest possible level, monthly consumption per adult equivalent in Nigeria, Tanzania, and Uganda would rise by 2.9%, 1.4%, and 10.7%, respectively, over their current levels. These kinds of gains are significant, especially considering that households with female-managed plots or parcels already consume less than households with males (Mukasa and Salami, 2016). The smaller percentage of female-managed plots (17.1% versus 34.9% in Tanzania) more than makes up for Nigeria's less favorable production gains, which result in higher consumption gains despite Tanzania's lower production gains.

Lastly, the study also shows how gender equality in agricultural production could reduce poverty. Once gender productivity gaps are closed, it is anticipated that approximately 1.2% of Nigeria's impoverished households with female-managed plots or parcels will escape poverty. In absolute terms, this amounts to two of the 171 households in the sample of female managers that currently live below the poverty line. These potential gains in poverty would reach 4% in Tanzania and 13% in Uganda, which corresponds to 6 out of 123 and 14 out of 108 current impoverished households in those two nations, respectively. This implies that reallocating productive resources between male and female managers 6 could greatly improve the welfare conditions of many impoverished people at the national level. These immediate results of gender equality in agriculture would eventually lead to additional possible advantages. These indirect effects would include improved child nutrition, health, and educational attainment in households with female-managed plots, as well as increases in the bargaining power of female managers and improvements in their social status as their earnings rise (FAO, 2011). Table 4.1 shows the statistics obtained in a study on EU nations.

Table 4.1. *Female- vs. Male-Operated Farms in the EU Countries, Var % 2016–2005*

Countries	Female Operated Farms	Male Operated Farms	Total Farms
Belgium	−31.01	−27.97	−28.42
Bulgaria	−46.83	−65.42	−62.13
Czechia	−54.74	−33.71	−37.23
Denmark	−53.69	−29.45	−32.19
Germany	−20.87	−29.96	−29.18
Estonia	−46.10	−36.13	−39.82
Ireland	−1.65	4.37	3.68
Greece	−10.47	−20.32	−17.84
Spain	3.86	−16.66	−12.76
France	−20.53	−19.48	−19.70
Italy	−25.16	−37.04	−33.72
Cyprus	−18.15	−23.84	−22.63
Latvia	−47.92	−43.63	−45.64
Lithuania	−37.22	−43.06	−40.58
Luxembourg	9.68	−23.83	−19.59
Hungary	−30.66	−42.69	−39.84
Malta	−38.46	−13.86	−15.88
Netherlands	−47.36	−30.85	−31.96
Austria	−29.36	−20.90	−23.79
Poland	−47.54	−40.92	−43.04
Portugal	−2.41	−26.05	−20.24
Romania	−6.85	−24.88	−19.66
Slovenia	−30.00	−2.12	−9.42
Slovakia	−63.31	−62.37	−62.55
Finland	−19.95	−30.71	−29.59

Countries	Female Operated Farms	Male Operated Farms	Total Farms
Sweden	12.60	−20.79	−16.98
United Kingdom	−45.83	−33.24	−35.51
Total	−22.84	−30.82	−28.72

Source: own calculations based on Eurostat and Word Bank data; Fanelli RM. Bridging the Gender Gap in the Agricultural Sector: Evidence from European Union Countries. Social Sciences. 2022; 11(3):105. https://doi.org/10.3390/socsci11030105.

However, the fight against gender productivity bias would likely be gradual and yield gaps would only be gradually reduced given the magnitude of structural challenges that SSA countries face. The approximate production, consumption, and poverty gains from reducing the current gender productivity disparities at different levels are reported. Since the approximate production gains in that scenario would be of the order of 0.3% in Nigeria, 1% in Tanzania, and approximately 1.5% in Uganda, reducing the current yield gap by just 10% would have very little impact.

Then, as we progressively close the existing gaps at accelerating rates, the gains keep growing. Insofar as consumption gains at that policy target are insufficient to initiate any transition out of poverty, no poverty gains would be anticipated in Nigeria with a 10% gap reduction. The poverty gains in Tanzania and Uganda are already positive at that point, although Tanzania's gain is only about 1%. Gains in poverty are not always growing as we close the gender gap, in contrast to gains in production and consumption. Therefore, in Nigeria, poverty gains do not improve between 40% and 90% of gender gap reduction, stagnating at about 0%–6%. This is because there are no poverty gains if less than 30% of the current gender productivity differentials are reduced, even though they later turn positive. Before beginning an upward trend, they stay at 2% in Tanzania between 10% and 60%. This suggests that significant work must be done to reduce gender productivity bias and achieve observable, beneficial welfare improvements. If governments wish to lift a sizable percentage of their nations' female managers out of poverty, they will need to implement ambitious programs to combat this gender bias (Table 4.2).

Table 4.2. *Gender Inequalities*

Glopan Food Systems Compo-nents	**Sub-Themes**	**Document Number**																	
Agri-cultural production		1	2	3	4	5	6	7	8	9	10	11	12	13	14	15	16	17	18	Total
	Extension			✓	✓	✓	✓	✓												5
	Land			✓	✓	✓	✓	✓	✓			✓		✓	✓	✓			✓	11
	Water						✓		✓					✓						3
Markets and trade	Trade					✓	✓	✓		✓									✓	5
Consumer purchasing power	Employment		✓	✓	✓	✓	✓	✓					✓			✓			✓	9
	Social protec-tion	✓	✓	✓		✓			✓					✓					✓	7
Food transforma-tion and consumer demand	Marketing			✓	✓	✓	✓									✓				5

Glopan Food Systems Compo-nents	**Sub-Themes**	**Document Number**																	
Agri-cultural production		1	2	3	4	5	6	7	8	9	10	11	12	13	14	15	16	17	18	Total
	Processing				✓		✓									✓				3
Diet quality	Diet			✓	✓	✓	✓		✓	✓	✓			✓		✓	✓		✓	11
	Maternal and child health			✓	✓		✓	✓						✓			✓		✓	7
	Water Health and Sanitation (WASH)		✓		✓	✓		✓											✓	5
Cross-cutting	Agribusiness					✓							✓			✓		✓	✓	5
	Capacity strengthening			✓	✓	✓	✓	✓		✓							✓		✓	8
	Financial resources			✓	✓	✓	✓	✓	✓					✓		✓			✓	9
	Infrastructure					✓						✓							✓	4
	Resource Management					✓	✓												✓	3

Glopan Food Systems Components	**Sub-Themes**	**Document Number**																	
Agricultural production		1	2	3	4	5	6	7	8	9	10	11	12	13	14	15	16	17	18	Total
	Technology			✓	✓	✓				✓				✓					✓	6
Social systems	Unpaid work			✓		✓		✓											✓	4
	Shared family responsibility				✓	✓		✓											✓	4
	Food culture				✓			✓	✓											3
Governance	Gender mainstreaming					✓	✓	✓				✓			✓	✓			✓	7
	Legislation		✓	✓	✓	✓	✓	✓	✓		✓	✓			✓	✓			✓	12
	Participation			✓	✓	✓	✓	✓	✓			✓				✓		✓	✓	10
Total		1	4	13	16	20	15	14	9	4	3	3	2	7	3	11	4	3	19	

Source.: Mkandawire E, Mentz-Coetzee M, Mangheni MN, Barusi E. Enhancing the Glopan Food Systems Framework by Integrating Gender: Relevance for Women in African Agriculture. Sustainability. 2021; 13(15):8564. https://doi.org/10.3390/su13158564.

4.2. GENDER INEQUALITIES IN AGRICULTURAL LABOR

Global economies still rely heavily on agriculture, especially in developing nations where a sizable section of the labor force is employed in this sector. In agricultural production, women are essential to both economic stability and food security. Gender inequality in agricultural labor still exists despite their substantial contributions, showing up as differences in recognition, decision-making authority, and resource access. A more equitable agricultural sector and the accomplishment of more general socioeconomic development objectives depend on an understanding of these disparities.

4.2.1. The Role of Women in Agriculture

In many regions of the world, especially in South Asia and sub-Saharan Africa, women make up almost half of the agricultural workforce. Planting, weeding, harvesting, processing, and marketing produce are just a few of the agricultural tasks they engage in. Along with managing food production in the home and sustaining family incomes, women frequently balance several responsibilities as caregivers, housewives, and agricultural laborers. However, in agricultural policy and practice, their contributions are usually disregarded and underappreciated.

4.2.2. Contribution to Food Security

Given that they are frequently in charge of making sure their families have access to enough wholesome food, women's participation in agriculture is essential for food security. Studies show that when women have equal access to resources, agricultural productivity can rise dramatically, improving community food security. For example, research indicates that reducing the gender gap in agricultural productivity could result in a 100–150 million reduction in the world's hunger rate.

4.2.3. Barriers to Gender Equality in Agriculture

Despite playing a crucial role, women in agriculture encounter many obstacles that limit their ability to be economically empowered and productive. These obstacles fall into many important categories.

4.2.3.1. Access to Land

One of the biggest obstacles facing women in agriculture is access to land. In many cultures, men own the majority of the land, and women are frequently denied the legal right to own land. Women may have limitations on how they can use land or be forced to rely on male family members to make decisions for them, even if they do have access to it. Women's capacity to invest in their agricultural output and attain long-term financial stability is hampered by this lack of secure land tenure.

4.2.3.2. Financial Resources

Financial resource accessibility is yet another significant obstacle. Women frequently lack access to financial services and credit, which are necessary for purchasing agricultural inputs like equipment, seeds, and fertilizer. The lack of financial products specifically designed for women may result from traditional banking systems' preference for male farmers. Though they are not always accessible and frequently have exorbitant interest rates, microfinance programs have surfaced as a viable remedy.

4.2.3.3. Education and Training

To increase agricultural productivity, education, and training are essential. But compared to men, women usually have lower educational attainment, which limits their understanding of contemporary farming methods and best practices. Women have fewer opportunities and frequently face discrimination when it comes to accessing training programs. Furthermore, cultural expectations might deter women from attending training sessions or looking for outside-the-home educational opportunities.

4.2.3.4. Technology and Resources

Technology access is crucial to raising agricultural productivity and practices. Women's efficiency and effectiveness may be hampered by their frequent lack of access to contemporary agricultural equipment, tools, and technologies. In addition to cultural biases that favor male farmers in implementing innovative practices, a lack of training on how to use these tools further exacerbates the gender gap in technology access.

4.2.3.5. Labor Dynamics

Labor dynamics in agriculture are also influenced by gender norms and expectations. Women's labor is frequently devalued and categorized as "unpaid labor," even though it makes a substantial contribution to agricultural production. In many communities, men may handle more profitable activities like marketing and selling produce, while women may be involved in labor-intensive tasks like weeding and harvesting. This division of labor restricts women's agency in agricultural decision-making and perpetuates traditional gender roles.

4.2.3.6. Decision-Making Power

When it comes to household management and agricultural practices, women frequently lack decision-making authority. Men typically control financial resources and make important decisions about crop selection, resource allocation, and sales, even when they make a substantial contribution to agricultural production. In the home and in the larger community, this disparity perpetuates gender inequality and threatens women's autonomy (World Bank, 2012, 2015).

4.2.4. The Impact of Gender Inequalities

Gender disparities in agricultural work have an effect on more than just individual women and families. The effects of these discrepancies on food security, economic growth, and agricultural productivity are extensive. The following are some of the main effects.

4.2.4.1. Reduced Agricultural Productivity

Agricultural production is inefficient as a result of gender inequality. Women's productivity declines when they lack access to training, technology, credit, and land. Equal access to resources has been shown to considerably boost agricultural yields, enhancing both food security and economic resilience.

4.2.4.2. Economic Inequality

Inequalities in the economy as a whole are exacerbated by gender differences in agricultural labor. Because women have less access to resources and decision-making authority, their income levels are lower, which feeds poverty cycles. Overall development is hampered by this economic disadvantage, which impacts not only women but also their families and communities.

4.2.4.3. Food Insecurity

Food insecurity is made worse by the underappreciation of women's contributions to agriculture. Communities struggle to achieve food security when systemic barriers prevent women from realizing their full agricultural potential. Improving nutritional outcomes and decreasing hunger require women's empowerment in agriculture.

4.2.4.4. Social and Cultural Consequences

In addition, gender disparities in agricultural work reinforce negative cultural and social norms. Women's limited opportunities to participate in economic activities outside the home are a result of their marginalization in agriculture, which perpetuates traditional gender roles. Gender relations and societal perceptions of women's empowerment may be impacted for some time by this dynamic.

4.2.5. Strategies for Promoting Gender Equality in Agriculture

A multifaceted strategy that includes community engagement, capacity building, and policy reforms is needed to address gender disparities in agricultural labor. Effective tactics include the following.

4.2.5.1. Legal Reforms

It is crucial to put legal changes into place that protect women's land rights. Women can invest in agriculture and raise their economic status by having equal access to land ownership and inheritance. Organizations and governments alike ought to support laws that advance gender parity in land ownership.

4.2.5.2. Access to Financial Services

Improving women's financial services accessibility is essential to their economic empowerment. Microfinance programs that offer low-interest loans and financial literacy training should be customized to women's unique needs. Women can also manage financial risks by having greater access to insurance and savings accounts.

4.2.5.3. Education and Training Programs

Women's productivity in agriculture can be greatly increased by funding educational and training initiatives. Women farmers can become more capable and self-assured through customized training programs that cater to their unique needs, such as modern farming methods and business management abilities.

4.2.5.4. Technology and Innovation

Encouraging innovation and technology access is essential to empowering women in agriculture. Women should have easier access to contemporary agricultural technologies and tools, as well as instruction on how to use them, thanks to governments and organizations. Women may become more productive and efficient as a result.

4.2.5.5. Community Engagement

Changing social norms and attitudes requires involving communities in conversations about gender equality in agriculture. Campaigns to raise awareness of women's contributions to agriculture can aid in changing attitudes and advancing gender-inclusive methods.

4.2.5.6. Support Networks

Women in agriculture can gain access to resources, information, and mentorship by creating support networks. Women's organizations and farmer cooperatives can improve cooperation, encourage the exchange of knowledge, and give women the confidence to stand up for their rights.

A major obstacle to attaining sustainable development, food security, and economic empowerment is gender inequality in agricultural labor. Fostering an inclusive agricultural sector that values the contributions of all people, regardless of gender, requires acknowledging and resolving these disparities. The full potential of women in agriculture can be realized by societies through the implementation of focused strategies to advance gender equality, which will ultimately benefit economies and communities overall. In addition to being a question of social justice, ensuring that women have equal access to resources, decision-making authority, and recognition is also a vital first step in creating resilient and sustainable food systems around the world (World Bank, 2012, 2015).

4.3. ACCESS TO LAND, RESOURCES, AND DECISION-MAKING FOR WOMEN IN AGRICULTURAL WORK

Although agriculture is an essential industry that supports livelihoods and food security globally, women still encounter major obstacles when trying to obtain essential resources like land, financial capital, technology, and decision-making authority. These obstacles restrict women's economic empowerment, productivity, and overall agricultural development contribution. In addition to attaining gender equality, addressing these disparities is crucial for raising agricultural productivity, boosting food security, and advancing sustainable development.

4.3.1. The Importance of Access to Land

4.3.1.1. Land as a Fundamental Resource

An essential resource for agricultural production is land. It offers the framework for farming, allowing people to cultivate crops, rear livestock, and make money. Access to land is especially important for women since it has a direct impact on their capacity to engage in agriculture.

4.3.1.2. Gender Disparities in Land Ownership

When it comes to land ownership, women are disproportionately disadvantaged worldwide. Women's rights to inherit, own, or control land are restricted in many areas by cultural norms and legal frameworks. In developing nations, women only make up around 13% of agricultural landowners, according to the Food and Agriculture Organization (FAO). This discrepancy raises their susceptibility to poverty, limits their ability to make decisions regarding land use, and limits their income potential.

4.3.1.3. Implications of Land Inequality

The lack of secure land tenure for women has far-reaching implications:

- **Economic Disempowerment:** Without ownership or secure access to land, women are less likely to invest in improvements, such as better seeds or irrigation systems, which can enhance productivity.
- **Food Security:** Women play a critical role in food production, and their inability to access land can lead to reduced agricultural output, impacting food security for families and communities.
- **Social Status:** Land ownership is often linked to social status and influence. Women without land may find themselves marginalized in both community decision-making processes and household dynamics.

4.3.1.4. Legal Frameworks and Reforms

Strong legal frameworks that support women's land rights are necessary to address land inequality. Governments must pass legislation guaranteeing women's equal rights to inheritance and land ownership. Initiatives for land reform have advanced the recognition of women's rights in some areas, although execution is still inconsistent. Legal aid programs and awareness-raising initiatives can support women in asserting their rights.

4.3.2. Access to Financial Resources

4.3.2.1. Financial Inclusion and Economic Empowerment

For women in agriculture to invest in their farms and raise their standard of living, they must have access to financial resources. However, women frequently encounter major obstacles when trying to obtain insurance, savings, and credit.

4.3.2.2. Gender Bias in Financial Institutions

Financial institutions usually favor male borrowers due to gender bias. Women frequently lack the networks, credit history, and collateral needed to obtain loans. The gender gap in financial inclusion caused by this systemic bias restricts women's access to training, technology, and agricultural input investments.

4.3.2.3. Microfinance and Empowerment

One tool that has shown promise in expanding women's access to financial resources is microfinance. Women can start or grow their agricultural businesses with the help of microfinance institutions that offer small loans without the need for conventional collateral. To guarantee that women can successfully manage their finances, microfinance must be combined with financial literacy training, even though it can make it easier for them to obtain funds.

4.3.2.4. Building Supportive Financial Ecosystems

To promote women's access to financial resources, it is essential to create supportive financial ecosystems that include:

- **Tailored Financial Products:** Financial institutions should develop products specifically designed for women farmers, considering their unique needs and circumstances.
- **Training Programs:** Providing training on financial management, budgeting, and investment strategies can empower women to utilize financial resources effectively.
- **Community-Based Financing:** Women's cooperatives and community savings groups can provide alternative financing mechanisms, fostering collective support and resource sharing.

4.3.3. Access to Technology and Resources

4.3.3.1. The Role of Technology in Agriculture

Improving agricultural sustainability and productivity requires having access to resources and modern technology. Crop yields can be greatly increased and labor intensity decreased with the use of technologies like better seeds, fertilizers, irrigation systems, and machinery.

4.3.3.2. Gender Gap in Technology Adoption

The adoption of agricultural technology varies by gender, despite the possible advantages. Due to a lack of knowledge, training, and funding, women frequently encounter obstacles when attempting to adopt new technologies.

Cultural norms may also deter women from taking part in training programs or using particular technologies.

4.3.3.3. Bridging the Technology Gap

To bridge the technology gap, it is essential to:

- **Promote Gender-Sensitive Agricultural Extension Services:** Agricultural extension programs should prioritize reaching women, providing them with information and training on modern agricultural practices and technologies.
- **Encourage Female Role Models:** Highlighting the success stories of women who have effectively adopted new technologies can inspire others to do the same.
- **Facilitate Access to Resources:** Organizations should work to ensure that women have access to essential agricultural resources, including seeds, fertilizers, and tools, through targeted programs that prioritize female farmers.

4.3.3.4. Innovative Approaches

Bridging the information gap can be greatly aided by creative strategies like mobile technology. By providing them with useful information that improves their decision-making, mobile applications that offer weather forecasts, market prices, and best farming practices can empower women.

4.3.4. Decision-Making Power

4.3.4.1. The Importance of Agency

In agriculture, decision-making authority is essential to women's empowerment. Women are better able to choose crops, manage their farms following their needs and preferences, and allocate resources when they have a voice in agricultural decisions.

4.3.4.2. Barriers to Participation

Women cannot frequently make decisions in their homes and communities, despite their contributions. Women's participation in formal and informal decision-making processes may be limited by cultural norms and traditional gender roles. This lack of agency can restrict women's ability to influence agricultural policies and practices and reinforce gender inequality.

4.3.4.3. Strategies for Empowerment

To enhance women's decision-making power, several strategies can be implemented:

- **Promote Women's Leadership:** Encouraging women to take on leadership roles in agricultural cooperatives, community organizations, and decision-making bodies can amplify their voices and influence.
- **Engage Men as Allies:** Involving men in discussions about gender equality and women's empowerment can help shift cultural norms and promote shared decision-making within households.
- **Capacity Building:** Providing training and resources that enhance women's leadership skills, negotiation abilities, and knowledge of agricultural practices can empower them to participate effectively in decision-making.

4.3.4.4. Community-Based Approaches

Inclusive decision-making can be promoted by community-based methods. Women's needs and viewpoints are taken into account in agricultural development projects when they are included in participatory planning processes. Women's collective voice can also be strengthened by establishing safe spaces for them to talk about problems and exchange stories (World Bank, 2012, 2015).

4.3.5. Challenges and Future Directions

4.3.5.1. Implementation Gaps

Gender inequality has been addressed, but there are still large implementation gaps. There may be legal frameworks, but their accessibility and enforcement may be restricted. To guarantee that policies result in significant changes on the ground, ongoing advocacy and observation are crucial.

4.3.5.2. Intersectionality

It is essential to comprehend how gender intersects with other variables like age, ethnicity, and socioeconomic status. Programs designed to advance gender equality in agriculture can be more successful if interventions are specifically designed to address the particular difficulties faced by various groups of women.

4.3.5.3. Collaborative Efforts

Governments, non-governmental organizations, and community groups must work together to create an environment that supports women in agriculture.

Partnerships involving multiple stakeholders can use resources, skills, and knowledge to promote significant change.

The empowerment of women in agricultural work is contingent upon their ability to access land, resources, and decision-making authority. To promote gender equality and increase agricultural productivity, the structural obstacles preventing women from accessing these essential components must be addressed. Societies can unleash the full potential of women in agriculture by supporting legislative changes, increasing financial inclusion, facilitating technology access, and giving women more decision-making authority. In addition to helping women personally, ensuring that they have fair access to resources and land also promotes sustainable farming methods, food security, and overall economic growth. In addition to being morally required, achieving gender equality in agriculture is a critical first step in creating robust and sustainable food systems for coming generations (World Bank, 2012, 2015).

4.4. ADDRESSING GENDER WAGE GAPS AND IMPROVING WORKING CONDITIONS FOR WOMEN

Many economies rely heavily on agriculture, which employs millions of people worldwide, particularly in developing nations. Women are vital to this industry, contributing greatly to agricultural labor and food production. Women frequently experience significant pay disparities and unfavorable working conditions despite their significant contributions. To advance gender equality, increase agricultural productivity, and accomplish more general socioeconomic development objectives, these issues must be addressed.

4.4.1. Understanding Gender Wage Gaps in Agriculture

4.4.1.1. The Scale of Wage Disparities

In agriculture, gender wage disparities are widespread and complex. For comparable work, women are frequently paid much less than men. According to research, depending on the area and particular farming methods, women in agriculture may make anywhere from 60%–80% of what men do. Numerous factors, such as discrimination, labor market segmentation, and sociocultural norms, all have an impact on these disparities.

4.4.1.2. Factors Contributing to Wage Gaps

Several key factors contribute to the wage gap between men and women in agriculture:

- **Occupational Segregation:** Women often work in lower-paying, less secure roles, such as seasonal labor or informal positions. They are frequently concentrated on tasks like weeding and harvesting, while men may take on more lucrative roles like management or marketing.
- **Lack of Negotiation Power:** Women often lack the bargaining power to negotiate fair wages due to cultural norms that prioritize male authority in economic decisions. This power imbalance can lead to the acceptance of lower wages.
- **Discrimination:** Gender discrimination in hiring, promotion, and pay persists in agricultural labor markets. Employers may undervalue women's contributions, leading to systemic inequities in compensation.

4.4.1.3. Impacts of Wage Gaps

The implications of gender wage gaps extend beyond individual women and their families. They affect economic development and food security:

- **Poverty:** Lower wages for women contribute to higher poverty rates among female-headed households. Women's limited income reduces their ability to invest in education, health, and nutrition.
- **Productivity:** When women are not compensated fairly, it can lead to decreased motivation and productivity. Addressing wage disparities can enhance overall agricultural output.
- **Social Inequality:** Persistent wage gaps reinforce societal inequalities and hinder progress toward gender equality. This perpetuates cycles of discrimination and marginalization.

4.4.2. Improving Working Conditions for Women in Agriculture

4.4.2.1. Current Challenges in Working Conditions

Working conditions for women in agriculture are often harsh and unsafe. Many women work long hours in physically demanding jobs, frequently without adequate protection or support. Key issues include:

- **Occupational Health Risks:** Women are often exposed to hazardous working conditions, including pesticide exposure, strenuous labor, and inadequate safety measures. This can lead to serious health problems and injuries.

- **Informal Employment:** A significant portion of women's agricultural work is informal, lacking protections such as contracts, benefits, and labor rights. This precarious status leaves women vulnerable to exploitation and abuse.
- **Lack of Support Services:** Women often lack access to childcare, healthcare, and transportation, making it challenging to balance work and family responsibilities. This can hinder their ability to engage fully in agricultural activities.

4.4.2.2. Addressing Working Conditions

Improving working conditions for women in agriculture requires a comprehensive approach that encompasses policy reforms, training, and community support:

- **Regulatory Frameworks:** Governments should enact and enforce labor laws that protect the rights of women in agriculture. This includes regulations on minimum wage, safe working conditions, and the right to organize.
- **Health and Safety Training:** Providing training on occupational health and safety can empower women to recognize and mitigate risks in their work environments. Access to personal protective equipment (PPE) is also essential.
- **Support Services:** Establishing support services, such as childcare facilities and transportation, can help women balance work and family responsibilities, making it easier for them to engage in agricultural activities.

4.4.2.3. Empowering Women Through Training and Education

Education and training are crucial for improving working conditions and addressing wage gaps:

- **Skills Development:** Providing women with access to vocational training and skill development programs can enhance their productivity and earning potential.

Training in areas such as leadership, negotiation, and business management can empower women to advocate for fair wages and better working conditions.

- **Financial Literacy:** Equipping women with financial literacy skills can help them understand their rights, negotiate wages, and manage their finances more effectively.
- **Awareness Campaigns:** Raising awareness about women's rights in agricultural work and the importance of fair wages can foster a culture of equality and support within communities.

4.4.3. The Role of Policy and Governance

4.4.3.1. Government Intervention

Effective government policies are essential for addressing gender wage gaps and improving working conditions:

- **Legislation:** Enacting and enforcing legislation that mandates equal pay and protects labor rights is crucial for creating an equitable work environment.
- **Monitoring and Evaluation:** Establishing mechanisms to monitor and evaluate the implementation of labor laws can ensure accountability and transparency.

4.4.3.2. Multi-Stakeholder Approaches

Engaging multiple stakeholders, including governments, NGOs, private sector actors, and community organizations, is vital for fostering an inclusive approach to addressing gender inequalities in agriculture.

- **Public-Private Partnerships:** Collaborations between public and private sectors can leverage resources and expertise to implement programs that support women in agriculture.
- **Community Engagement:** Involving communities in the development and implementation of initiatives can foster a sense of ownership and ensure that programs are culturally appropriate and effective.

4.4.4. Challenges and Barriers to Progress

4.4.4.1. Cultural Norms and Resistance

Gender inequality in agriculture is frequently sustained by ingrained cultural norms and stereotypes. Persistent advocacy and education are necessary to change societal perceptions of women's roles in agriculture.

4.4.4.2. Economic Constraints

Programs designed to address wage disparities and improve working conditions may not be able to be implemented due to a lack of funds and resources. Investments in gender-responsive projects must be given top priority by governments and organizations.

4.4.4.3. Data Gaps

Attempts to address gender inequality may be hampered by a lack of disaggregated data on wages, working conditions, and gender dynamics in agriculture. Data collection and analysis are essential for guiding interventions and policies.

4.4.5. The Way Forward

4.4.5.1. Comprehensive Gender Policies

It is crucial to create comprehensive gender policies that address working conditions and wage disparities in agriculture. These regulations ought to cover every facet of women's rights in the agriculture industry, from labor protections to land ownership.

4.4.5.2. Capacity Building for Women

Women can be empowered to negotiate fair wages and fight for their rights by funding capacity-building initiatives. Training in financial literacy, negotiation techniques, and leadership are a few examples of this.

4.4.5.3. Advocacy and Awareness

To promote societal change, it is essential to increase awareness of gender wage disparities and unfavorable working conditions. Campaigns for gender equality have the power to inspire communities and foster support.

4.4.5.4. Strengthening Cooperatives

Women's collective bargaining power, access to resources, and ability to advocate for improved pay and working conditions can all be increased by empowering them through agricultural cooperatives.

Promoting gender equality and increasing agricultural productivity require addressing gender wage gaps and improving working conditions for women employed in agriculture. By putting specific policies into place, encouraging community involvement, and empowering women via education and training, we can build a more just agricultural industry that recognizes the contributions of all employees.

In addition to helping women, ensuring safe working conditions and fair wages also benefits families, communities, and economies. Addressing these gender disparities in agriculture must continue to be a top priority as the world works to meet the SDGs. In addition to being a matter of justice, empowering women in agriculture is a means of fostering resilience and sustainable development for coming generations (World Bank, 2012, 2015).

4.5. ANALYZING EFFORTS TO IMPROVE WOMEN'S RIGHTS AND PARTICIPATION IN AGRICULTURAL WORK

Women are essential to agriculture, particularly in developing nations where they make major contributions to rural economies and food production. Nevertheless, despite their efforts, women's rights and involvement in agricultural labor are restricted by structural obstacles. The need for gender equality in the agricultural sector has spurred efforts to increase women's rights and participation in recent years. This study looks at many projects, laws, and activities meant to improve women's rights and involvement in agriculture.

Due to ingrained gender norms and cultural prejudices, women have historically been marginalized in the agricultural industry. Gender-based agricultural roles have been established in many societies, with men controlling resources and making decisions while women are restricted to low- or no-paying labor. Economic development has been hampered and poverty cycles have been sustained by this systemic inequality.

Governments, non-governmental organizations, and international organizations have implemented programs to empower women in agriculture as a result of their recognition of these disparities.

4.5.1. Policy Frameworks

4.5.1.1. International Agreements

Gender equality in all fields, including agriculture, has been emphasized by international frameworks like the SDGs and the Convention on the Elimination of All Forms of Discrimination Against Women (CEDAW). These agreements stress the necessity of policies that support women's rights and participation and demand the abolition of discrimination against women.

4.5.1.2. National Policies

National policies aimed at promoting gender equality in agriculture have been developed in many nations.

The implementation of progressive land tenure reforms in Rwanda, for example, has greatly increased women's involvement in agricultural production by ensuring their rights to own and inherit land. Women play a vital role in ensuring food security, and these policies seek to increase their economic empowerment.

4.5.2. Community-Based Initiatives

4.5.2.1. Women's Cooperatives

Women's cooperatives have become a potent instrument for advancing women's participation in agriculture and their rights. Women can pool resources, get training, and participate in collective bargaining through these cooperatives. Women who band together can access markets, increase their bargaining power, and bargain for lower prices for their goods.

Women's self-help groups (SHGs) in India, for instance, have effectively empowered women by offering support networks, training, and financial services. These organizations have made it easier for women to obtain resources and credit, allowing them to invest in their farming businesses and raise their standard of living.

4.5.2.2. Local Advocacy and Capacity Building

Community-based organizations are frequently essential in promoting women's rights and offering leadership, financial literacy, and agricultural practices training. Women can be empowered to take on leadership roles in their communities and fight for their rights through programs that emphasize capacity building.

NGOs have organized training workshops in many locations, giving women the tools they need to actively engage in agricultural and resource management decision-making. Women now have more self-assurance and agency as a result of these initiatives, which empowers them to stand up for their rights and have an impact on agricultural policies.

4.5.3. Access to Resources

4.5.3.1. Land Rights

Improving women's involvement in agriculture requires securing their land rights. To guarantee equal rights for women, numerous initiatives concentrate on changing land tenure systems. Laws have been put in place to support women's access to land in nations like Ghana. Women farmers now enjoy greater economic security and higher agricultural productivity as a result of these reforms.

4.5.3.2. Financial Inclusion

Financial resources are essential for empowering women in agriculture. Programs for microfinance have become a crucial tactic to deal with this problem. These initiatives give women access to small loans so they can purchase tools, seeds, and fertilizer, among other agricultural inputs.

Microfinance organizations frequently modify their offerings to accommodate women's needs by providing flexible repayment plans and reduced interest rates. One successful example is the Grameen Bank in Bangladesh, which has given millions of women support and financial resources, thereby empowering them.

4.5.2.3. Technology and Training

To increase agricultural productivity, technology access is crucial. The productivity and decision-making abilities of women can be greatly increased by initiatives to give them access to contemporary agricultural equipment, knowledge, and training.

Initiatives that use mobile technology to inform women about markets, weather, and best practices have demonstrated the potential to enhance agricultural results. For example, programs in Africa have used mobile applications to spread important agricultural knowledge, empowering women to manage crops with knowledge.

4.5.4. Addressing Social Norms and Cultural Barriers

4.5.4.1. Advocacy Campaigns

Comprehensive advocacy efforts are needed to address the social and cultural barriers that restrict women's rights and involvement in agriculture. Dismantling discriminatory practices and advancing gender equality can be aided by campaigns that try to alter societal perceptions of gender roles.

Campaigns to promote women's rights and increase awareness of their contributions to agriculture have been started by organizations and grassroots movements. These advertisements frequently dispel stereotypes and spur change by showcasing prosperous female farmers and their contributions to regional economies.

4.5.4.2. Engaging Men and Boys

Promoting an inclusive atmosphere that upholds women's rights requires involving boys and men in conversations about gender equality. Cultural norms can be changed and shared responsibility in agricultural work encouraged by programs that include men as allies in the fight for gender equality.

Education programs aimed at men and boys can encourage fair decision-making in the home and respect for the contributions of women. These initiatives have the potential to improve the conditions for women to succeed in agriculture.

4.5.5. Monitoring and Evaluation

4.5.5.1. Data Collection

Understanding the unique requirements and difficulties faced by women in agriculture requires the collection of gender-disaggregated data. Programs and policies can be informed by this data, guaranteeing that gender disparities are adequately addressed.

Governments and organizations must spend money on data collection and research to track advancements in women's rights and agricultural participation. Analyzing the results of initiatives can reveal important information and point out areas that need work.

4.5.5.2. Accountability Mechanisms

Putting in place accountability systems is essential to making sure that pledges to promote gender equality are carried out. Holding organizations and governments responsible for their commitments can be achieved by keeping an eye on the execution of policies and initiatives meant to advance women's rights in agriculture. To ensure that women's opinions are heard during decision-making processes, civil society organizations can be extremely effective in promoting accountability and transparency.

Recognizing the vital role women play in food production and rural economies has fueled efforts to advance women's rights and participation in agricultural work. The issue of gender inequality in agriculture has advanced significantly thanks to a combination of advocacy campaigns, community-based projects, and legislative frameworks. But there are still obstacles to overcome, and continued dedication is required to guarantee that women have equal access to opportunities, resources, and decision-making.

Gender equality in agriculture must be further promoted in the future via community involvement, capacity-building programs, and inclusive policies. We can establish a more equitable agricultural sector that benefits not only women but also families, communities, and economies overall by empowering women and removing the structural obstacles they encounter. In addition to being a question of justice, achieving gender equality in agriculture is a step toward long-term growth and future generations' food security (World Bank, 2012, 2015).

4.6. CHALLENGES FACED BY WOMEN IN AGRICULTURE

The difficulties faced by women in agriculture are a more severe and upsetting version of the "progressive collection of issues" that other oppressed groups in society face. These concerns include land ownership, protection of tenure, land quality in situations where land ownership is assured, and land management about agriculture and the infrastructure it requires. Given the current gender roles that women must fulfill, particularly concerning household management in their reproductive roles—fuel wood selection, fodder selection, livestock tending in general, food protection needs, etc.—some changes in land ownership and farming practices have a greater positive or negative impact on women than on men. Their dependence on agriculture, common lands, forests, and water is even more pronounced. The land ownership status of farm households influences how women participate in farm production. Women work in a variety of roles, from supervisors to laborers without land. According to estimates, women contribute between 50% and 60% of all labor in agricultural production on average, and even more in some regions. Due to their lower pay than men, girls are given preference in the cotton industry. They are also typically easier to manage and put in more hours and effort. Land planning, planting, sowing, and fertilizer application are examples of tasks that are typically completed by men and usually done within a set amount of time. However, weeding is one of the repetitive, everyday tasks that women perform from the time the seed is sown until it is harvested. (Verma et al., 2016).

24.6.1. Land Access

About 1% of the land is owned by women, who also produce 80% of the crops. This demonstrates the lack of ownership security and, typically, the 1% of men who own smaller land parcels to less fertile men.

4.6.2. Lack of Social Status

In society, women who work in agriculture are not given the recognition they deserve. They are not receiving the social standing they are due in India. In this situation, female farm workers should be given social status because the agriculture industry and agriculture would not be complete without them.

4.6.3. Health

Another significant problem that women farm workers face is health issues. They operate in extremely dire conditions in rural areas. To raise money, they have even sacrificed their well-being. The unsanitary working conditions for women farm workers are a result of their lack of access to adequate food, rest, and sanitation facilities.

4.6.4. Overworked

Women are the foundation of every family, but in rural areas, their responsibilities have expanded beyond simple housework; they must now work outside the home to support their families financially. Farmworkers frequently juggle work and family obligations; they are mentally and physically exhausted.

4.6.5. Discrimination

Due to discrimination, women farmers continue to face more obstacles than men because they have less access to information, technology, property, inputs, and credit. Additionally, a greater percentage of them were illiterate and engaged in subsistence farming without being up to date with modern technologies, and their varied tasks limited their time and mobility. It is emphasized that people must be trained, educated, and have access to inputs and services to produce cash crops. Training facilities for farmers, which are designed to offer residential instruction on technical subjects, served as an example of this male bias. Women have been unable to attend many trainings because, like most other places where farmers receive training, facilities specifically designed for them are disregarded. Furthermore, men have historically and currently made up the majority of the workforce in the government's extension services because there are likely insufficiently qualified women to fill these roles at the field level.

4.6.6. Lack of Infrastructure

Women farmers would be more accessible and able to sell more of their produce on schedule if they had access to logistics and transportation.

4.6.7. Less Market Opportunities

Women farmers' access to markets is limited by a lack of market research and information. Women can only shop at local markets because prices there are typically lower than in cities.

4.6.8. Access to Finance

Another significant obstacle that women in agriculture face is limited access to financing. Because farmers' production activities are typically seasonal in nature and there is a significant lag between when they incur costs and when they can profit from their produce, credit is a very helpful resource for them.

Several researchers have identified many reasons why women farmers are still not able to access credit easily, some of the most relevant ones include;

- Lack of collateral requirements;

- High transaction cost;
- Limited education and mobility;
- Socio-cultural impediments;
- Irregularity of employment; and
- The nature of women's businesses limits their ability to obtain credit.

4.6.9. Access to Agricultural Inputs

Accessing essential agricultural inputs like better seedlings, fertilizer, pesticides, machinery, etc., can be challenging for these women, frequently due to their precarious financial circumstances. According to women farmers, the high cost of improved inputs on the open market prevents them from using them.

When combined, these elements limit women's productivity by limiting their access to information about the input and output markets.

4.6.10. Gender Division of Labor in Agriculture

The widespread practice of gender division of labor in agriculture has frequently been perceived by some as a constraint for the majority of women. There has long been a rigid division of labor by gender in agriculture in the majority of Africa. Additionally, subsistence crops are referred to as "female crops," whereas cash crops and export crops are frequently referred to as "male crops."

The primary rationale behind this classification is that women prefer to cultivate subsistence crops for domestic use because they are in charge of providing for the family. Conversely, men in agriculture are expected to grow cash and export crops that will increase the family's income because they are the breadwinners of their households.

4.6.11. Participation in Farm Management, Agricultural Policy Deficiencies and Other Socio-Economic Factors

Like in most patriarchal societies, women's decision-making in agriculture is influenced by a variety of factors, including socioeconomic status. While it's clear that rural women's low economic standing limits their ability to fully engage in agricultural production, other factors like their designated roles as housewives, caregivers, and childbearing also play a part. Although women are a vital component of the farming system, policymakers have largely ignored this gender dimension, which has had a negative impact on agriculture and gender equity. The fact that men are more likely than women to hold higher education levels in rural areas and are therefore more likely to be selected for such government-level positions may be one factor contributing to the low levels of female representation in policy formulation.

4.6.12. Safety

For these women, the most significant problem is that their disorganized structure makes them invisible. For the women who work in rural areas, this results in insecurity.

Women are not only an essential component of agriculture, but they are also a key factor in its success. It is impossible to overstate their significance in agriculture. Because of their diligence and hard work, female farmers have played a crucial role in feeding the world. Agriculture wouldn't exist without them, and without agriculture, there wouldn't be any food for everyone.

In addition to being hampered by their gender, women farmers also had to deal with the difficulties that all small-scale farmers face. Women farmers are playing a key role in small-scale agriculture, which is two to four times more effective than other sectors at reducing poverty and hunger. Empowering women farmers is more crucial than ever to end world hunger and ensure climate change resilience (World Bank, 2012, 2015).

REFERENCES

1. African Development Bank (AfDB), Organization for Economic Co-operation and Development (OECD), World Bank (WB), & World Economic Forum (WEF). (2015). *Africa Competitiveness Report 2015.* Retrieved from: http://www3.weforum.org/docs/WEF_ACR_2015/Africa_Competitiveness_Report_2015.pdf (accessed on 17 December 2024).
2. Aguilar, A., Carranza, E., Goldstein, M., Kilic, T., & Oseni, G. (2014). Decomposition of gender differentials in agricultural productivity in Ethiopia (Policy Research Working Paper No. 6764). The World Bank.
3. Akresh, R. (2005). Understanding pareto inefficient intrahousehold allocations (Discussion Paper No. 1858). IZA.
4. Alene, A. D., Manyong, V. M., Omanya, G. O., Mignouna, H. D., Bokanga, M., & Odhiambo, G. D. (2008). Economic efficiency and supply response of women as farm managers: Comparative evidence from Western Kenya. *World Development, 36*(7), 1247–1260.
5. Ali, D., Bowen, D., Deininger, K., & Duponchel, M. (2015). Investigating the gender gap in agricultural productivity: Evidence from Uganda (Policy Research Working Paper No. 7262). The World Bank.
6. Backiny-Yetna, P., & McGee, K. (2015). Gender differentials and agricultural productivity in Niger (Policy Research Working Paper No. 7199). The World Bank.
7. Barrett, C. B. (1996). On price risk and the inverse farm size productivity relationship. *Journal of Development Economics, 51*, 193–215.
8. Barrett, C. B., Sherlund, S. M., & Adesina, A. A. (2008). Shadow wages, allocative inefficiency, and labor supply in smallholder agriculture. *Agricultural Economics, 38*(1), 21–34.
9. Croppenstedt, A., Goldstein, M., & Rosas, N. (2013). Gender and agriculture: Inefficiencies, segregation, and low productivity traps (Policy Research Working Paper No. 6370). The World Bank.
10. Food and Agriculture Organization (FAO). (2011). *The State of Food and Agriculture: Women in Agriculture–Closing the Gender Gap for Development.* Rome, Italy: FAO. Retrieved from: http://www.fao.org/docrep/013/i2050e/i2050e.pdf (accessed on 17 December 2024).
11. FAO. (2011). *Women in Agriculture: Closing the Gender Gap for Development.* Retrieved from: http://www.fao.org/publications/sofa/2010-11/en/ (accessed on 17 December 2024).
12. Fox, J. (1990). Describing univariate distributions. In J. Fox & J. S. Long (Eds.), *Modern Methods of Data Analysis* (pp. 58–125). Newbury Park, CA: Sage.

13. Gilbert, R. A., Sakala, W. D., & Benson, T. D. (2002). Gender analysis of a nationwide cropping system trial survey in Malawi. *African Studies Quarterly, 6*(1).
14. Goldstein, M., & Udry, C. (2008). The profits of power: Land rights and agricultural investment in Ghana. *Journal of Political Economy, 116*(6), 981–1022.
15. Horrell, S., & Krishnan, P. (2009). Poverty and productivity in female-headed households in Zimbabwe. *Journal of Development Studies, 43*(8), 1351–1380.
16. Kilic, T., Palacios-Lopez, A., & Goldstein, M. (2013). Caught in a productivity trap: A distributional perspective on gender differences in Malawian agriculture (Policy Research Working Paper No. 6381). The World Bank.
17. Lalitha, V., & Kumar, T. V. (2016). Farm management in rural areas: Role of women in decision-making process. *Social Sciences International Research Journal, 2*, 63–65.
18. Meenakshi, J. R., & Panneer, S. (2020). Occupational health of agricultural women workers in India. *Indian Journal of Community Medicine, 45*, 546–549.
19. Mukasa, A. N., & Salami, A. O. (2016). Gender productivity differentials among smallholder farmers in Africa: A cross-country comparison (Working Paper Series No. 231). African Development Bank.
20. Oladeebo, J. O., & Fajuyigbe, A. A. (2007). Technical efficiency of men and women upland rice farmers in Osun State, Nigeria. *Journal of Human Ecology, 22*(2), 93–100.
21. Oseni, G., Corral, P., Goldstein, M., & Winters, P. (2013). Explaining gender differentials in agricultural production in Nigeria. *Background Paper: Levelling the Field–Improving Opportunities for Women Farmers in Africa.* World Bank & ONE Report.
22. Palacios-López, A., & López, R. (2015). The gender gap in agricultural productivity: The role of market imperfections. *The Journal of Development Studies, 51*(9), 1175–1192.
23. Patil, B., & Babu, V. S. (2018). Role of women in agriculture. *International Journal of Applied Research, 4*(12), 109–114.
24. Peterman, A., Quisumbing, A., Behrman, J., & Nkonya, E. (2011). Understanding the complexities surrounding gender differences in agricultural productivity in Nigeria and Uganda. *Journal of Development Studies, 47*, 1482–1509.

25. Saito, K., Mekonnen, H., & Spurling, D. (1994). Raising the productivity of women farmers in sub-Saharan Africa. *World Bank Discussion Papers: Africa Technical Department Series* (No. 230). Washington, DC: World Bank.
26. Singh, S. K. (2018). Role of women in agriculture sector. *International Journal of Scientific Current Research and Development.*
27. Timothy, A. T., & Adeoti, A. L. (2006). Gender inequalities and economic efficiency: New evidence from cassava-based farm holdings in rural south-western Nigeria. *African Development Review, 18*(3), 428–443.
28. Tiruneh, A., Testfaye, T., Mwangi, W., & Verkuijl, H. (2001). Gender differentials in agricultural production and decision-making among smallholders in Ada, Lume, and Gimbichu woredas of the central highlands of Ethiopia. *International Maize and Wheat Improvement Center and Ethiopian Agricultural Research Organization: Mexico, DF, and Addis Ababa.*
29. Udry, C. (1996). Gender, agricultural production, and the theory of the household. *Journal of Political Economy, 104*(5), 1010–1046.
30. Vargas Hill, R., & Vigneri, M. (2011). Mainstreaming gender sensitivity in cash crop markets supply chains (Working Paper No. 11–08, ESA). FAO.
31. Verma, M., Singh, S., & Verma, V. K. (2016). Problems of women farmers in agriculture. *Journal of Agroecology and Natural Resource Management, 3*(1), 10–13.
32. World Bank. (2012). *World Development Report 2012: Gender Equality and Development.* Washington, DC: The World Bank.
33. World Bank. (2015). *World Development Indicators 2015.* Washington, DC: The World Bank. https://doi.org/10.1596/978-1-4648-0440-3.
34. Yadav, S. (2017). Role of rural women in agricultural activities in Jaipur district of Rajasthan (M.Sc. Thesis). Sri Karan Narendra Agriculture University, Jobber.

CHAPTER 5

OCCUPATIONAL SAFETY AND HEALTH IN AGRICULTURE

Contents

INTRODUCTION

The multidisciplinary field of occupational safety and health (OSH) is concerned with the welfare, safety, and health of employees in the workplace. It entails foreseeing, recognizing, evaluating, and managing workplace risks that could endanger employees' health and welfare. Enhancing working environments and conditions is another goal of OSH. Agriculture workers are frequently at risk for lung conditions, noise-induced hearing loss, skin conditions, work-related injuries, and some types of cancer linked to chemical use or extended sun exposure. Agricultural machinery use is a common cause of injuries on industrialized farms. Rollover protection structures reduce the risk of injury if a tractor rolls over, preventing the most frequent cause of fatal agricultural injuries in the United States. Workers who are exposed to pesticides or other chemicals used in farming may suffer from illnesses or birth defects, making them potentially harmful to their health (Garzia et al., 2018). Agriculture is a common source of occupational illnesses and injuries among younger workers because it is an industry where families, including children, frequently work side by side. Young farm workers are frequently fatally injured in accidents involving machinery, motor vehicles, and drowning.

Increased prevalence rates of many occupational exposures that may have a detrimental effect on health were discovered in the forestry, fishing, and agriculture sectors by the 2010 NHIS-OHS. Frequently, these laborers put in long hours. Of those working in these industries, 37% reported working more than 48 hours per week, and 24% reported working more than 60 hours per week (Kumaraveloo et al., 2018). 85% of all workers in these industries worked outside regularly, compared to 25% of all US workers. Furthermore, compared to 25% of all US workers, 53% were regularly exposed to vapors, gas, dust, or fumes.

5.1. OCCUPATIONAL HEALTH HAZARDS IN AGRICULTURAL WORK

Anything that endangers a worker, a machine, or the environment is considered an occupational hazard. The potential for harm in the form of illness or injury, property damage, environmental damage at work, or a combination of these are the sources or circumstances of occupational hazards. Women who work in agriculture may be subject to the same risks and dangers as men. They also face other risks, especially those related to reproductive health (such as biological agents and pesticides). Women may also be more susceptible to upper limb and neck disorders related to their jobs. Women's exposure to injury-causing situations may be influenced by sociocultural, economic, and physical factors. They also possess information and solutions for medication and injury prevention. When women seek medical attention worldwide, they are less aware of the severity of the injury and less aware of the risks associated with their jobs. The physical strain was increased by differences in size and stature, and women may be more susceptible to ergonomic-related injuries due to low maximal oxygen uptake.

5.1.1. Types of Health Hazards

Anything that can be harmful if left unchecked is considered a hazard. It is an unanticipated, unplanned, or uncontrollable event that typically has unfavorable outcomes. It also alludes to the possible hazards to the health and safety of people who work both indoors and outdoors. Farm women are more likely to experience this since they participate in both domestic and agricultural tasks. Both indoor and outdoor environments are exposed to them. The following lists the various kinds of hazards.

5.1.1.1. Physical Hazards

Workplace physical elements that can negatively impact health include noise, vibration, dim lighting, ionizing, and non-ionizing radiation, and microclimatic conditions. One of the most common occupational health consequences in both developed and developing nations is noise-induced hearing loss. Health can be impacted by coming into contact with poisonous and wild animals, such as insects, spiders, scorpions, snakes, and some wild mammals.

5.1.1.2. Chemical Hazards

In contemporary settings, there are approximately 100,000 distinct chemical products in use, and the number is constantly increasing. Exposure differs greatly. Health effects include cancer, reproductive disorders, pesticide poisoning, dermal, and respiratory allergies, dermatomes, and harm to the liver and central nervous system (from exposure to solvents). Menstrual dysfunction, such as dysmenorrhea, irregular cycles, and spontaneous abortions, has been reported to be more common in women exposed to toluene. When exposed to concentrations higher than the threshold limit value (TLV), toxic, corrosive, allergenic, and carcinogenic chemicals act locally, through inhalation, and through ingestion.

5.1.1.3. Biological Hazards

At work, employees may come into contact with parasitic agents and infections. Agricultural workers and those handling animal products are likely to be exposed to biological hazards.

5.1.1.4. Mechanical Hazards

Numerous musculoskeletal conditions related to the agricultural industry have been documented. Women typically have smaller stature, less physical strength, an 11% lower vital capacity, and hemoglobin levels that are below normal. 20% less; they have more body fat; and their skin area is greater than their circulating volume. They are more tolerant of cold and less tolerant of heat. They are therefore more vulnerable to mechanical risks. These include carpel tunnel syndrome, repetitive motion disorders, shoulder and tendon disorders, cumulative trauma disorders, and back and neck pain. Some of the most common workplace hazards in both developed and developing nations are mechanical hazards, unshielded machinery, unsafe workplace structures, and dangerous tools. About 30% of workers in developed nations and 50%–70% of workers in developing nations may be subjected to a high physical workload, poor ergonomic conditions, a lot of lifting and moving of heavy objects, or repetitive manual labor. These may result in musculoskeletal conditions and injuries. These conditions result in financial losses and are the primary cause of both temporary and permanent work disabilities. The mechanical risks in the industry are centered on moving parts, protruding machinery, and similar elements. Mechanical causes account for around 10% of industrial accidents.

5.1.1.5. Psychosocial Hazards

One of the main issues from a gender perspective is occupational stress. Over the past 10 years, stress brought on by work and time constraints has increased. Negative psychological effects can also result from shift work, seasonal work, monotonous work, work requiring continuous focus, and irregular working hours. Sleep issues, burnout syndromes, depression, and high blood pressure have all been linked to psychological stress and overload. Concerns regarding workplace stress are raised by social conditions of work, such as gender distribution, job segregation, and equality. Farmwomen engage in domestic tasks in addition to farming. These result from the employee's inability to adjust to a strange psychosocial setting. Among the psychosocial elements that can jeopardize employees' physical and mental well-being are frustration, a lack of job satisfaction, insecurity, strained interpersonal relationships, and emotional stress.

5.1.1.6. Ergonomical Hazards

Ergonomics encompasses the task, the organization, the tool, the workstation, and the surroundings. It aims to lower the number of musculoskeletal disorders (MSDs) that employees develop at work. MSDs are injuries and illnesses that affect muscles, nerves, tendons, ligaments, joints or spinal disks. MSDs are frequently characterized by painful joints, numbness in the hands, waist, forearms, shoulders, knees, and feet, as well as pain in the neck or back. Inflammations or swelling are typical. Awkward posture, repetitive motion, forceful exertion, static posture, and an excessive range of motion are risk factors.

5.1.2. Health Hazards in Agriculture

Agricultural workers are exposed to a wide range of environmental hazards that could be detrimental to their health and well-being, possibly more than any other occupational group. The prevalence of respiratory conditions, noise-induced hearing loss, skin conditions, some types of cancer, chemical toxicity, and heat-related illnesses is higher among farmers and farmworkers. These possible risks can be reduced or eliminated with the help of preventative measures.

5.1.2.1. Farm Noise

Farm equipment and tool noise can result in irreversible hearing loss. Although hearing loss may initially be transient, repeated exposure will cause irreversible harm. Over several years, the damage may develop gradually and go undetected until it is too late. Certain sounds, like gunfire, are so loud that they can result in immediate and irreversible harm. For an eight-hour day, the acceptable noise exposure level is 90 dB(A). For peak noise, like a gunshot, the exposure standard is 140 dB.

5.1.2.1.1. Spot the Hazard

Some early warning signs of hearing loss include:

- Ringing in the ears after work;
- Difficulty understanding a normal conversation;
- Turning up the volume on radio or television when others appear to hear adequately; and
- Failing to hear background noises, such as a ringing telephone or doorbell.

Typical farm noises that can damage hearing include:

- Tractor (95–100dB(A);
- Header (88–90dB(A);
- Orchard sprayer (85–100dB(A);
- Angle grinder (95–105dB(A);
- Bench grinder (90–95dB(A);

- Chainsaw (105–120dB(A);
- Pig shed at feed time (95–105dB(A); and
- Shotgun (over 140dB(lin)).

5.1.2.1.2. Assess the Risk

Your hearing may be in danger if you have to shout loud enough for someone a meter away to hear you. Wearing protective hearing equipment is required if noise cannot be eliminated or reduced at its source and if there is no other means of keeping people away from harmful noise exposure. Some farmers employ a noise consultant to take noise readings, assess hearing risks and recommend preventive measures.

5.1.2.1.3. Make the Changes

You can reduce noise at its source by:

- Using quieter machines;
- Reducing and modifying noise;
- Maintaining machines; and
- Using low revs when appropriate.

You can protect people from loud noise exposure by:

- Limiting time spent;
- Insulation of noisy machines;
- Scheduling noisy work with a few workers; and
- Job rotation.

5.1.2.1.4. Protective Equipment

- Hearing protection;
- Ear muffs;
- Lowering the SLC80 sound level conversion) figure;
- Ear plugs;
- Maintaining noisy parts; and
- Regular cleaning.

5.1.2.1.5. Remember

Hearing aids are not very helpful once hearing is lost; once lost, it cannot be recovered. They are able to increase the volume of speech, but not its clarity.

5.1.2.2. Heat Stress

When the body accumulates more heat than it can manage, heat stress results. The risk of heat stress is increased by high temperatures, high humidity, sunlight, and demanding work schedules. When feasible, make use of fans, ventilation systems, and shade. Sometimes a canopy or tarp can provide shade for a work area. In addition to wearing cooling vests—clothes with ice or frozen gel inserts—make sure you drink lots of water before, during, and after work. Give yourself time to get used to the workload and the heat. Heat stress is less likely to affect people who are accustomed to working in hot conditions. For a few days in a row, work in the heat for about two hours each day to get used to it. Then, over the following few days, progressively increase the amount of time and workload. It is advised to give yourself at least seven days to adjust. Workers may naturally adapt if the warm weather comes gradually. It has long been known that one of the most important factors affecting quality of life is good health. One of the most important resources to safeguard is the health of farm workers. Your quality of life can be greatly improved by taking the suggested preventative actions to safeguard your health.

5.1.2.3. Stress on the Farm

5.1.2.3.1. What is Stress?

A person's response to something viewed as a threat or challenge is stress. It is the emotional stress and strain that oneself and other people place on one's physical and mental health. The body starts preparing for action when it is under stress. Although it requires more energy, doing this makes one stronger and more aware.

5.1.2.4. How is Stress Harmful?

5.1.2.4.1. Health

The body starts to do more of some things and less of others when it is under stress. While digestion slows down or even stops, blood circulation increases. Major health issues like ulcers and heart disease could result from this. Other minor but significant health issues include headaches, poor digestion, and insomnia.

5.1.2.4.2. Relationships With Other People

Most people forget about other people when they are under stress because they are too preoccupied with their own issues. They start venting their emotions on friends and family at the same time. Stress soon affects not just the individual but the entire family as well.

5.1.2.4.3. Efficiency in the Workplace

Stress can temporarily improve an individual's productivity. However, a person will deteriorate over time, growing weaker and more easily fatigued. Poor management choices could arise from a lack of focus. This can be particularly risky when using machinery.

5.1.2.4.4. More Stress

Stress will spread like wildfire. The issues it brings about with work, family, and personal health will all turn into new ones. When stress is not managed, it can spiral out of control (WHO, 1990; Mostafalou and Abdollahi, 2017).

5.1.2.5. Steps to Take to Control Stress

- Analyze how you feel.
- List the stressors.
- Check the effects of stress on you.
- Check the level of stress now vs. earlier.

5.1.2.6. Learn How to Manage Stress

- Talking about problems is a good way to relieve stress. Choose someone you can be honest with, and then share your problems and discuss solutions with them.
- Learn to identify stressors. These could include stomach issues, tightness in the shoulders and neck, or behavioral or interpersonal changes. A sophisticated system in the body alerts the user when stress levels are too high.
- Examine the list of things that stress you out and consider how serious each one is. Choose variables like prices and the weather that are beyond anyone's control. Next, assess the source of your stress. Is it a small matter or something beyond your control?
- When tackling a significant issue, attempt to deconstruct it into smaller components. Choose one task and focus on finishing it if the barn needs to be repaired. Proceed to the next task after finishing the previous one.
- Schedule the time realistically.
- Short breaks are to be taken.
- Learn to relax.
- Develop other interests.
- Take help like counselling.

5.1.2.7. Take Care of Your Self

Tips from the American Heart Association:

- Regular exercise.
- Eat a well-balanced diet. Sleep and rest.
- Balance work and play.

Acknowledge the things you cannot alter. Seek out the best in both people and circumstances. Keep in mind that nobody is flawless. Recognize that you have no control over the time and money constraints brought on by weather, crop prices, and market demand. (WHO, 1990; Mostafalou and Abdollahi, 2017).

5.1.3. General Health for Farm Workers

5.1.3.1. Know Your Physical Limits

- A common cause of farm accidents is going beyond one's personal limits. The risk of an accident or illness increases when you work in extremely hot or cold conditions or try to do tasks that are beyond your physical capabilities.
- Prepare for a secure day. This entails getting enough sleep, eating the right food, and dressing appropriately for the climate and job. To combat fatigue and prolong your energy, take breaks from your work. When you've had enough, stop.
- Seek assistance if lifting or carrying something will be difficult. Make sure you possess the strength, skill, and stamina needed for the job or activity to complete it successfully and safely. Look for the least demanding method. When feasible, use motor power instead of muscle power. Arrange your tasks to utilize as much of your available energy as possible.
- When determining what and how much you can do safely, take your age and health into account. Be prepared to reassign tasks and activities that become unsafe due to age or health issues. Engage in regular exercise to maintain your agility and to improve your cardiovascular fitness and muscle tone.

5.2. EXPOSURE ROUTES AND HEALTH RISKS ASSOCIATED WITH PESTICIDE APPLICATION

Pesticides are crucial to the development of agriculture because they can increase the affordable yield and quality of food while lowering the loss of agricultural products. The development of pesticides increased during World War II (1939–1945) due to the pressing need to improve food production and control insect-borne diseases. From the 1940s onward, the increased use of synthetic crop protection chemicals allowed for an even greater increase in food production (Carvalho et al., 2017). Furthermore, from 0–2 million tons in the 1950s to over 5 million tons in 2000, the global production of pesticides has grown at an annual rate of roughly 11%. Only 1% of all pesticides were successfully used to control insect pests on target plants, although three billion kilograms of pesticides are consumed annually worldwide. The majority of the remaining pesticides enter or reach environmental media and non-target plants. Pesticide contamination has consequently had a negative influence on human health and greatly contaminated the environment.

To minimize spray solution loss, minimize pesticide residues in the environment, and prevent harmful effects on human health from over spraying and residues, it is crucial to have the right equipment for applying pesticides. People are exposed to pesticides both directly at work and indirectly through potentially pesticide-contaminated environmental media like the air, water, soil, and food chain. The three main ways that pesticides typically enter the human body are through the skin, mouth, and respiratory system. A WHO and UNEP report states that exposure to pesticides causes 200,000 fatalities and three million poisonings globally (Yadav et al., 2015).

Acute and long-term toxicity to humans from pesticides is still a major issue, despite some steps being taken to lessen their negative effects on the environment and human health. Over the next 10 years, it is anticipated that the risk of pesticide exposure will rise globally, particularly in developing nations. The need to use fewer pesticides has been extensively discussed, but less attention has been paid to how using the right application techniques can lower exposure to pesticides. Thus, it is necessary to talk about the ways in which pesticides are exposed and the associated health issues. To help policymakers develop strategies regarding the establishment of appropriate pesticide application techniques and methods to minimize pesticide exposure and reduce the health effects on both applicators and communities, this section aims to review the most recent pesticide application methods, the main routes of pesticide exposure and the associated human health effects, the fundamental framework of health risk characterization, and the benefits and drawbacks of various risk characterization techniques.

5.2.1. Pesticide Application Methods

The properties of the pesticides, the target site's features, and the target pests' habits are taken into account when choosing pesticide application techniques. Band spraying, broadcast spraying, foliar, drench, soil injection, space treatment, and spot treatment are the most widely used pesticide application techniques. Band spraying and broadcast spraying are two popular techniques for treating crop areas consistently. Band spraying is a technique used to eradicate all vegetation along fence rows and borders using non-selective herbicides. Broadcast spraying involves applying pesticides evenly over a sizable turf grass area, either by hand or with the aid of motorized equipment. Important variables about the effectiveness of these application techniques include spray pressure, walking speed, nozzle tip, and height. Additionally, the Kentucky Pesticide Education Program (2016) demonstrates that the spray pressure is determined by hand pumping. The right use of pesticides is made possible by choosing a nozzle tip for the volume of mixed spray and a suitable constant walking speed.

When treating dispersed clumps of brush or weeds, spot, and space treatments are frequently utilized. Without treating the entire turf area, these application techniques are effective ways to address particular trouble spots. Larger areas are not suitable for manual sprayers, which are made for spot and space treatment applications. They are easy to use, maintain, and store, and they are reasonably priced when compared to other sprayers. Lawn care sprays are also applied with adjustable spray guns (Xiongkui et al., 2017).

Non-spray pesticides can be applied in a variety of ways, such as foliar, drench, and soil injection techniques. Soil injection involves applying pesticides directly to the soil rather than onto a growing plant, the foliar application involves applying pesticides directly to the leafy parts of a plant, and soaking involves applying pesticides specifically to the root system and allowing them to pass through a plant. The purpose of the wiper application is to kill specific weeds by wiping a non-selective herbicide onto the plant. These methods eliminate drift and enable more effective placement and use of pesticides, particularly against certain borers.

5.2.1.1. Hydraulic Sprayers

Hand-pumped backpack sprayers and powered units with multiple nozzle booms are examples of hydraulic sprayers. The spray mix is always atomized at the nozzle using pressure from a pump, compressed gas, or air. For large trees, adequate spray coverage requires high-pressure pumps.

5.2.1.2. Backpack Sprayers

A tank, a pump, a spray wand, and a nozzle make up the basic backpack sprayer. It works well for spot spraying, treating small areas, and covering difficult-to-reach places. Broadcast, band, space, and spot treatment are the primary spray options.

5.2.1.3. Basal Trunk Sprayers

Basal trunk spraying is a quick and simple method of applying pesticides that only needs a garden sprayer and no additional tools. Using a water-soluble pesticide, the lower five feet of a tree trunk are thoroughly wetted. The tree's vascular system distributes and absorbs pesticides through the bark. When farmers apply the chemicals correctly, they do not pollute the soil, so this application method does not damage the tree.

5.2.1.4. Aerial Sprayers

In large, inaccessible areas, such as highlands and special areas where pests and crop diseases occur frequently and repeatedly, or in places that are unsuitable for ground-based spraying, helicopters or fixed-wing aircraft are utilized to apply pesticides to crops and plantations.

Pesticides have the potential to enter the environment when they are applied to a target plant using various application techniques. Pesticides may go through processes like transfer (or movement) and degradation once they enter the environment (Figure 5.1). Adsorption, leaching, volatilization, spray drift, and runoff are some of the transfer processes that cause pesticides to move from the target site to other environmental media or non-target plants. Food contamination, soil pollution, water pollution, and air pollution are all results of improper pesticide use, management, and behavior in the environment. As a result, farmers and locals are exposed to pesticides through contaminated food, soil, water, and air.

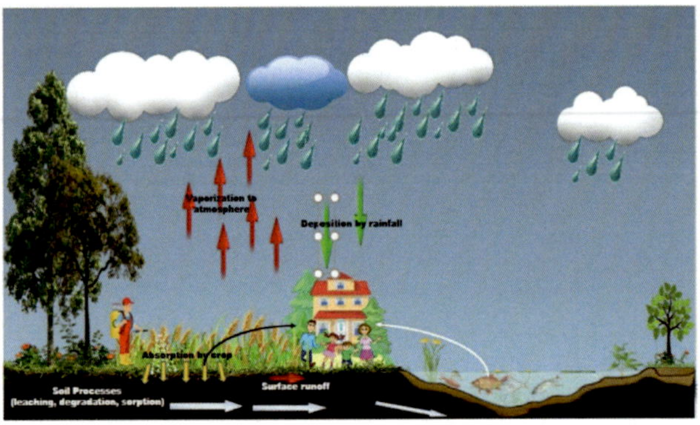

Figure 5.1. *Pesticide behavior in the natural environment in a crop field.*

Source: Tudi M, Li H, Li H, Wang L, Lyu J, Yang L, Tong S, Yu QJ, Ruan HD, Atabila A, et al. Exposure Routes and Health Risks Associated with Pesticide Application. Toxics. *2022; 10(6):335. https://doi.org/10.3390/toxics10060335.*

5.2.2. Routes of Pesticide Exposure

In addition to being directly exposed to pesticides through various application techniques used in the workplace, in agriculture, and in the home, people are also indirectly exposed to pesticides through contaminated environmental media such as food, water, soil, and air. The level of pesticide toxicity is determined by these various exposure types. The skin, mouth, and respiratory systems are the primary routes by which pesticides enter the human body (Figure 5.2).

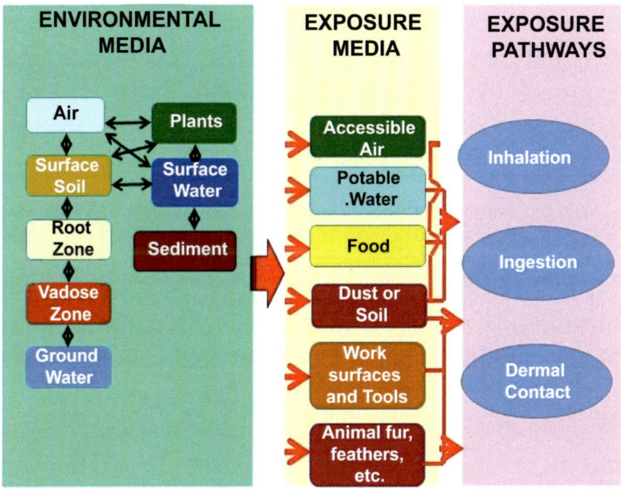

Figure 5.2. *Routes of exposure to pesticides.*

Source: Tudi M, Li H, Li H, Wang L, Lyu J, Yang L, Tong S, Yu QJ, Ruan HD, Atabila A, et al. Exposure Routes and Health Risks Associated with Pesticide Application. Toxics. *2022; 10(6):335. https://doi.org/10.3390/toxics10060335.*

For farmers exposed to pesticides through splashing, spilling, or spray drift, dermal exposure is the most frequent and efficient route of exposure, particularly when the farmers use the pesticides for household or agricultural purposes. For instance, Wang et al. used absorbent paper patches to conduct a risk assessment of employees who were directly exposed to Trinexapac-ethyl (TE) topically. Results of the exposure intensity (EI) and potential dermal exposure (PDE) for worker body sections revealed that the EI of each body section varied from 7.22–73.0 $pg \cdot cm^{-2}$ for workers using manual sprayers, hand-held power sprayers, and mixing/loading equipment. The hands of manual sprayers had the highest EI of TE, whereas the upper arms of mixing and loading employees had the lowest EI of TE. The maximum exposure values were 32% for the thighs of manual sprayers, 40% for the chests and backs of hand-held power sprayers, and 29% for the chests and backs of mixing/loading workers. Taufeeq et al. conducted a risk assessment for human health and looked into the contamination of OCPs in River Barandu. The findings indicated that the levels of ΣOCP in water ranged from 0.340–0.935 μg/L, while those in sediments ranged from 32.9–98.8 ng/g. The most common pesticides in both river matrices were heptachlor and hexachlorocyclohexanes (HCHs). For the local populations, the lifetime carcinogenic and non-carcinogenic health risks linked to skin contact with OCP-contaminated river water were deemed insignificant. Han et al. evaluated the risk of pesticide exposure for individuals engaged in the seed-coating procedure and identified dermal exposure. The study's findings showed that the type of work performed and the area of a worker's body exposed affected the levels of chemical exposure. While handling carbofuran presented a possible risk to human health, handling tebuconazole during seed coating carried a low risk.

Oral exposure is the most frequent poisoning route and results in serious health issues. Oral exposure to pesticides can occur when those who manufacture or use them do not wash their hands before eating or smoking, and it can also happen to the general public when they eat pesticide-contaminated food.

Workers' health is at risk, particularly their nose, throat, and lung tissues if they breathe in significant quantities of pesticides from the air, water, and soil. Respiratory exposure (by inhalation or breathing) is caused by the volatile components of pesticides. Yoshida et al. for instance, examined the role of the inhalation pathway in Japanese children's exposure to pyrethroids in the home using all exposure pathways. Subjects between the ages of 6 and 15 ($n = 132$) had their urine collected as soon as they woke up, and samples of airborne pyrethroids were taken in their bedrooms for a full day. Profluthrin (median 15%) ≈ transfluthrin (14%) > metofluthrin (1%) > bifenthrin (0.1%) was the order in which the contribution rates of the amounts absorbed by inhalation compared to the amounts absorbed via all exposure pathways tended to decrease. The most well-known pyrethroid to pollute indoor air was transfluthrin, which had negative health effects on residents who were exposed to it (WHO, 1990; Mostafalou and Abdollahi, 2017).

5.2.3. Human Health Effects Related to Pesticide Exposure

There is a significant reliance on pesticide applications to meet the enormous demand for food production by an expanding population because of the significant role that pesticides play in agricultural development. This has a negative impact on human health globally and stresses the environment. Workers, formulators, sprayers, mixers, loaders, and agricultural farm workers are among the high-risk groups that are directly exposed to pesticides. Due to the risky nature of the processes involved, contact with hazardous materials and situations increases significantly during manufacturing and formulation.

Additionally, workers are exposed to pesticides directly through their hands and through the airborne chemicals they inhale. Because they must handle a variety of hazardous chemicals in industrial settings, such as pesticides, raw materials, toxic solvents, and inert carriers, workers are also more vulnerable. Indirect pesticide exposure to humans occurs through tainted soil, water, air, and food chains. There are numerous direct and indirect ways for humans to be exposed to pesticides, and each has a unique impact on health (Figure 5.3).

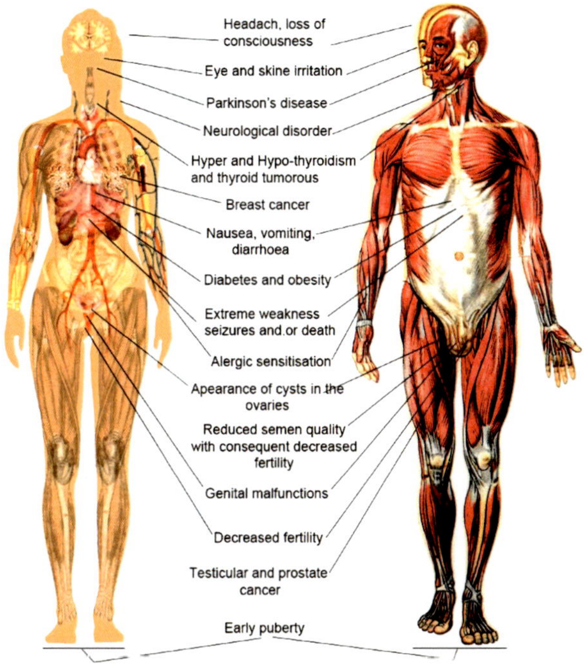

Figure 5.3. *Health problems due to pesticide exposure.*

Source: Tudi M, Li H, Li H, Wang L, Lyu J, Yang L, Tong S, Yu QJ, Ruan HD, Atabila A, et al. Exposure Routes and Health Risks Associated with Pesticide Application. Toxics. *2022; 10(6):335. https://doi.org/10.3390/toxics10060335.*

Both direct and indirect pesticide exposure can cause acute toxic effects, such as mass food poisoning, suicide attempts, industrial chemical accidents, occupational exposure in agriculture, and several severe chronic diseases, such as cancer, asthma, diabetes, Parkinson's disease, leukemia, and cognitive impairment. There have been numerous reports of farmers, rural laborers, and their families becoming intoxicated while applying pesticides for agricultural purposes. Additionally, reports of poisoning and the negative health effects of chemicals from the environment and tainted food have surfaced. An estimated 355,000 people worldwide are killed by unintentional poisonings every year, according to the WHO, and poisonings are closely linked to excessive exposure and improper use of toxic pesticides.

5.2.3.1. Acute Toxic Effects

Within a few minutes to several hours following pesticide poisoning, acute toxic effects manifest. Both the central nervous system and peripheral muscarinic and nicotinic receptors are affected by poisoning. Nausea, vomiting, diarrhea, cramping in the abdomen, incontinence, miosis, lacrimation, salivation, bradycardia, bronchorrhea, hypotension, fasciculation, muscle paralysis, dizziness, confusion, seizures, coma, and respiratory failure are some of the symptoms of a cholinergic crisis. When exposed to pesticides, these effects might manifest right away. Death is also a possibility if life-threatening complications are not appropriately and promptly treated.

A few case studies of the acute toxic effects of pesticides on humans and animals are compiled in the following. Mishra et al. evaluated the harmful effects of CPF on the biochemical changes (acetylcholinesterase (AChE) activity and antioxidant markers) in the brain, the histopathological changes in pseudobranchial neurosecretory cells (PNSCs) of a neuroendocrine system of the gill region, the optic tectum (OT), and the cerebellum, and the related locomotory behavioral alterations in air-breathing catfish *Heteropneustes fossilis*. The study's findings suggested that brief acute exposure to CPF may cause their locomotory and swimming performance to decline, their neurosecretory activity of PNSCs to become dysregulated, and their brain's biochemical activity to change. The acute toxic effects of Thmx on A were examined by Uçkun et al. Acetylcholinesterase, carboxylesterase, glutathione S-transferase, glutathione, superoxide dismutase, glutathione peroxidase, glutathione reductase, and adenosine triphosphatases are some of the biomarkers used to identify Leptodactylus. Thmx's 96-hour LC_{50} value as an active ingredient was 8.95 mg/L. Crayfish are in danger in areas where Thmx is used because of its extremely toxic effects on them.

Kwon et al. used a silicone O-ring as a reservoir in a passive dosing format and assessed its suitability for determining PCP's effects on *Daphnia magna*. According to the study's findings, a dependable exposure concentration could be maintained during the experiment because the partition coefficient of PCP between methanol and a test medium (log KMeOH: ISO) was 2.1 Similar EC_{50} values of 576 and 485 μg/L for 24 hours and 362 and 374 μg/L for 48 hours, respectively, were demonstrated by solvent spiking and passive dosing in the acute toxicity tests. These values overlap with those of earlier research. The acute lethal effects of a 1-hour pulse exposure to specific insecticides on adult and juvenile (less than 72-hour-old) crimson-spotted rainbowfish, *Melanotaenia duboulayi* (Castlenau), were determined by Brown et al. With 24-hour pulse exposure LC_{50} values of 27 and 15 ug/L, respectively, the study's findings showed that juveniles were toxically exposed to temephos and pirimiphos-methyl. With a lethal dose ratio (pulse exposure LC_{50} temephos/pulse exposure LC_{50} pirimiphos-methyl) of 1.8 (95% CL 0.5Ð 6.4), pirimiphos-methyl was the most toxic of the two OPs. For a 15 cm deep water body, these pulse exposure LC_{50} values corresponded to 40% and 4.5% of the estimated environmental concentrations (EECs), respectively.

The harmful effects of beta-cypermethrin (BCP), bensulfuron-methyl (BSM), and prometryne (PM) on *Pseudomonas putida* (P.) were examined by Chen et al. using a series of calorimetric experiments. With an IC_{50} of 19–24 µg/mL against *P. putida*, the study's findings showed that BSM was the most toxic. Putida. With an IC_{50} of 27.86 µg/mL, PM demonstrated moderate virulence, while BCP had the lowest toxicity, measuring 39.64 µg/mL. Abss et al. talked about how 18 pesticides could inhibit the selective activities of all the major xenobiotic-metabolizing enzymes, including CYP1A11/2, 2A6, 2B6, 2C8, 2C9, 2C19, 2D6, 2E1, and 3A4. The study's findings showed that the IC_{50} values for profennofos, fenitrothion, and chlorpyrifos were 4 µm, CYP2B6 had IC_{50} values of chlorpyrifos and fenitrothion were 2.5 µm, CYP2CB had fenitrothion at 4.3 µm, CYP2C9 had fenitrothion and malathion at 4.8 and 2.5 µm, respectively, and CYP2D6 had IC_{50}values for chlorpyrifos and phenthoate were 3 µm.

In two distinct test systems, Lo et al. examined the impact of endocrine disruptors with antiandrogenic effects on 5a-reductase activity. The findings showed that the organotin compounds DBT (dibutylin), TBT (tributyltin), and triphenyltin (TPT) had nearly identical effects on enzyme activity in both systems, with IC_{50} values ranging between 2.7 and 11.2 µm. As acetylcholinesterase (AChE) inhibitors, Das et al. calculated the IC_{50} concentrations of the pesticides monocrotophos, chlorpyrifos, profenofos, and acephate. The study's findings showed that the IC_{50} values for RBC-AChE were 0.12, 0.25, 0.35, and 4.0 µm for monocrotophos, profenofos, acephate, and chlorpyrifos, respectively. Zhang et al. investigated the connection between neurodegenerative diseases, liver conditions, and butyrylcholinesterase (BChE); the findings showed that tacrine had accurate IC_{50} values for BchE (8.6 nm). Imidacloprid and glyphosate had the level of inhibition concentration (IC_{50}) values of 0.023 and 0.025 mm, respectively, according to Abdel-Halim et al.'s discussion of the ability to induce *in vitro* cytotoxic and oxidative stress in normal human cells using an MTT test. Forsythe et al. screened a panel of glyphosate and thallium using bioengineered 3D human liver and cardiac organoids, and they talked about how the organoids responded to these substances. The findings showed that the glyphosate IC_{50} was 13.5 µm.

5.2.3.2. Chronic Disease

Numerous studies conducted on laboratory animals provide evidence for the possibility of long-term health problems associated with pesticide exposure. However, not all health issues have available epidemiological data. Pesticide exposure has been linked to many chronic illnesses and conditions, such as cancer, poor reproductive outcomes, male sterility, peripheral neuropathies, neurobehavioral disorders, weakened immune systems, and allergic sensitization reactions, especially skin-related ones (Nankongnab et al., 2020). Furthermore, the majority of the pesticides analyzed have an impact on male reproductive systems, leading to aberrant sperm morphology, DNA damage, and sperm damage. Chronic diseases, for instance, are caused by cumulative inhalation of cholinesterase activity brought on by prolonged, low-dose exposure to organophosphorus compounds. Low concentrations of glyphosate were tested by Antonine et al.; the drop in Clusterin mRNA levels indicated that glyphosate targets Sertoli cell integrity. The reduction in germ cell counts starting on day 14 demonstrated the long-term effects of glyphosate at 50 nm, 500 nm, or 5 μm. Meltzer et al. sought to assess the relationship between chronic health outcomes and self-reported exposure to the Ringwood Mines/Landfill Superfund Site among Ringwood, New Jersey, residents and members of the Ramapough Lunaape Turtle Clan nation. The study's findings showed a strong correlation between health problems and self-reported exposure to Superfund sites among Native American residents of Ringwood. The findings also demonstrated a significant correlation between the prevalence of asthma and bronchitis and self-reported exposure to Superfund sites. An immature mouse Sertoli cell line (TM4) was used by Vanlaeys et al. to test glyphostate alone, glyphosate-based herbide formulations, and POEA at concentrations ranging from environmental to agricultural-use levels. The findings showed that glyphosate-based herbicide formulations cause TM4 mitochondrial dysfunction, disruption of cell detoxification systems, lipid droplet accumulation, and mortality at sub-agricultural doses. Additionally, the results demonstrated that formulants—particularly those found in Glyphogan—are more harmful than glyphosate. Séralini et al. talked about a 90-day feeding study that Monsanto carried out using the same rat strain and assessing the animals' biochemical parameters to achieve the commercial release of this GMO. According to the findings, liver congestion and necrosis were 2–5 times more common in the treated males than in the control groups. Nephropathies that were marked and severe were typically 1–3 times more severe. All treatment groups in females had mortality rates that were two to three times higher, and deaths happened sooner. Mammary tumors were more than four times as common in men as in women.

A summary of research on how different exposure routes affect chronic disease is provided below. Fang et al. found pesticide residues in 300 samples they gathered from China's eight primary growing regions. Pesticides' acute and chronic intake risks were evaluated. Additionally, a predetermined ranking matrix was used to rank the intake risk for each pesticide that was detected. 175 of the 300 samples had one or more pesticide residues, according to the results. The maximum residue limit was exceeded by carbofuran, one of the 25 pesticides that were identified. For these 25 pesticides, the assessed chronic and acute intake risks fell between 0 and 1.80 and between 0.05 and 28.0, respectively. Individual pesticides intake risks were ranked. The most dangerous pesticides were Avermectin, Triazophos, Chlorpyrifos, Dimethoate Oxygen, and Carbofuran. Tudi et al. talked about the possible long-term health risks of pymetrozine in soil and water in common rice-growing regions of China. The findings indicated that adults in both regions had higher lifetime non-cancer risks from soil exposure than from dermal contact with paddy water. Adults in both areas who were exposed to soil had lower potential lifetime non-cancer risks than those who were exposed to soil through ingestion. When Pan

et al. talked about the organophosphorus pesticides (OPPs) applied to agricultural soils in China's Yangtze River Delta, they discovered that dimethoate was the main ingredient, followed by methyl parathion and parathion. 70–80% of the risks were caused by soil ingestion, which was the main exposure pathway for OPP. Children were comparatively more at risk for non-cancer risks than adults were. In surface soil samples from four major Nepalese cities, Yadav et al. examined the presence, distributions, and profiles of specific OCP chemicals. The findings indicated that the primary exposure pathway for the OPPs used in Nepalese soil concerning cancer risk was soil ingestion.

There may be a higher number of mild but unreported cases of intoxication and acute conditions than the three million cases of severe acute poisoning. Nevertheless, there are fewer instances of chronic effects than acute ones. Medical and rehabilitation services are desperately needed due to the high levels of acute and chronic morbidity. Future epidemiological research is necessary to precisely estimate the number of less severe health effects that would contribute to the overall disease burden. The risk of contracting pesticide-related diseases is higher for children, expectant mothers, elderly people, and employees who are directly exposed to pesticides.

5.2.4. The Impact of Pesticide Application Methods on Exposure and Health Risks Associated with Pesticide Use

Choosing the proper equipment for the right pesticide application is crucial for application efficiency, minimizing spray solution loss, lowering environmental pesticide residues, and lowering risks to farmers and local residents. The effects of various spray tools and formulations on pyraclostrobin persistence in *R.*

roxburghii were investigated by Han et al. According to the results, the following order was found: suspension concentrate (SC) (6.78 d) > wettable powder (WP) (5.64 d) > water dispersible granule (WG) (4.69 d), gaston gasoline piggyback agricultural sprayer (5.38 d) > manual agricultural backpack sprayer (3.37 d) > knapsack multi-function electric sprayer (2.91 d). Konthonbut et al. examined the degree of occupational exposure and talked about paraquat exposure among backpack sprayers in Thailand.

The study's findings showed that using battery-operated backpack sprayers and standing upwind significantly decreased inhalation exposure. The impact of varying application speeds and nozzle types on the target area coverage and uniformity of UAV applications was examined by Hunter et al. (2019); the study's findings showed that AIXR nozzles offered the best coverage of all the nozzles tested and could lower the chance of off-target movement.

The results of the study showed that unmanned aerial sprayers had advantages over traditional spraying machinery, such as better penetration and more uniform liquid deposition distribution. Li et al. discussed the various operating parameters that could provide a better reference for determining field operating parameters and technical references for field pesticide application. This helped to establish a uniform standard for pesticide application technology and provide a technical reference for field spray operations.

According to Illyassou et al.'s assessment of the possible dermal exposure to pesticides linked to the use of backpack and hand-held sprayers, operators of hand-held sprayers had higher exposure levels than those of backpack sprayers. Mahaboonpeeti et al. evaluated the possible exposure and risk levels related to pesticides for farmers in Thailand who either used a backpack sprayer with a battery-operated pump or a backpack sprayer with a two-stroke gasoline motor and fan.

The study's findings indicated that the estimated total body alachlor exposures of applicators using the battery-operated pump backpack sprayer (15.50 μg/h) were significantly lower than those using the two-stroke engine/fan backpack sprayer (219.48 μg/h). Tractor/boom pesticide application reduced overall population occupational exposure, according to Lozier et al.'s evaluation of occupational inhalation exposure to atrazine during pesticide application in a developing nation. Inhalation exposure can be reduced or eliminated by keeping an eye on boom nozzles from a distance rather than on the rear of a tractor or a boom. Pump backpack sprayers may lessen their exposure to inhalation by applying herbicides using flat spray nozzles (WHO, 1990; Mostafalou and Abdollahi, 2017).

5.2.5. General Health Problems Associated with Pesticide Exposure in Developing Countries

The major issue of pesticide contamination in the environment and direct and indirect human exposure to pesticides remains a major concern globally, even though the WHO and some developed countries have taken certain steps to lessen the harmful effects of pesticides.

People in developing nations and high-risk groups in other nations bear a disproportionate amount of the burden, even though no segment of the general population is totally immune to pesticide exposure and its potentially dangerous health effects. For instance, the use of pesticides results in approximately 750,000 chronic illnesses, 220,000 fatalities, and three million poisonings annually worldwide, with the majority of these incidents taking place in developing nations. Another example is the estimated USD 44.4 billion in 2005 cost of diseases and harm caused by pesticides in sub-Saharan Africa, which was predicted to rise to USD 90 billion by 2020. According to a WHO (1990) report, Sichuan Province in China, which has an agricultural area of 6.5 million hectares, is home to about 87 million people. 10 million people were exposed to the 4 kg of pesticides used per hectare, with approximately 12% of them becoming poisoned. Furthermore, things have been changing quickly in developing nations, and some new crop and product types need more pesticides because of climate change and the population's exponential growth. As a result, the proportion of individuals exposed to pesticides has been rising.

In developing nations, exposure to pesticides is frequently linked to health issues. The majority of farmers in developing nations continue to use pesticides incorrectly and at an increasing rate, which has major negative health effects despite the facts and evidence of harm. The major health issues associated with pesticide exposure in developing nations are caused by inadequately trained organizations that oversee and assess the production and distribution of these chemicals. Furthermore, the import and export of pesticides are not adequately regulated in developing nations due to a lack of stringent laws and regulations. For instance, Atreya et al. demonstrated that since farmers in Nepal still purchase extremely hazardous, out-of-date pesticides, it is imperative and required to establish stringent regulations to regulate farming practices. Additionally, some extremely harmful pesticides that are prohibited in developed nations are still in use there.

The primary causes of general health issues in developing nations are excessive pesticide use by farmers, risky practices, and a lack of knowledge about pesticide use. Furthermore, farmers in developing nations are increasingly under pressure to use pesticides to preserve their subsistence livelihoods due to their large populations. Additionally, because they lack access to adequate education and training, farmers in developing nations lack a thorough understanding of chemical toxicity and safe application techniques for chemicals used in household and agricultural pest and disease control. Additionally, they lack the authority to regulate internal policies and external factors like trade and market liberalization.

The inability of developing countries' public health systems to effectively address pesticide-related health issues is another significant factor contributing to the general health issues in these countries. The fact that there are numerous pesticide types in use, each requiring a different case management protocol, exacerbates the situation. Previous research predicts that in developing nations, the use of pesticides will double over the next 10 years. Intentional and unintentional acute poisoning cases are expected to rise as agricultural practices continue to evolve. While the use of insecticidal carbamates and insecticidal organophosphorus compounds is growing, the use of organochlorine pesticides will decrease. Acute intoxication risks will rise unless the majority of toxic pesticides are used less frequently. Over the next 10 years, there may be an increase in the number of people working in high-risk occupations due to the growth of cash crops and plantation-style farming in developing nations.

In conclusion, when pesticides are sprayed on target plants, applicators and communities are exposed directly through a variety of application techniques. They are also indirectly exposed to pesticides through environmental contamination through ingestion, inhalation, and oral routes, which can result in acute toxic effects and long-term illnesses. As a result, it is essential to evaluate pesticide exposure and the risks to human health (WHO, 1990; Mostafalou and Abdollahi, 2017).

5.2.6. Methods of Health Risk Assessment Regarding Pesticide Application

5.2.6.1. Deterministic Risk Assessment

In environmental and public health contexts, the deterministic risk assessment method is frequently employed. It assesses risk using a single metric. Deterministic risk assessment has both benefits and drawbacks. The approach has the benefit of being simple to comprehend and use for evaluating exposure and health risks. Deterministic risk assessment, however, only accounts for health risk and average exposure. This type of approach ignores the unpredictability of risk and exposure. The deterministic approach, for instance, only displays the risk if the average exposure value and risk index are higher or lower than a predetermined dose or standard of risk level. This occurs when 20% of the population may have exposure levels above the reference dose and standard of risk level, but the average exposure and health risk for a particular chemical type are below these levels. Researchers may conclude that there is no risk based on a deterministic risk assessment, in which case no risk-management measures may be taken.

This paragraph summarizes earlier research that used deterministic risk assessment to evaluate pesticide exposure and health risk levels. Based on animal-based foods gathered from Shanghai, China markets, Lei et al. evaluated health risks associated with exposure to organochlorine pesticides, polychlorinated biphenyls, polybrominated diphenyl ethers, polycyclic aromatic hydrocarbons, and toxic trace elements (mercury, chromium, cadmium, lead, and arsenic). The findings indicated that there would not be a substantial non-cancer health risk for Shanghai residents because the combined hazard quotient values for multiple contaminants via single or multiple food consumption were less than 1. Methylmercury was linked to the highest possible non-cancer risk among the contaminants examined. Nonetheless, inorganic arsenic was the primary cause of the combined cancer risk from several contaminants in the majority of foods, which was higher than the recognized risk level of 10^{-6}. In samples of marine edible fish taken from two significant nearshore fishing locations in Shantou Harbor and Haimen Bay in eastern Guangdong Province, China, Shi et al. measured the pesticide level and assessed the health risk level. The study's findings demonstrated that lifetime cancer risk is still a possibility in the studied area and that daily fish consumption in this area can be extremely concerning. Siriwat et al. examined the factors linked to children's exposure to chlorpyrifos in agricultural areas of Sakon Nakhon Province, Northeastern Thailand, as well as the health risks associated with skin exposure to the chemical. According to the study's findings, 73.1% of the children tested had residues of chlorpyrifos on their skin. According to the health risk assessment of non-cancerous effects, children's dermal chlorpyrifos exposure levels ranged from 1.46–10.5 mg/kg/day, with an acceptable HQ of 0.03.

In the Kurose River and its catchment area in Japan, Chidya et al. used an integrated approach that combined pesticide monitoring and risk assessment. The findings indicated that cyanazine had the highest detection level (64%), followed by simetryn (58%) and diazinon (57%), across all sample sites (n = 12). All of the pesticides, according to the HQ estimates, were below the threshold value of 1, meaning they did not pose any serious health risks to people. Using the HQ and HI (hazard index) methods, Wu et al. investigated the pesticide residue in honeysuckle in China and assessed the possible health risks for consumers. The study's findings revealed that the acute hazard quotient (HQa) of carbofuran was 1.54 for fetuses, infants, and pregnant or nursing women, suggesting that it posed a potential acute health risk. The acute hazard index (HIa) of insecticides in honeysuckle for children and the specific population was 1.34 and 3.36, respectively, in the cumulative risk assessment, indicating that they may have presented acute cumulative health risks. Wang et al. evaluated the seasonal and regional distribution of 17 polycyclic aromatic hydrocarbons (PAHs) in surface waters from four main water functional regions of Baiyangdian Lake. The findings indicated that the carcinogenic risk was generally acceptable and that the human health risk posed by PAHs in Baiyangdian Lake's surface water did not represent a potential non-carcinogenic risk to local residents, although the potential lifetime carcinogenic risks for infants in rural residential regions should raise concerns. Comparing urban and rural residential areas to other functional regions, the cumulative non-carcinogenic and carcinogenic risks were higher in the former.

In 48 composite fruit and vegetable samples (carrot, cucumber, tomato, and watermelon), Odewale et al. found residues of dichlorodiphenyltrichloroethanes (DDTs) and hexachl HCHs. They also measured the cancer benchmark concentration (CBC), hazard quotient (HQ) and index (HI), hazard ratio (HR), and estimated average daily intake (EADI). According to the findings, the non-carcinogenic health risk of α-HCH in watermelon and tomatoes had an HQ > 1, suggesting that there may be a systemic health risk for children who consume these foods. Only α-HCH and γ-HCH in children and α-HCH in adults had an HR > 1 for tomato and watermelon, according to the carcinogenic health risk, suggesting that eating them could pose a risk for cancer. Mercadante et al. calculated the health risks related to drinking water for the adult population in Lombardy and found the concentration levels of pesticides in groundwater. The findings showed that although the environmental quality standard was exceeded by roughly 1%–5% of the determinations, there was no evidence of a possible health risk to the general adult population from consuming contaminated groundwater. Simas et al. looked into the characteristics of work in banana production in a part of Brazil's Ribeira Valley as well as the prevalence of ventilatory dysfunction in workers. Mild obstructive disorder (13.3%), mild mixed disorder (3.3%), and moderate obstructive disorder (10.0%) were found to be the most common. A correlation between ventilatory dysfunction and the labor required for banana farming was established, and it was determined that one-third of the workers suffered from some form of ventilatory dysfunction.

5.2.6.2. Probabilistic Risk Assessment

From the lowest to the highest risk, the probabilistic risk assessment method calculates the exposure and risk distribution for a variety of populations. The use of probabilistic risk assessment has three key benefits. First, the percentage of the population that surpasses the designated reference dose—including the acceptable daily intake or risk-specific dose levels—can be ascertained using probabilistic risk assessment. The probabilistic risk assessment, for instance, can identify that 20% of the population is at risk if 20% of the population is exposed to a particular type of chemical (Duan et al., 2021). Second, cost-benefit analyses may make use of the outcomes of probabilistic risk assessments. By repeatedly running a probabilistic risk assessment model, different risk-management scenarios, such as eliminating contamination and recalling tainted foods, can be produced. As a result, these risk scenarios can be decreased by implementing the required financial actions. Thirdly, uncertainty and unreliability can be estimated through probabilistic risk assessment.

5.2.6.2.1. Risk Characterization and Quantification Using the Probabilistic Approach

Comparisons of the cumulative probability curves for exposure and effects can be used to characterize and quantify risk using the probabilistic approach. A qualitative indicator of the relative level of health risk is provided by the proximity or overlap between the exposure and effect curves, which are located on opposite sides of each other. A higher risk is indicated by a closer overlap between the two curves. On the other hand, if the curves are widely separated and do not overlap, the risk is minimal and should be insignificant.

A HQ must be computed to produce a quantitative comparison. The exposure dose (or concentration) at 95% cumulative probability on the exposure curve divided by that at 5% cumulative probability on the effect curve is known as $HQ_{95/5}$. A risk probability of 1/20, or 5%, is equivalent to a reference point of $HQ_{95/5} = 1$. Likewise, a reference $HQ_{50/50} = 1$ denotes a 50% risk probability. Additionally, this scenario would show that over the same dose intervals, the two cumulative probability curves substantially overlap. Despite being a straightforward technique for comparing and quantifying risk, the HQ method fails to adequately take into account the statistical distributions of the cumulative probability curves. When exposure and effect data distributions deviate from a normal distribution or have different slopes, the HQ might not accurately reflect the true risk characteristics. Albert et al. (2018), for instance, examined the absorbed dose levels and ensuing health risks linked to applicators' skin exposure to chlorpyrifos on Ghanaian rice farms. With HQ50 values ranging from 1–5, the results showed that the median exposed group was at high risk of experiencing acute adverse health effects as a result of exposure to chlorpyrifos. Furthermore, the 5% highly exposed group had HQ_{95} values of 2.7×10^{-9}, which indicated a higher risk of acute adverse health effects. With HQ_{95} values ranging from 1.2–2, the chronic exposure guideline values also indicated that the 5% highly exposed group might suffer negative effects from chlorpyrious exposure. Using a probabilistic approach, Phung et al. assessed the exposure of Vietnamese rice farmers to chlorpyrifos; the study's findings showed that the baseline exposure level, which varied from 0.03–1.98 mg/kg/day, was below the chronic guideline level established by both national and international organizations. In the 95%, the post-application exposure level, which varied from 0–35–94 mg/kg/day, was higher than the majority of the acute recommendations.

5.2.6.2.2. Monte Carlo Health Risk Assessment

Every variable and parameter used in risk assessment is treated as a probability distribution when the Monte Carlo technique is applied. Using randomly selected values of variables and parameters that span their range of variability and replicate the assumed distribution density, the hazard quotient and the degree of cancer risk are computed 10,000 times or more. A probability distribution for a particular risk is the output of a Monte Carlo health risk assessment. Furthermore, the outcomes of the Monte Carlo health risk assessment help comprehend the uncertainties associated with biological factors, interpersonal variability, and the dynamic nature of contaminated media. However, the method's application is limited because Monte Carlo simulation necessitates two cumulative probability distributions.

Numerous Monte Carlo health risk assessments have been carried out in connection with the risk to human health posed by pesticide exposure. Faezeh et al. (2021) found that six walnut cultivars from five different Iranian regions had concentrations of 18 organophosphorus, carbamate, pyrethroid, and nicotinoid pesticides. HIs and Monte Carlo simulations were used to assess the risks to human health. Based on exposure through walnut consumption, the 95% of human HIs was calculated to be 1.68, indicating a moderate level of concern for human consumers. According to sensitivity analysis done during the MCS, the most important parameters were concentrated and fell between 0.71 and 0.97. The findings indicated that eating walnuts was generally safe.

By exposing human and terrestrial species to soil and agricultural products in Turkish agricultural areas, Korucu et al. assessed the environmental risks of p,p′-DDT and its metabolites (DDX). The intake values for five distinct exposure groups, four distinct bird species, and four distinct mammal species were determined using the Monte Carlo simulation method. The findings indicated that humans were at a high risk of developing cancer. Additionally, it was found that there was a considerable risk of reproductive toxicity for both mammals and birds. Exposure to DDX-contaminated soils and ingestion of plants cultivated on these soils were the primary causes of the risk. Using the Monte Carlo approach, Eslami et al. assessed the hazard quotient and total hazard quotient (THQ) in a human health risk assessment for pesticides in fruit grown in Iran. The findings indicated that the THQ in adults based on fruit consumption was estimated to be 7.8%, while the THQ in children was 36.7%. As a result, neither adults nor children were at risk for health problems.

Using sophisticated probabilistic techniques, Atabila et al. evaluated the health risks of chlorpyrifos exposure among applicators on Ghanaian rice farms. The findings indicated that the odds of experiencing negative health effects from chlorpyrifos under chronic exposure scenarios ranged from 1%–8%, while those for acute exposure scenarios ranged from 31%–34%. Therefore, exposure to chlorpyrifos among Chana applicators has both acute and long-term effects on human health.

Using the Monte Carlo simulation method, Pirsaheb et al. evaluated the health risks of pyrethroid and organophosphorus pesticides for both adults and children after finding residues of these pesticides in flour and breads that were gathered from local markets in Kermanshah Province, Iran. According to the study's findings, detectable amounts of malathion and deltamethrin were found in 11% and 15% of the total samples, respectively. Because of the high rates of insecticide consumption, about 85% of pesticide residue detections were found in tropical and mild climates. The 95% of the THQ value was less than 1 for both malathion and deltamethrin in both adults and children (i.e., an acceptable level); consequently, no non-carcinogenic health risk was found for bread consumers in Kermanshah Province as a result of deltamethrin and malathion.

5.2.6.2.3. Overall Risk Probability Health Risk Assessment

Comparing the relative risks of different compounds is made simple by the overall risk probability (ORP) health risk assessment method, which offers a straightforward probability measure. Additionally, because the statistical distributions of the exposure and effect curves are appropriately taken into account, it is a multiple-point method. Furthermore, the combined effects of independent or interacting factors can be measured using the ORP method. But to determine the ORP, this approach necessitates building and integrating an exposure exceedance curve, which could take more processing power than single-point approaches.

For instance, Atabila et al. used the overall risk probability risk assessment method to evaluate the exposure and health risks of chlorpyrifos in Ghana. The findings showed the likelihood that applicators would experience negative effects from acute exposure from occupational application (31%), chronic exposure from occupational application (4%), and chronic background exposure (3%) to chlorpyrifos. In an overall risk assessment, Phung et al. evaluated the exposure and health risks posed by chlorpyrifos for Vietnamese rice farmers. The findings indicated the likelihood of negative effects among applicators as a result of acute exposure from occupational application (29%), chronic exposure from occupational application (2%), and chronic background exposure (1%).

5.2.7. Symptoms of Pesticide Poisoning

Pesticide poisoning can cause anything from a slight rash to a coma or even death. Different kinds of symptoms are caused by different chemical classes or families. Additionally, people differ in how sensitive they are to various concentrations of these substances. When exposed to something that could cause serious illness in others, some people might not react. Users and handlers of pesticides need to be aware of the typical symptoms of pesticide poisoning due to possible health risks.

In general, pesticide poisoning symptoms can be classified as either systemic or topical. The irritant qualities of the pesticide (either the active or inert ingredient) or the victim's allergic reaction cause topical effects, which typically appear at the site of pesticide contact. It is widely acknowledged that the most frequently reported topical effect linked to pesticide exposure is dermatitis or inflammation of the skin. Dermatitis symptoms include blisters, rashes, and/or skin reddening.

When exposed to pesticide sprays, some people have a tendency to cough, wheeze, or sneeze. The potent smell and irritating properties of petroleum distillates used as carriers in pesticide products cause reactions in certain people. One sign is a raw, scratchy feeling in the eyes, nose mucous membranes, and even the delicate mouth and back of the throat linings. Usually, this symptom goes away a few minutes after the person is taken out of the irritant's reach. On the other hand, a reaction to a pesticide product that results in severe acute respiratory symptoms in addition to sneezing and coughing is more likely to be a true hypersensitivity or allergic reaction. A genuine allergic reaction can cause respiratory discomfort that frequently resembles asthma as well as skin and eye reddening and itching.

Topical effects and systemic effects are very different. Due to the pesticide's absorption and distribution throughout the body, they frequently happen far from the initial site of contact. Nausea, vomiting, exhaustion, headaches, and intestinal problems are common systemic effects. In cases of advanced poisoning, the victim may have breathing difficulties, convulsions, a coma, and changes in heart rate, all of which could be fatal (WHO, 1990; Mostafalou and Abdollahi, 2017).

5.2.7.1. Seeking Medical Attention

Keep yourself and others on the lookout for the early warning signs and symptoms of pesticide poisoning. These can happen up to 24 hours after exposure, but they usually happen right away. At the very least, let someone know if you are experiencing symptoms but are not sure if they are caused by pesticides in case they worsen. However, you should get medical help right away if symptoms start to show up after coming into contact with pesticides.Bring the pesticide container to the phone if it's safe to do so. However, note the product name, active ingredient or ingredients and percentage, and EPA registration number if the pesticide container is contaminated. Information for medical personnel, including active ingredients, an antidote, and the manufacturer's emergency phone number, is provided on the product label. Take advantage of the Material Safety Data Sheet as well, if it is available, as it includes extra information for medical professionals.

You should bring the entire pesticide container, including the label if you have to visit the hospital or doctor's office. Make sure the container is tightly sealed and, if at all possible, place it inside a plastic bag to prevent fumes from being inhaled or the contents from spilling. You should never put the pesticide container in your car's enclosed passenger area.

5.2.8. Harmful Effects of Some Pesticide Families

5.2.8.1. Fungicides

Although fungicides are generally thought to have a low acute toxicity to humans, they can irritate the skin and eyes. Sneezing, coughing, and irritation of the throat can result from breathing in spray mist or dust from these pesticides. Long-term exposure to fungicides at lower concentrations can have negative health effects. The majority of human fungicide poisoning cases have resulted from eating seed grain. Nowadays, fungicide treatments include a vivid dye to make it obvious that the seed has been treated, preventing these kinds of poisonings. The symptoms and indicators of acute exposure to widely used fungicides are compiled in Table 5.1.

Table 5.1. *The Effects of Pesticides on DNA*

Pesticide	Pesticide group	Mode of interaction	References
Chloridazon or Pyrazon	Organochlorine herbicide	Intercalation via GC region	Ahmadi et al., 2011
Fenitrothion	Organophosphorus insecticide	Partially intercalation via NO_2 and the C Form conformation	Ahmadi et al., 2013
Permethrin, deltamethrin	Synthetic pyrethroid insecticides	Groove binding and partial intercalation	Ahmadi and Ghanbari, 2014

Methyl Thiophanate	Fungicide	Non-intercalative groove binding via AT region	Saquib et al., 2010
Propyzamide	Herbicide	Intercalation via AT region	Zhang et al., 2015
Edifenphos	Organophosphate pesticide	Electrostatic binding minor groove binding via AT region	Ahmad and Ahmad, 2018
Tau-fluvalinate, flumethrin	Synthetic pyrethroid pesticide	Hydrogen bonding and Van der Waals forces, minor groove binding via AT region	Tao et al., 2016
Dinitramine	Herbicide	Hydrophobic interactions, major groove binding	Daneshmehr et al., 2016
Resmethrin	Synthetic pyrethroid insecticides	Hydrogen bonds and Van der Waals forces, groove binding via GC region	Tao et al., 2015
Pendimethalin	Herbicide	Intercalation via GC region	Ahmad et al., 2016
Organophosphates	Pesticide	DNA methylation	Paul et al., 2018
Organophosphate, pyrethroids	Fumigation insecticide	DNA methylation	Paredes-Céspedes et al., 2019
Endosulfan	Pesticide	DNA hypomethylation	Mahna et al., 2021
Glyphosate	Pesticide	DNA hypermethylation	Mahna et al., 2021
Diazinon	Pesticide	DNA hypermethylation	Mahna et al., 2021
Fonofos, parathion, terbufos	Pesticide	DNA hypermethylation	Mahna et al., 2021

Source: Pathak VM, Verma VK, Rawat BS, Kaur B, Babu N, Sharma A, Dewali S, Yadav M, Kumari R, Singh S, Mohapatra A, Pandey V, Rana N and Cunill JM (2022) Current status of pesticide effects on environment, human health and it's eco-friendly management as bioremediation: A comprehensive review. Front. Microbiol. 13:962619. doi: 10.3389/fmicb.2022.962619.

5.2.8.2. Herbicides

Because human physiology differs greatly from that of plants, herbicides generally have low acute toxicity to humans. There are some exceptions, though, as many of them are strong acids, amines, esters, and phenols that can irritate the skin. Coughing and a burning feeling in the chest and nasal passages can result from inhaling spray mist. Dizziness can occasionally result from prolonged inhalation. Usually, ingestion results in muscle twitches, diarrhea, vomiting, and a burning feeling in the stomach. The symptoms and indicators of acute exposure to widely used herbicides are compiled in Table 5.2.

Table 5.2. *Effects of Pesticides*

Type of pesticide	Type of pests	Functions	Pests and disease management	References
Aldicarb	Nematicides	Inhibit nematodes (plants parasites)	Damage tissue *via* oxidative stress, and also binds and inhibits acetylcholinesterase (AChE) (controlling acetylcholine neurotransmitter)	Yadav and Devi, 2017; Hassaan and El Nemr, 2020
	Insecticides	Inhibit insects and other arthropods also		
Atrazine	Herbicides	Destroy weeds and other plants, photosystem-II (PSII)–inhibiting	Use to control grasses and broadleaf weeds in sorghum, corn, and sugar cane crops	Gupta and Crissman, 2013; Yadav and Devi, 2017
Avitrol	Avicides	Chemicals that lethal to small seed-eating birds	Used for population management of certain birds (crows, gulls, cowbirds, blackbirds, starlings, grackles, pigeons, sparrows, red-winged blackbirds)	Yadav and Devi, 2017; NIPHM, 2018; Hassaan and El Nemr, 2020
Azoxystrobin	Fungicides	Kill fungi (blights, rusts, molds, and mildews), azoxystrobin act fungal mitochondrion, binds to cytochrome bc_1 complex and inhibit electron transport thorough oxidative phosphorylation.	Uses to kill Oomycetes, Ascomycetes, Deuteromycetes, BasidiomycetesAnd it controls disease like apple scab rusts, rice blast, powdery and downey mildew.	Yadav and Devi, 2017; Hassaan and El Nemr, 2020

Benzoxazin	Ovicides	Prevention of mites and insects egg growth	In pest managmeent	Yadav and Devi, 2017; Hassaan and El Nemr, 2020
Bifenazate	Acaricides	Control spiders and mites that feed on plants and animals by altering their growth and development. Target site of Bifenazate is mitochondrial, particularly the Q_o site of that encoded for cytochrome b	Bifenazate uses as an acaricide on strawberry, flowering plants, and nursery ornamentals	Van Nieuwenhuyse et al., 2012; Hassaan and El Nemr, 2020; Authority et al., 2021
Boric acid	Desiccants	Act on plants by drying their tissues	Use to bed bug control	Hassaan and El Nemr, 2020; United States Environmental Protection Agency [US-EPA], 2022
Copper complexes	Bactericides	Prevent bacteria with greater doses, copper works as a broad-spectrum biocide by interfering with nucleic acids, disrupting enzyme active sites, interfering with the energy transport system, cell membranes integrity disrupted	Copper complexes are used to prevent infection of seedlings from plant pathogens by seed treatment	Yadav and Devi, 2017; Hassaan and El Nemr, 2020
Copper sulfate	Algaecides	Control or kill growth of algae	Alter the algal growth and photosynthesis	Lamichhane et al., 2018; Hassaan and El Nemr, 2020
Dichlorobenzene	Moth balls	Inhibit molds and moth larvae and prevent cloths damage	Commonly used to control moths, molds, and mildew	Yadav and Devi, 2017; Eastmond and Balakrishnan, 2010

Fipronil	Termiticides	Fipronil inhibits termites by acting as a GABA antagonist and leads to excessive CNS excitation and causes death	Used in seed coatings and granular soil treatments to control unwanted arthropods in many kinds of food, horticultural, and turf plants	Cannon and Ruha, 2013; Beasley, 2020; Hassaan and El Nemr, 2020
Methiocarb	Repellents	Repel pest vertebrates and invertebrates by its taste or smell	Use as a seedling bird repellant and also effective against frit fly larvae.	Finch et al., 2014; Yadav and Devi, 2017
Methoprene	Larvicides	Prevents larvae growth	Uses as mosquito larvicide, also effective against horn flies, mushroom flies in compost, dipteran pests of livestocks, nuisance flies, highly selectivity for insects and no acute toxicity is expected in humans	Ramaseshadri et al., 2012; Monteiro and Jurado, 2014; Yadav and Devi, 2017
Metaldehyde	Molluscicides	Prevent mollusk's (snail's) usually disturbing growth of plants or crops	Use in vegetables and gardens, to kill slugs, snails, other garden pests	Yadav and Devi, 2017; Hassaan and El Nemr, 2020
Rotenone	Piscicides	Toxic and act on fishes	Uses in fisheries and fish management strategies (where unbalanced population of fish)	Gupta, 2014; Hassaan and El Nemr, 2020
Scytovirin	Virucides	Acts against viruses	Control of viral infections and diseases	Hassaan and El Nemr, 2020
Tebuthiuron	Silvicides	Specific to woody vegetation and act on it	Uses to manage the undesirable plants or unwanted forest species and apply to eliminate trees and brush or "entire forest"	Yadav and Devi, 2017
Trifluromethyl nitrophenol (TFM)	Lampricides	Target larvae of lampreys by uncoupling mitochondrial oxidative phosphorylation and ATP production reduces which ultimately leads to death	TFM used to control invasive sea lamprey (*Petromyzon marinus*)	Birceanu and Wilkie, 2018; Hassaan and El Nemr, 2020

Source: Pathak VM, Verma VK, Rawat BS, Kaur B, Babu N, Sharma A, Dewali S, Yadav M, Kumari R, Singh S, Mohapatra A, Pandey V, Rana N and Cunill JM (2022) Current status of pesticide effects on environment, human health and it's eco-friendly management as bioremediation: A comprehensive review. Front. Microbiol. 13:962619. doi: 10.3389/fmicb.2022.962619.

5.2.8.3. Insecticides

The majority of pesticide poisonings in the US are caused by insecticides. The most severe pesticide poisonings are typically caused by acute exposure to insecticides that contain carbamates and organophosphates. Chlorpyrifos, diazinon, dimethoate, disulfoton, malathion, methyl parathion, and ethyl parathion are examples of organophosphate insecticides. Carbaryl, carbofuran, methomyl, and oxamyl are examples of carbamate compounds. Cholinesterase is inhibited by organophosphates and carbamates, which disrupts the nervous system. These substances can poison all living things that have cholinesterase in their nervous systems, including fish, birds, insects, humans, and other mammals.

One must comprehend the functioning of the nervous system to comprehend the effects of organophosphate and carbamate insecticides. The body's most intricate system is the nervous system, which includes the brain. It is made up of millions of cells that communicate with one another. This intricate network of cells carries messages or electrical impulses, also known as stimuli. There is a space or synapse between nerve cells or neurons instead of physical contact. To maintain the message's flow throughout the entire network, the impulses must bridge or cross the synapse between nerve cells.

Acetylcholine is released to transmit the message to the subsequent cell when an impulse reaches the synapse. The main substance in charge of nerve impulse transmission across the synapse between two neurons is acetylcholine. The enzyme cholinesterase breaks down the acetylcholine after the impulse has passed through the synapse. The synapse is then free to receive a new transmission after this has taken place.

Acetylcholine accumulates in the body as a result of organophosphate and carbamate insecticides' inhibition of cholinesterase activity. Uncontrolled nerve transmissions between nerve cells are caused by an increase in acetylcholine. Acetylcholine builds up in the nervous system, causing impulses to continuously travel across synapses, poisoning the system.

Systemic and topical symptoms are possible outcomes of organophosphate or carbamate poisoning. For instance, topical symptoms like pupil constriction, blurred vision, headaches from the eyebrows, and extreme eye irritation and reddening can result from direct exposure to the eye. Acetylcholine buildup at the nerve endings is nearly solely responsible for the symptoms and indicators of systemic poisoning.

The severity of the poisoning and the entry point determine when symptoms appear. If the material has been consumed, gastric symptoms like diarrhea, vomiting, nausea, and cramping start to show up early. Similarly, if the material has been inhaled, the first symptoms include headache, dizziness, salivation, and excessive secretions that make breathing difficult. Respiratory failure may arise from the involvement of the respiratory muscles. If the pesticide is absorbed through the skin, symptoms related to the stomach, intestines, and respiratory system typically manifest simultaneously. Convulsions may be the initial sign of poisoning in children.

The victim is pale, perspiring, and frothing at the mouth in cases of advanced poisoning. The pupils are unable to react to light and are constricted. Additional symptoms include coma, convulsions, mental disorientation, muscle weakness, and/or heart rate fluctuations. If treatment is not received, the victim may die (WHO, 1990; Mostafalou and Abdollahi, 2017).

5.2.8.4. Cholinesterase Testing

Those who frequently handle carbamates and organophosphates ought to think about getting cholinesterase tests done regularly. The impact of exposure to carbamate and organophosphate insecticides is measured by the blood cholinesterase test. A baseline needs to be set for every individual because cholinesterase levels can differ significantly between people. Actually, a tiny portion of people have low levels of cholinesterase that are genetically determined. For these individuals, even a small amount of exposure to cholinesterase inhibitors can be extremely dangerous. Always conduct baseline testing at least 30 days after the most recent exposure, or during the season when pesticides are not being used. Two tests that are conducted at least 72 hours apart but within 14 days of one another are frequently necessary to establish a baseline value. A third test is frequently advised if there is a 20% discrepancy between the test results.

When using carbamate and organophosphate insecticides, cholinesterase tests can be performed again and compared to the baseline level. Routine cholinesterase monitoring is intended to help a doctor identify instances of excessive exposure to carbamates and organophosphates. A worker should be retested right away if a laboratory test reveals a cholinesterase drop that is 30% below the predetermined baseline. A pesticide handler or agricultural worker should be shielded from additional exposure to organophosphate and carbamate insecticides until their cholinesterase levels return to the preexposure baseline range if a second test reveals a decrease in cholinesterase. The frequency of this testing program can be determined with assistance from your primary care physician.

5.2.9. Exposure and Preventative Measures

As previously stated, the risk or hazard associated with using a pesticide is determined by the product's toxicity as well as the degree of exposure to it

(Hazard = Toxicity x Exposure). Even low-toxicity products can be hazardous if your exposure level is high, though it is ideal to use them whenever possible. However, if the exposure level is low, the hazard will also be low, regardless of the toxicity of the product. Wear the personal protective equipment (PPE) as directed on the product label at all times to minimize exposure and safeguard your health. The general PPE recommendations to guard against the four points of entry are as follows.

5.2.9.1. Dermal

Dermal exposures account for over 95% of all exposures. When equipment is being cleaned or repaired, or when there is a splash, spill, or drift, dermal absorption may take place. To prevent the majority of dermal exposures, put on unlined gloves that can withstand chemicals. For most pesticides, wearing long sleeves, long pants, gloves, and appropriate footwear provides the bare minimum of dermal protection. Consider donning unlined rubber boots, a waterproof hat, and coveralls for added protection. Additionally, it is advised to wear a rain suit or liquid-proof apron when handling highly toxic products or when mixing and pouring concentrates.

5.2.9.2. Inhalation

The respiratory (breathing) system is the quickest and most direct pathway for a lot of harmful chemicals to enter the circulatory system. Protection of the respiratory system is particularly crucial when inhaling pesticide dusts, powders, gases, vapors, or tiny spray droplets. Always adhere to the manufacturer's instructions and use the respirator as intended. Choose only equipment that has been authorized by the Mine Safety and Health Administration (MSHA) and the National Institute of Occupational Safety and Health (NIOSH).

5.2.9.3. Oral

The most common way that accidental oral exposure happens is when pesticides are removed from their original container and placed in an unlabeled bottle or food container. Regretfully, the most frequent victims in these circumstances are children. Keep the original label on the pesticide container and only store it in its original container. Keep the original label on the pesticide container and only store it in its original container. To keep out of children's reach, store on a high shelf or in a locked cabinet. Never siphon a pesticide from a tank or container or clear a spray line with your mouth. Before eating, drinking, or smoking after handling or working with pesticides, thoroughly wash your hands and face with soap and water.

5.2.9.4. *Eyes*

Because of their size, eyes can absorb a lot of chemicals and are extremely sensitive to many pesticides. Splashes, spills, drifts, and rubbing the eyes with contaminated hands or clothing can all cause serious eye exposure. If there is a possibility of getting pesticides in the eyes, it is best to wear a full-face shield or tight-fitting chemical splash goggles, particularly when handling dusts and pouring or mixing concentrates. To prevent chemicals from splashing or spilling on your face or protective clothing, keep the container below eye level when pouring (WHO, 1990; Mostafalou and Abdollahi, 2017).

REFERENCES

1. Abdel-Halim, K. Y., & Osman, S. R. (2020). Cytotoxicity and oxidative stress responses of imidacloprid and glyphosate in human prostate epithelial WPM-Y. 1 cell line. *Journal of Toxicology, 2020*, 4364650.

2. Afata, T. N., Mekonen, S., Shekelifa, M., & Tucho, G. T. (2022). Prevalence of pesticide use and occupational exposure among small-scale farmers in Western Ethiopia. *Environmental Health Insights, 16*, 11786302211072950.

3. Aktar, M. W., Sengupta, D., & Chowdhury, A. (2009). Impact of pesticide use in agriculture: Their benefits and hazards. *Interdisciplinary Toxicology, 2*(1), 1–12.

4. Ali, N., Kalsoom, K. S., Ihsanullah, R. I. U., & Muhammad, S. (2018). Human health risk assessment through consumption of organophosphate pesticide-contaminated water of Peshawar Basin, Pakistan. *Exposure and Health, 10*, 259–272.

5. An, X. H., Ji, X. F., Jiang, J. H., Wang, Y. H., Wu, C. X., & Zhao, X. P. (2015). Potential dermal exposure and risk assessment for applicators of chlorothalonil and chlorpyrifos in cucumber greenhouses in China. *Human and Ecological Risk Assessment, 21*(4), 972–985.

6. An, X. H., Wu, S. G., Guan, W. B., Lv, L., Liu, X. J., Zhang, W. P., Zhao, X. P., & Cai, L. M. (2018). Effects of different protective clothing for reducing body exposure to chlorothalonil during application in cucumber greenhouses. *Human and Ecological Risk Assessment, 24*(1), 14–25.

7. Anderson, S. E., & Meade, B. J. (2014). Potential health effects associated with dermal exposure to occupational chemicals. *Environmental Health Insights, 8*, EHI-S15258.

8. Bányiová, K., Čupr, P., & Kohoutek, J. (2015). An experimentally refined tool to assess the risks of human dermal exposure to herbicide chlortoluron. *Environmental Science and Pollution Research, 22*, 10713–10720.

9. Blanco, L. E., Aragon, A., Lundberg, I., Liden, C., Wesseling, C., & Nise, G. (2005). Determinants of dermal exposure among Nicaraguan subsistence farmers during pesticide applications with backpack sprayers. *Annals of Occupational Hygiene, 49*(1), 17–24.

10. Bootsikeaw, S., Kongtip, P., Nankongnab, N., Chantanakul, S., Sujirarat, D., Mahaboonpeeti, R., Khangkhun, P., & Woskie, S. (2021). Urinary glyphosate biomonitoring of sprayers in vegetable farms in Thailand. *Human and Ecological Risk Assessment, 27*(4), 1019–1036.

11. Bormann, J. L., Acipayam, A. S. F., & Maibach, H. I. (2021). Percutaneous absorption of chemicals from fabric (textile). *Journal of Applied Toxicology, 41*(3), 363–374.

12. Cao, L. D., Zhang, H. J., Li, F. M., Zhou, Z. L., Wang, W. L., Ma, D. K., Yang, L., Zhou, P. G., & Huang, Q. L. (2018). Potential dermal and inhalation exposure to imidacloprid and risk assessment among applicators during treatment in cotton fields in China. *Science of the Total Environment, 624*, 1195–1201.
13. Carvalho, F. P. (2017). Pesticides, environment, and food safety. *Food and Energy Security, 6*(1), 48–60.
14. Chang, C., Chen, M., Gao, J., Luo, J., Wu, K., Dong, T., Zhou, K., He, X., Hu, W., & Wu, W. (2017). Current pesticide profiles in blood serum of adults in Jiangsu Province of China and a comparison with other countries. *Environment International, 102*, 213–222.
15. Chawla, P., Kaushik, R., Swaraj, V. S., & Kumar, N. (2018). Organophosphorus pesticide residues in food and their colorimetric detection. *Environmental Nanotechnology, Monitoring and Management, 10*, 292–307.
16. Chen, D. W., Li, J. G., Zhao, Y. F., & Wu, Y. N. (2022). Human exposure to fipronil insecticide and the associated health risk. *Journal of Agricultural and Food Chemistry, 70*(1), 63–71.
17. Cui, L. L., Ge, J., Zhu, Y. D., Yang, Y. Y., & Wang, J. (2015). Concentrations, bioaccumulation, and human health risk assessment of organochlorine pesticides and heavy metals in edible fish from Wuhan, China. *Environmental Science and Pollution Research, 22*(20), 15866–15879.
18. Damalas, C. A., & Eleftherohorinos, I. G. (2011). Pesticide exposure, safety issues, and risk assessment indicators. *International Journal of Environmental Research and Public Health, 8*(5), 1402–1419.
19. Das, G. P., Jamil, K., & Rahman, M. (2006). Effect of four organophosphorus compounds on human blood acetylcholinesterase: *In vitro* studies. *Toxicology Mechanisms and Methods, 16*(9), 455–459.
20. Delcour, I., Spanoghe, P., & Uyttendaele, M. (2015). Literature review: Impact of climate change on pesticide use. *Food Research International, 68*, 7–15.
21. Demir, V., & Ergin, S. (2013). Occurrence and assessment of chemical contaminants in drinking water in Tunceli, Turkey. *Journal of Chemistry, 2013*, 238374.
22. Devi, N. L., Yadav, I. C., Qi, S. H., Chakraborty, P., & Dan, Y. (2014). Distribution and risk assessment of polychlorinated biphenyls (PCBs) in the remote air and soil of Manipur, India. *Environmental Earth Sciences, 72*(11), 3955–3967.

23. Devi, N. L., Yadav, I. C., Raha, P., Qi, S. H., & Dan, Y. (2015). Spatial distribution, source apportionment, and ecological risk assessment of residual organochlorine pesticides (OCPs) in the Himalayas. *Environmental Science and Pollution Research, 22*(21), 20154–20166.
24. Ding, G. D., & Bao, Y. X. (2014). Revisiting pesticide exposure and children's health: Focus on China. *Science of the Total Environment, 472*, 289–295.
25. Duan, Y., Ramilan, T., Luo, J. H., French, N., & Guan, N. (2021). Risk assessment approaches for evaluating cumulative exposures to multiple pesticide residues in agro-products using seasonal vegetable monitoring data from Hainan, China: A case study. *Environmental Monitoring and Assessment, 193*(1), 1–19.
26. Fang, L. P., Zhang, S. Q., Chen, Z. L., Du, H. X., Zhu, Q., Dong, Z., & Li, H. D. (2015). Risk assessment of pesticide residues in dietary intake of celery in China. *Regulatory Toxicology and Pharmacology, 73*, 578–586.
27. Fang, Y. Y., Nie, Z. Q., Yang, Y. M., Die, Q. Q., Liu, F., He, J., & Huang, Q. F. (2015). Human health risk assessment of pesticide residues in market-sold vegetables and fish in a northern metropolis of China. *Environmental Science and Pollution Research, 22*(8), 6135–6143.
28. Farooq, S., Eqani, S. A. M. A. S., Malik, R. N., Katsoyiannis, A., Zhang, G., Zhang, Y. L., Li, J., Xiang, L., Jones, K. C., & Shinwari, Z. K. (2011). Occurrence, fingerprinting, and ecological risk assessment of polycyclic aromatic hydrocarbons (PAHs) in the Chenab River, Pakistan. *Journal of Environmental Monitoring, 13*(9), 3207–3215.
29. Fenske, R., & Simcox, N. J. (1995). Agricultural workers. In B. S. Levy & D. H. Women (Eds.), *Occupational Health: Recognizing and Perceiving Work-Related Disease and Injury* (3rd ed., pp. 665–684). Boston: Little Brown and Co.
30. Gangstad, E. O. (1989). Practical management for rights-of-way herbicide application. In *Woody Brush Control* (p. 31). CRC Press: Boca Raton, FL, USA.
31. Garzia, N. A., Spinelli, J. J., Gotay, C. C., & Teschke, K. (2018). Literature review: Dermal monitoring data for pesticide exposure assessment of farm workers. *Journal of Agromedicine, 23*(3), 187–214. https://doi.org/10.1080/1059924x.2018.1448734.
32. Gil, E., Salcedo, R., Soler, A., Ortega, P., Llop, J., Campos, J., & Oliva, J. (2021). Relative efficiencies of experimental and conventional foliar sprayers and assessment of optimal LWA spray volumes in trellised wine grapes. *Pest Management Science, 77*(11), 2462–2476.

33. Gish, T., Isensee, A., Nash, R., & Helling, C. (1991). Impact of pesticides on shallow groundwater quality. *Transactions of the ASAE, 34*(6), 1745–1753.
34. Guodong, D., Pei, W., Ying, T., Jun, Z., Yu, G., Xiaojin, W., Rong, S., Guoquan, W., & Xiaoming, S. (2012). Organophosphate pesticide exposure and neurodevelopment in young Shanghai children. *Environmental Science and Technology, 46*(5), 2911–2917.
35. Han, R., Wu, Z., Huang, Z., Man, X., Teng, L., Wang, T., Liu, P., Wang, W., Zhao, X., Hao, J., et al. (2021). Tracking pesticide exposure to operating workers for risk assessment in seed coating with tebuconazole and carbofuran. *Pest Management Science, 77*(14), 2820–2825.
36. Hanafi, A., Hindy, M., & Ghani, S. A. (2016). Effect of spray application techniques on spray deposits and residues of bifenthrin in peas under field conditions. *Journal of Pesticide Science, 41*(1), 49–54.
37. Hernández, A. F., Parrón, T., Tsatsakis, A. M., Requena, M., Alarcón, R., & López-Guarnido, O. (2013). Toxic effects of pesticide mixtures at a molecular level: Their relevance to human health. *Toxicology, 307*, 136–145.
38. Hunter, J. E., Gannon, T. W., Richardson, R. J., Yelverton, F. H., & Leon, R. G. (2020). Coverage and drift potential associated with nozzle and speed selection for herbicide applications using an unmanned aerial sprayer. *Weed Technology, 34*(2), 235–240.
39. Islam, M. S., Rahman, M. R., Prodhan, M. D. H., Sarker, D., Rahman, M. M., & Uddin, M. K. (2021). Human health risk assessment of pesticide residues in pointed gourd collected from retail markets of Dhaka City, Bangladesh. *Accreditation and Quality Assurance, 26*(4), 201–210.
40. Khairy, M. A., Kolb, M., Mostafa, A. R., El-Fiky, A., & Bahadir, M. (2012). Risk posed by chlorinated organic compounds in Abu Qir Bay, East Alexandria, Egypt. *Environmental Science and Pollution Research, 19*(2), 794–811.
41. Kim, K. H., Kabir, E., & Jahan, S. A. (2017). Exposure to pesticides and the associated human health effects. *Science of the Total Environment, 575*, 525–535.
42. Konthonbut, P., Kongtip, P., Nankongnab, N., Tipayamongkholgul, M., Yoosook, W., & Woskie, S. (2020). Paraquat exposure of backpack sprayers in agricultural areas in Thailand. *Human Ecology Risk Assessment, 26*, 2798–2811.
43. Korucu, M. K., Elibol, P. S., & Isleyen, M. (2021). An environmental risk assessment for a DDX-contaminated agricultural area in Turkey:

Soil vs. plant or human vs. animal. *Environmental Science and Pollution Research, 28*, 50127–50140.

44. Krieger, R. I., Bernard, C. E., Dinoff, T. M., Fell, L., Osimitz, T. G., Ross, J. H., & Thongsinthusak, T. (2000). Biomonitoring and whole-body cotton dosimetry to estimate potential human dermal exposure to semivolatile chemicals. *Journal of Exposure Analysis and Environmental Epidemiology, 10*, 50–57.
45. Kumaraveloo, K. S., & Kolstrup, C. L. (2018). Agriculture and musculoskeletal disorders in low- and middle-income countries. *Journal of Agromedicine, 23*(3), 227–248. https://doi.org/10.1080/1059924x.2018.1458671.
46. Kwong, L. H., Ercumen, A., Pickering, A. J., Unicomb, L., Davis, J., Leckie, J. O., & Luby, S. P. (2021). Soil ingestion among young children in rural Bangladesh. *Journal of Exposure Science and Environmental Epidemiology, 31*, 82–93.
47. Lang, C., Tao, S., Wang, X. J., Zhang, G., & Fu, J. M. (2008). Modeling polycyclic aromatic hydrocarbon composition profiles of sources and receptors in the Pearl River Delta, China. *Environmental Toxicology and Chemistry, 27*, 4–9.
48. Lei, B. L., Zhang, K. Q., An, J., Zhang, X. Y., & Yu, Y. X. (2015). Human health risk assessment of multiple contaminants due to consumption of animal-based foods available in the markets of Shanghai, China. *Environmental Science and Pollution Research, 22*, 4434–4446.
49. Li, H., He, Y. J., Qin, C. B., Liu, D. Q., & Zhang, K. F. (2019). Ecological analysis on spray performance of multi-rotor unmanned aerial sprayer in soybean field. *Ekoloji, 28*, 4573–4579.
50. Liu, S. Y., Jin, Q., Ren, R., & Zhu, G. N. (2021). Risk assessment of endocrine-disrupting pesticides exposure through consumption of *Carassius auratus* collected from Qiantang River, China. *Human Ecology Risk Assessment, 27*, 865–875.
51. Liu, Y., Pan, X., & Li, J. (2015). A 1961–2010 record of fertilizer use, pesticide application, and cereal yields: A review. *Agronomy for Sustainable Development, 35*, 83–93.
52. Lozier, M. J., Montoya, J. F. L., Del Rosario, A., Martinez, E. P., Fuortes, L., Cook, T. M., & Sanderson, W. T. (2013). Personal air sampling and risks of inhalation exposure during atrazine application in Honduras. *International Archives of Occupational and Environmental Health, 86*, 479–488.

53. Lu, P., Li, Q. F., Liu, H. M., Feng, Z. Z., Yan, X., Hong, Q., & Li, S. P. (2013). Biodegradation of chlorpyrifos and 3,5,6-trichloro-2-pyridinol by *Cupriavidus* sp. DT-1. *Bioresource Technology, 127*, 337–342.
54. Macfarlane, E., Carey, R., Keegel, T., El-Zaemay, S., & Fritschi, L. (2013). Dermal exposure associated with occupational end use of pesticides and the role of protective measures. *Safety and Health at Work, 4*, 136–141.
55. Mahaboonpeeti, R., Kongtip, P., Nankongnab, N., Tipayamongkholgul, M., Bunngamchairat, A., Yoosook, W., & Woskie, S. (2018). Evaluation of dermal exposure to the herbicide alachlor among vegetable farmers in Thailand. *Annals of Work Exposures and Health, 62*, 1147–1158.
56. Matthews, G. (2008). *Pesticide Application Methods*. John Wiley & Sons: Hoboken, NJ, USA.
57. Mostafalou, S., & Abdollahi, M. (2017). Pesticides: An update of human exposure and toxicity. *Archives of Toxicology, 91*, 549–599.
58. Msibi, S. S., Chen, C. Y., Chang, C. P., Chen, C. J., Chiang, S. Y., & Wu, K. Y. (2021). High pesticide inhalation exposure from multiple spraying sources amongst applicators in Eswatini, Southern Africa. *Pest Management Science, 77*, 4303–4312.
59. Nankongnab, N., Kongtip, P., Tipayamongkholgul, M., Bunngamchairat, A., Sitthisak, S., & Woskie, S. (2020). Difference in accidents, health symptoms, and ergonomic problems between conventional farmers using pesticides and organic farmers. *Journal of Agromedicine, 25*, 158–165.
60. Nayak, J., Singh, S. P., & Moharana, G. (2012). Occupational health hazards of farm women at their workplace in Bhopal district. *Final Report*, DRWA, Bhubaneswar.
61. Odewale, G. O., Sosan, M. B., Oyekunle, J. A. O., & Adeleye, A. O. (2021). Human health risk assessment of dichlorodiphenyltrichloroethane (DDT) and hexachlorocyclohexane (HCH) pesticide residues in fruits and vegetables in Nigeria. *Environmental Science and Pollution Research, 28*, 33133–33145.
62. Pan, L., Sun, J., Li, Z., Zhan, Y., Xu, S., & Zhu, L. (2018). Organophosphate pesticide in agricultural soils from the Yangtze River Delta of China: Concentration, distribution, and risk assessment. *Environmental Science and Pollution Research, 25*, 4–11.
63. Pedroso, T. M. A., Benvindo-Souza, M., Nascimento, F. D., Woch, J., Dos Reis, F. G., & Silva, D. D. E. (2022). Cancer and occupational exposure to pesticides: A bibliometric study of the past 10 years. *Environmental Science and Pollution Research, 29*, 17464–17475.

64. Phung, D. T., Connell, D., Miller, G., Rutherford, S., & Chu, C. (2013). Needs assessment for reducing pesticide risk: A case study with farmers in Vietnam. *Journal of Agromedicine, 18*, 293–303.

65. Praveen, P. L. R. J., Rao, A. R. M., & Rao, V. (2005). Decision making pattern of rural women in farm-related activities. *Agriculture Extension Review, November-December*, 3–5.

66. Ratanachina, J., Amaral, A., De Matteis, S., Cullinan, P., & Burney, P. (2022). Farming, pesticide exposure, and respiratory health: A cross-sectional study in Thailand. *Occupational and Environmental Medicine, 79*, 38–45.

67. Saeed, M. F., Shaheen, M., Ahmad, I., Zakir, A., Nadeem, M., Chishti, A. A., Shahid, M., Bakhsh, K., & Damalas, C. A. (2017). Pesticide exposure in the local community of Vehari District in Pakistan: An assessment of knowledge and residues in human blood. *Science of the Total Environment, 587*, 137–144.

68. Sampath, S., Shanmugam, G., Selvaraj, K. K., & Ramaswamy, B. R. (2015). Spatio-temporal distribution of polycyclic aromatic hydrocarbons (PAHs) in atmospheric air of Tamil Nadu, India, and human health risk assessment. *Environmental Forensics, 16*, 76–87.

69. Sapbamrer, R., Hongsibsong, S., Naksata, M., & Naksata, W. (2021). Insecticide filtration efficiency of respiratory protective equipment commonly worn by farmers in Thailand. *International Journal of Environmental Research and Public Health, 18*, 2624.

70. Schneider, T., Vermeulen, R., Brouwer, D. H., Cherrie, J. W., Kromhout, H., & Fogh, C. L. (1999). Conceptual model for assessment of dermal exposure. *Occupational and Environmental Medicine, 56*, 765–773.

71. Scholtz, M., & Bidleman, T. (2007). Modelling of the long-term fate of pesticide residues in agricultural soils and their surface exchange with the atmosphere: Part II. Projected long-term fate of pesticide residues. *Science of the Total Environment, 377*, 61–80.

72. Séralini, G. E., Clair, E., Mesnage, R., Gress, S., Defarge, N., Malatesta, M., Hennequin, D., & De Vendômois, J. S. (2014). Republished study: Long-term toxicity of a Roundup herbicide and a Roundup-tolerant genetically modified maize. *Environmental Sciences Europe, 26*, 1–17.

73. Shi, J. C., Li, Y. L., Liang, H., Zheng, G. J., Wu, Y. L., & Liu, W. H. (2013). OCPS and PCBs in marine edible fish and human health risk assessment in the eastern Guangdong, China. *Archives of Environmental Contamination and Toxicology, 64*, 632–642.

74. Simas, J. M. M., Yamauchi, L. Y., & De Alencar, M. D. B. (2021). Risk factors associated with respiratory health and banana farming. *Archives of Environmental and Occupational Health, 76*, 181–187.
75. Singh, T., & Singh, D. K. (2017). Phytoremediation of organochlorine pesticides: Concept, method, and recent developments. *International Journal of Phytoremediation, 19*, 834–843.
76. Siriwat, S., Nganchamung, T., Ponrachom, C., Siriwong, W., & Robson, M. G. (2021). Health risk assessment of dermal exposure to chlorpyrifos among children in agricultural areas in Sakon Nakhon Province, Thailand. *Human Ecology Risk Assessment, 27*, 2277–2287.
77. Strassemeyer, J., Daehmlow, D., Dominic, A., Lorenz, S., & Golla, B. (2017). Synops-Web, an online tool for environmental risk assessment to evaluate pesticide strategies on field level. *Crop Protection, 97*, 28–44.
78. Sumner, H. R., & Herzog, G. A. (2000). Assessing the effectiveness of air-assisted and hydraulic sprayers in cotton via leaf bioassay. *Journal of Cotton Science, 4*, 79–83.
79. Taufeeq, A., Baqar, M., Sharif, F., Mumtaz, M., Ullah, S., Aslam, S., Qadir, A., Majid, M., & Jun, H. (2021). Assessment of organochlorine pesticides and health risk in tobacco farming associated with River Barandu of Pakistan. *Environmental Science and Pollution Research, 28*, 38774–38791.
80. Tefera, Y. M., Gaskin, S., Thredgold, L., & Pisaniello, D. (2022). The role of formulation co-ingredients in skin and glove barrier protection against organophosphate insecticides. *Pest Management Science, 78*, 177–183.
81. Thouvenin, I., Bouneb, F., & Mercier, T. (2016). Operator dermal exposure and individual protection provided by personal protective equipment during application using a backpack sprayer in vineyards. *Journal of Consumer Protection and Food Safety, 11*, 325–336.
82. Tsui, M. T., & Chu, L. (2003). Aquatic toxicity of glyphosate-based formulations: Comparison between different organisms and the effects of environmental factors. *Chemosphere, 52*, 1189–1197.
83. Tudi, M., Atabila, A., Ruan, H. D., Wang, L., Lyu, J., Tong, S., Yu, Q. J., Sadler, R., Phung, D. T., & Connell, D. (2022). Natural dynamics and residues of pymetrozine for typical rice-growing areas of China. *Ecotoxicology and Environmental Safety, 232*, 113230.
84. Tudi, M., Ruan, H. D., Wang, L., Lyu, J., Sadler, R., Connell, D., Chu, C., & Phung, D. T. (2021). Agriculture development, pesticide application and its impact on the environment. *International Journal of Environmental Research and Public Health, 18*, 1112.

85. Tudi, M., Wang, L., Ruan, H. D., Tong, S., Atabila, A., Sadler, R., Yu, Q. J., Connell, D., & Phung, D. T. (2022). Environmental monitoring and potential health risk assessment from pymetrozine exposure among communities in typical rice-growing areas of China. *Environmental Science and Pollution Research*, 1–14.

86. Tyagi, H., Gautam, T., & Prashar, P. (2015). Survey of pesticide use patterns and farmers' perceptions: A case study from cauliflower and tomato cultivating areas of district Faridabad, Haryana, India. *International Journal of Medipharm Research, 1*, 139–146.

87. Vercruysse, F., Drieghe, S., Steurbaut, W., & Dejonckheere, W. (1999). Exposure assessment of professional pesticide users during treatment of potato fields. *Pesticide Science, 55*, 467–473.

88. Wang, J. Y., Wei, D., Deng, Y. J., Feng, W. Z., Gao, Q., Shi, Y. H., & Xiao, J. J. (2021). Bioaccessibility and health risk of neonicotinoids in apple and pear samples as affected by *in vitro* digestion. *Quality Assurance and Safety of Crops & Foods, 13*, 74–81.

89. Wang, X., Murison, J., Wang, J., Leong, G., Wu, Z. C., & Li, Q. X. (2021). Dermal exposure assessment to trinexapac-ethyl: A case study of workers in golf courses in Hawaii, USA. *Environmental Science and Pollution Research, 28*, 1072–1076.

90. Wang, Z. N., Meng, Y. X., Mei, X. D., Ning, J., Ma, X. D., & She, D. M. (2020). Assessment of handler exposure to pesticides from stretcher-type power sprayers in orchards. *Applied Sciences, 10*, 8684.

91. Weirich Neto, P. H., Meer, R. W., Vink, P., Justino, A., & Garcia, L. C. (2014). Automatic and manual spray bar sections control. *Engineering in Agriculture, 34*, 1201–1209.

92. World Health Organization. (1990). *Public Health Impact of Pesticides Used in Agriculture*. WHO.

93. Xiongkui, H., Bonds, J., Herbst, A., & Langenakens, J. (2017). Recent development of unmanned aerial vehicle for plant protection in East Asia. *International Journal of Agricultural and Biological Engineering, 10*, 18–30.

94. Yadav, I. C., Devi, N. L., Li, J., Zhang, G., & Shakya, P. R. (2016). Occurrence, profile, and spatial distribution of organochlorine pesticides in soil of Nepal: Implication for source apportionment and health risk assessment. *Science of the Total Environment, 573*, 1598–1606.

95. Yadav, I. C., Devi, N. L., Syed, J. H., Cheng, Z., Li, J., Zhang, G., & Jones, K. C. (2015). Current status of persistent organic pesticides residues in

air, water, and soil, and their possible effect on neighboring countries: A comprehensive review of India. *Science of the Total Environment, 511*, 123–137.

96. Yan, X. J., Zhou, Y. Y., Liu, X. H., Yang, D. B., & Yuan, H. Z. (2021). Minimizing occupational exposure to pesticide and increasing control efficacy of pests by unmanned aerial vehicle application on cowpea. *Applied Sciences, 11*, 9579.
97. Yang, C. C., & Deng, J. F. (2007). Intermediate syndrome following organophosphate insecticide poisoning. *Journal of the Chinese Medical Association, 70*, 467–472.
98. Yohannes, Y. B., Ikenaka, Y., Saengtienchai, A., Watanabe, K. P., Nakayam, S. M. M., & Ishizuka, M. (2014). Concentrations and human health risk assessment of organochlorine pesticides in edible fish species from a rift valley lake—Lake Ziway, Ethiopia. *Ecotoxicology and Environmental Safety, 106*, 95–101.
99. Yoshida, T., Mimura, M., & Sakon, N. (2021). Estimating household exposure to pyrethroids and the relative contribution of inhalation pathway in a sample of Japanese children. *Environmental Science and Pollution Research, 28*, 19310–19324.
100. Yuan, G. L., Sun, Y., Qin, J. X., Li, J., & Wang, G. H. (2014). Chiral signature of alpha-HCH and O, P′-DDT in the soil and grass of the central Tibetan Plateau, China. *Science of the Total Environment, 500*, 147–154.
101. Zhang, Q., Fu, C., Guo, X., Gao, J., Zhang, P., & Ding, C. (2021). Fluorescent determination of butyrylcholinesterase activity and its application in biological imaging and pesticide residue detection. *ACS Sensors, 6*, 1138–1146.
102. Zhao, Z. H., Zhang, L., Cai, Y. J., & Chen, Y. W. (2014). Distribution of polycyclic aromatic hydrocarbon (PAH) residues in several tissues of edible fishes from the largest freshwater lake in China, Poyang Lake, and associated human health risk assessment. *Ecotoxicology and Environmental Safety, 104*, 323–331.

CHAPTER 6

FAIR WAGES AND LIVELIHOODS FOR AGRICULTURAL WORKERS

Contents

INTRODUCTION

To promote resilient and equitable rural economies, agricultural workers must be paid fairly and have sustainable means of subsistence. Despite playing a vital role in the production of food, agricultural workers frequently endure appalling working conditions, low pay, and restricted access to social protection, especially in developing nations. Setting living standards that cover necessities like housing, food, healthcare, and education is necessary to guarantee fair wages. To improve livelihoods beyond wages, a comprehensive strategy is required, one that grants access to resources, technology, and training to increase productivity and diversify sources of income. To stop exploitation, labor rights and protections—such as the ability to enter into contracts, secure working conditions, and the ability to organize—are also essential. Consumers, businesses, and governments all have a part to play in advancing fair trade policies that put agricultural workers' welfare first. Raising living standards can be facilitated by certification schemes, ethical supply chains, and laws that assist farmworkers and small-scale farmers. In the end, providing agricultural workers with sustainable livelihoods and fair wages not only raises their standard of living but also enhances global economic stability and food security.

6.1. WAGED AGRICULTURAL WORKERS

The men and women who work in the glasshouses, livestock units, orchards, crop fields, and primary processing facilities that create the world's food and fibers are known as wage agricultural workers. They work on large industrialized farms and plantations, as well as small and medium-sized farms. They differ from farmers in that they are waged workers since they do not own or rent the land they work on or the tools and equipment they use. These employees are not a uniform group. They fall into a variety of categories due to the wide range of their terms and conditions of employment, including full-time, permanent agricultural workers, seasonal, temporary, or casual workers, migrant agricultural workers, piece-rate workers, and workers who receive "in-kind" compensation. The employed workforce also includes a large number of indigenous agricultural workers. Workers in agriculture receive some form of "wage," which could be in the form of cash, in-kind compensation, or a mix of these. Whether they work for a farmer, a farming or plantation company, a labor contractor, or a subcontractor, they are all engaged in an employment relationship.

The nature of the workforce reflects the seasonal variations in the demand for agricultural labor. Planting and harvesting typically require very long workdays, with shorter workdays during off-peak periods. Field work can last from sunrise to sunset during peak hours, plus transportation time to and from the fields. Because of the length of the working day and the intensity of the work, there is little opportunity for rest periods. Payment systems have the potential to make this worse. There is a strong financial incentive for casual, temporary, or seasonal workers to increase their working hours to increase their earnings because minimum wages for agricultural workers generally are low or nonexistent, and many of them are paid at least partially on a piece-work basis, such as per kilogram of crop picked, row weeded, or hectare sprayed.

By its very nature, a lot of agricultural labor is physically taxing, requiring extended standing, bending, stooping, and performing repetitive motions in awkward body positions. Fatigue, ill-designed equipment, challenging terrain, weather exposure, and poor general health all raise the chance of accidents. Even though technological advancements have made agricultural labor less physically taxing, they have also brought about new hazards, particularly those related to the use of highly advanced machinery and extensive chemical use that frequently occurs without the proper safety precautions, knowledge, or training. It should come as no surprise that there are many serious and fatal accidents as well as illnesses. However, when it comes to access to health care, workers' compensation, long-term disability insurance, and survivors' benefits, agricultural workers are among the least protected.

6.1.1. Full Time, Permanent Workers

Compared to other waged agricultural workers, full-time, permanent employees enjoy greater job security, comparatively higher pay, better housing, and better health and work benefits. This does not imply, however, that these workers receive high compensation in comparison to the national average. In comparison to cities, rural areas typically have lower real and cash wages, as well as longer workweeks.

A large number of full-time agricultural workers are poor or below. The least common type of contracts are permanent ones, and in the majority of nations, their proportion of all agricultural employment has been falling. Known as the "casualization" or "flexibilization" of employment, there is a shift away from full-time work toward seasonal and casual work with little to no social protection.

6.1.2. Casual, Temporary, and Seasonal Workers

In the majority of developing and some developed nations, waged agricultural workers are employed on a seasonal basis, frequently on a casual or temporary basis. People who are employed and paid at the end of each workday or on a task basis are said to be doing casual work. Temporary workers are those who are employed for a set but constrained amount of time. The majority of temporary, seasonal, and casual workers do not receive any kind of unemployment insurance, social security, paid holidays, or maternity or sick leave. In fact, these same benefits are not available to many full-time waged agricultural earners. A growing proportion of women are working as temporary or casual employees. Furthermore, even when there is continuous employment, jobs are frequently classified as temporary or casual. Individual workers are frequently rotated to prevent them from receiving the benefits of permanent employment status.

For instance, statistics from the Ugandan National Union of Plantation and Agricultural Workers confirm that women make up the majority of casual workers in the cut flower sector. More than 70% of this workforce works in casual jobs without benefits like annual leave or job security. The majority of the female employees work in the grading and harvesting departments. A casual worker receives 1,500 Uganda shillings (75 cents US) per day, while a full-time employee receives a monthly total package of 70,000 Uganda shillings ($35 US). Unpredictable weather patterns, erratic produce market demand, and labor laws requiring seasonal and permanent employees to receive benefits like notice pay, leave allowances, and medical care are some of the factors driving the trend toward casual and temporary labor.

Unions in the plantation industry in many nations, including Kenya, have attempted to include clauses in collective bargaining agreements that restrict the amount of time that workers may be employed on a seasonal, casual, or temporary basis to combat this trend toward temporary and casual labor.

6.1.3. Migrant Workers

There are migrant workers in every kind of employment arrangement, whether they are full-time, seasonal, temporary, or casual. They could be foreign workers or migrant laborers from other regions of the nation. Migrant workers are always at a significant disadvantage when it comes to compensation, social protection, housing, and health care, regardless of where they are from.

In many nations, agriculture has become an export-oriented industry that is largely reliant on migrant labor as a result of globalization and structural adjustment. According to a 2003 trade union consultation on irregular migration and human trafficking in Europe, the big supermarket chains' influence basically drives farmers to produce at extremely low prices. In response to the pressure, farmers lower labor costs, which puts more of the burden on the workers. The national workforce does not find low pay and status for hard work appealing. As a result, farmers depend more and more on migrant labor. In some places, like the Spanish region of Almeria, over 90% of agricultural laborers are foreign-born. One of the main effects of increasing work flexibility, casual employment, low wages, unfavorable working conditions, and poverty is labor migration.

Globally, this labor mobility is extremely important. Although only the family's head is officially employed, entire families frequently make up the migrant labor force in practice. Children of seasonal and migrant workers often work alongside their parents in many countries, but they are not paid. Children must work for migrant and seasonal workers to earn a living wage because much of the labor is paid on a piece-rate basis.

6.1.4. Indigenous Rural Workers

In many nations, indigenous people make up a portion of the agricultural workforce. They frequently become agricultural laborers after being evicted from their own lands, frequently with subpar pay and working conditions. For instance, the living and working conditions of indigenous workers and their families are often worse than those of workers from what is known as the dominant society, including non-indigenous rural workers, in many Latin American countries. Access to education, social security, trade union membership, wages, and occupational health and safety all exhibit these disparities.

6.1.5. Lack of Standardization of Definitions of Categories of Workers

These waged worker categories frequently overlap. The precise computation of the workforce is impacted by the absence of definitional standardization. For instance, farmers in South Africa's Western Cape who hired women made a distinction between seasonal, temporary, and permanent workers. However, there were significant differences in how permanent employees were defined; many farmers considered seasonal workers to be casual workers, and vice versa. In actuality, there wasn't much uniformity within a sector or region, or even between farms.

6.1.6. Wage-dependent Small Farmers

To augment their meager earnings, many small farmers spend a portion of the year working on a farm or plantation. They rely on waged labor as a consistent source of income each year. They are actually a combination of wage worker and farmer. For instance, a 1998 study of the agricultural labor market in Mexico calculated that up to 40.8 million farmers, or 78.3% of the rural labor force, were employed as waged workers for at least a portion of the year. Despite being a well-known phenomenon, wage-dependent smallholders are rarely included in agricultural programs aimed at empowering small farmers. Programs and strategies for eradicating poverty rarely take into consideration the need to raise smallholders' wages and enhance working conditions.

6.1.7. Developing Links between Waged Agricultural Workers and Small Farmers

Small farmers and waged agricultural workers face many of the same challenges in sustainable agriculture and rural development, including food security. Their roles and needs are also very similar, particularly in cases where farmers are frequently wage-dependent. Similar interests include, for instance, enhancing the environmental, safety, and health conditions on farms and plantations as well as the freedom of association and organization. The line separating waged agricultural workers from the rest of the rural working population is becoming increasingly blurred as rural employment becomes more casualized. Consequently, waged agricultural workers have emerged as potential allies of marginalized rural populations, including small-scale subsistence farmers, sharecroppers, and tenants, the unemployed, and those without land. They frequently come from the same home, participate in comparable activities, and share the same surroundings. As their interests align and shared approaches are developed to address shared issues, relationships based on solidarity and mutual support can grow, despite the occasional tensions and conflicts between these various actors.

The example of a sugar plantation and processing plant in Kenya run by a multinational corporation (MNE) serves to highlight the possible connections. Despite having 3,200 permanent workers on its own nucleus plantation, this sugar company contracts with, purchases from, and processes sugar from a network of 65,000 out growers, i.e., small farmers, collectively referred to as their own business. The outgrower network and the sugar company are completely dependent on one another.

6.1.8. The "Employment Relationship"

The various types of waged employment involve employment relationships that are frequently complicated and challenging to comprehend, particularly when labor is provided through labor contractors or casual labor is used. A person known as the employee and another known as the employer, to whom the employee provides labor or services under specific conditions in exchange for compensation, are legally linked by the concept of an employment relationship. Although the idea of an employment relationship is universal across legal systems and customs, each nation has different requirements, rights, and benefits related to it. An increasing number of workers who are actually employees but lack the protection of an employment relationship are a result of changes in the labor market's structure, how work is organized, and the inadequate application of the law. In economies with less formalization, this type of false self-employment is more prevalent. However, this phenomenon is also becoming more prevalent in many nations with developed labor markets. While some of these advancements have been around for decades, others are brand new.

Work patterns are evolving both inside and outside the confines of the employment relationship as a result of changes in the labor market's structure and how work is organized. Sometimes it's difficult to tell if a worker is truly self-employed or an employee. When an employer disguises an employee's true legal status by treating them as someone else, this is known as "disguised employment." Among the most common ways to conceal the employment relationship are false self-employment, false subcontracting, the creation of fictitious cooperatives, false service provision, and false company restructuring. Such actions may have the dual effect of denying workers labor protection and avoiding expenses such as taxes and social security contributions. Any time work is done or services are rendered in a way that raises a real and legitimate question about whether an employment relationship is in place, there is an ambiguous employment relationship. Even in situations where there is no intention to conceal the employment relationship, it is becoming increasingly difficult to discern between dependent and independent work. In this regard, it is recognized that the line separating independent contractors from employees has blurred in many places. Employee autonomy or increased independence is one of the features of some new work arrangements. Everyone involved benefits from the law's predictability and clarity. Both employers and employees should be aware of their status and, as a result, their legal rights and responsibilities. All parties involved in the labor market have an interest in making sure that the many different arrangements under which a worker performs work or renders services can be placed within a suitable legal framework. Fair labor market governance requires well-defined regulations.

6.2. WORKERS AND THEIR TRADE UNIONS IN AGRICULTURE AND THE FOOD INDUSTRY/FOOD CHAIN

The purpose of a trade union, also known as a labor union, is to defend and advance workers' rights and means of subsistence in the workplace. Patronage does not exist within a free trade union. It is not an organization founded by the elite or the government for the same reasons, nor is it a workers' organization that was openly or covertly formed by an employer to regulate the demands or actions of the workers. Trade unions strive to guarantee the job security of their members and improve their pay and working conditions through collective bargaining with employers. Additionally, unions collaborate with governments and lobby them to uphold and enhance labor laws, standards, and policies. Unions are funded in part by the dues paid by employees. At conferences and congresses, representatives chosen by the members democratically decide on union policies. Workers can also voice their opinions on broader societal, political, ethical, and economic issues through unions. The trade unions, federations, and confederations found in agriculture are included.

6.2.1. National Level

1. National agricultural trade unions with only agricultural workers on commercial farms and plantations as members. Typically, this kind of union occurs in developing nations where agriculture continues to be the main economic sector. Workers in all agricultural sectors may be organized by such unions, or they may only have members in particular crop sectors, such as tobacco, sugar, and tea. These unions are frequently the biggest workers' group in a nation. Large unions and plantation agriculture are frequently associated. Two such organizations are the Kenya Plantation and Agricultural Workers Union in Kenya and the National Union of Plantation and Agricultural Workers in Uganda.
2. In nations with comparatively small agricultural labor forces, national general workers' unions unite members from various commercial and industrial sector industries. Within these sizable general unions, waged agricultural workers are organized into sections or trade groups. Developed nations frequently have these kinds of unions. Examples include the Transport and General Workers in the UK, the Agricultural Workers' sections of Kommunal in Sweden, and SiD in Denmark.
3. Additionally, there are hybrid organizations known as agricultural trade unions that represent both independent farmers and wage earners. For instance, there are about 9 million members of CONTAG in Brazil (3 million waged workers and 6 million smallholders). Typically, these unions have distinct political and organizational structures to handle the two kinds of members. The two forms of membership in the same organization can occasionally have a difficult relationship, particularly when small farmer members also work for pay.
4. Whether they are general, agriculture-specific, or hybrid, national trade unions typically affiliate with or are part of a national trade union center. Unions that organize workers in various commercial and industrial sectors are represented by a national center. A nation may have more than one national center.

5. Some agricultural unions are organized on a federal state basis such as, for example, in India.
6. In some countries, labor laws confine unions to organizing in one company.

6.2.2. International and Regional Levels

The International Union of Food, Agricultural, Hotel, Restaurant, Catering, Tobacco, and Allied Workers' Associations (IUF) is one example of a global union (labor) federation to which national unions may affiliate on a sectoral basis. Workers on farms and plantations, in businesses that manufacture food and beverages, in lodging facilities, dining establishments, and in the tourism and catering industries worldwide are organized by the IUF, a network of 336 national affiliated unions. Bananas, cocoa, coffee, sugar, tea, and (cut) flowers are the six main crops that the IUF has singled out to build ties between unions along the food chain and unite workers at every level of production.

The International Confederation of Free Trade Unions (ICFTU), which is composed of national trade union centers, collaborates with global union federations. 148 million workers are members of the ICFTU, which has 233 affiliated organizations spread across 154 nations and territories on all five continents. Its three main regional organizations are ORIT for the Americas, AFRO for Africa, and APRO for Asia and the Pacific. Additionally, it keeps tight ties with the Global Union Federations and the European Trade Union Confederation (ETUC). There are also two other national center-based international trade union organizations. With more than 26 million members, the World Confederation of Labor (WCL) is an international trade union confederation that brings together 144 independent and democratic trade unions from 116 nations. Brussels, Belgium, is home to its headquarters. (WCL, 2004). To establish a new international trade union organization, the ICFTU and WCL are currently in merger negotiations. Prague is home to the World Federation of Trade Unions (WFTU).

6.2.3. Union Organization in Transnational Corporations in Agriculture and Along the Food Chain

International union federations like the IUF represent and organize workers in catering services, restaurants, hotels, tourist attractions, retail, and wholesale distribution, as well as in large multinational corporations involved in agriculture and the food chain. Through their international federations, national, and regional trade unions can play a significant role in collaborating with these MNEs and tracking their advancements toward sustainable agriculture, particularly with regard to labor standards.

6.2.4. Levels of Trade Union Organization in Agriculture

Generally speaking, most countries have low levels of trade union representation among agricultural workers, especially among women agricultural laborers and temporary workers. There is a dearth of complete data on agricultural trade union membership. Agricultural workers continue to be poorly organized for a variety of reasons. Some of these challenges are financial or practical, such as the inability to organize across wide geographic areas, the lack of transportation for organizers, the low membership dues that only provide the most basic union services, and so forth.

However, workers who want to exercise their fundamental human right to freedom of association by joining the union of their choice are frequently prevented from doing so by administrative and legal obstacles. For instance, Turkey's labor laws weren't updated to include agricultural workers until 2003. However, the public notary must approve each new application for union membership, and there is a significant fee associated with this process. This requirement hinders freedom of association and makes it expensive for unions, particularly low-income agricultural ones, to recruit members. For indigenous workers, who are frequently denied the legal ability to form or join such organizations, the situation is even more dire.

The principle of freedom of association, as stated in the ILO Right of Association (Agriculture) Convention, 1921, and the Freedom of Association and Protection of the Right to Organize Convention, 1948, enshrines the fundamental right of both employers and workers to form and join organizations of their own choosing to defend their interests and improve their conditions. The ILO International Declaration on Fundamental Principles and Rights at Work includes both of these conventions.

The goal of the 1975 ILO Rural Workers' Organizations Convention is to give agricultural workers access to the freedom of association principle. However, the ILO Committee on Freedom of Association has examined cases that demonstrate the prevalence of violations of freedom of association among agricultural workers. These infractions include everything from practical obstacles and regulatory limitations to physical attacks and the killing of trade union officials and leaders in rural regions. The Committee of Experts has regularly reviewed cases of forced and bonded labor involving agricultural workers in various nations and regions. Additionally, the ICFTU releases an annual report on global trade union rights abuses. Furthermore, waged workers engaged by labor contractors or subcontractors often face difficulties in registering as union members. This is causing growing concern, due to the increasing numbers of workers hired by these labor-supplying intermediaries.

6.3. CHARACTERISTICS OF AGRICULTURAL LABOR

The NSSO uses "agricultural labor households" to identify agricultural workers in India. To estimate the total number of agricultural laborers, the number of households must be multiplied by the average number of people living in an agricultural labor household. The estimated number of "agricultural workers" and "agricultural laborers," however, is directly provided by the census. A household is considered to be an "agricultural labor household" if agriculture accounts for more than half of its income. Because agricultural laborers have certain unique characteristics, policy planners have had to create and execute unique employment-generating programs to increase the number of working days they have throughout the year. The availability of work and income, the unorganized nature of the workforce, which contributes to their poor job quality, and the high rate of poverty among agricultural workers are all included in these characteristics.

6.3.1. Irregularity in Employment and Income

Agrarian economies must differentiate between the types of jobs in the agricultural and non-agricultural sectors. The National Sample Survey Organization (NSSO) has developed three methods to measure employment following the global consensus for this purpose, namely the current weekly status method, the current daily status method, and the standard status method. The other two approaches capture the nature of under-employment in agriculture by focusing on measuring the activities/work performed during "days within a week" and "hours within a day," whereas the usual status approach captures the numbers employed on a fairly stable basis (by taking into account persons employed for the majority of the larger reference period of one year). According to the most recent Census data, 46% of all agricultural workers were employed in 2001. With a daily wage of $4, an average agricultural worker was able to find employment for roughly 209 days of the year (i.e., an average of roughly $70 per person per month). At this rate, at least three members of an agricultural labor household must be able to find work for at least 200 days a year for the household to stay above the poverty line, since the poverty line expenditure level for such a household was set at about $180 per month in 2002–03. However, in practice, a large number of agricultural labor households live below the poverty line since they are denied many days of work.

6.3.2. Poverty

At the national level, the prevalence of rural poverty in India was higher in 2004–05 (28.3%) than in urban areas (25.7%). Specifically, households with "all agricultural workers" had a higher poverty ratio (31.1%). Additionally, the poverty rate for households with agricultural laborers was the highest among rural households involved in agriculture, at 46.4%. The corresponding ratio for cultivators was notably lower, at 21.5%. However, as land holdings increased, the poverty ratio for agricultural households gradually decreased.

The corresponding poverty ratios for various farmer classes were as follows: landless, at 22%; sub-marginal, at 20.2%; marginal, at 18.1%; small, at 14.8%; and medium and large, at 9.8%. Therefore, even though the average poverty ratio for farmers in both household classes (i.e., 15.2% of both landholders and landless people was in "agricultural labor households," it was more than three times higher. Therefore, a major determinant of poverty across farmer classes based on the size of their land holdings was the extent of "land holding." This emphasizes the significance of policy focus required for the successful implementation of land reform measures (to give cultivating farmers ownership rights) and to make it easier for farmers to obtain inputs, such as extension services, that are necessary for cultivation.

6.3.3. Unorganized Employment

The organized and unorganized sectors of the Indian workforce are conceptually separated. The former refers to the sense of job security that employees experience as a result of at least one or more statutory regulations (such as minimum wage payments, paid holidays, and work hours). supplied by the company. However, these jobs are extremely rare in India (approximately 7% of all jobs), and they are primarily offered to workers with some formal education or training in non-agricultural businesses. Due to their lack of literacy and resulting low skill levels, agricultural workers make up nearly all of the unorganized or informal sector workforce. This trait of agricultural workers directly contributes to their inability to negotiate for the defense of their legal rights. As a result, they are now unable to receive even the minimum wage that the Minimum Wages Act guarantees. Due to this feature of agricultural workers, the National Statistical Commission (2012) noted that, in the Indian context, roughly 52% of workers involved in agricultural activities (or roughly 65% of workers in rural areas with usual status and 7% of workers in urban areas with usual status) are not included in the NSSO's employment-unemployment surveys, which specifically focus on the informal sector. The Commission further notes that "the implication of this is a serious underestimation of the economic contribution of the large unorganized employment engaged in agricultural and allied activities to the estimates of national income of the country." In 2004–05, the estimated labor input share in the unorganized sector activities of agriculture and forestry was 99.9%, while in fishing it was 98.7%. Therefore, the country suffers both personal and financial loss when there are so many people working in agriculture in the unorganized or informal sector. Although it requires more capital investment, the degree of formalization in the workforce should theoretically decrease with development. For this to occur in agriculture, small and marginal farmers must have access to capital so they can cultivate their land productively and adopt increasingly advanced technologies. At the same time, policy initiatives must concentrate on increasing agricultural workers' ability to absorb technology by providing them with the necessary extension training.

6.3.4. Wage-Employment and Self-Employment

In agriculture, 28.8% of households are self-employed, according to the 2009–10 Employment-Unemployment Survey report. This is twice as much as the 13% that corresponds to non-agriculture. 41.2% of households are estimated to be wage-employed.

People who work in other people's farms or non-farm businesses in exchange for regular salaries or wages—that is, not based on daily or periodic work contract renewals—are considered to be engaged in wage-employment. A subset of wage workers is known as "casual wage laborers," who perform work for other farm or non-farm businesses (both domestic and commercial) in exchange for pay based on the conditions of a daily or periodic work contract. Road construction, bund construction, and other public works projects employ these casual wage workers. Meeting both the immediate need for income and the development of public infrastructure is essential to increasing the productivity of self-employed people in particular.

Conversely, self-employed people are those who run their own farm or non-farm business or work as independent contractors in a trade or profession. The key characteristics of self-employed people are their economic independence (with regard to market, scale of operation, and capital) and autonomy (about how, where, and when to produce).

The portion of their labor and the business's profit make up the fee or compensation that self-employed people receive. In other words, the sales or profits of the goods or services they produce determine all or most of their compensation.

Given that the vast majority of agricultural workers rely on daily job searches, the government has over the years put in place many unique employment promotion initiatives tailored to the requirements of rural agricultural workers. To lessen the amount of unemployment or underemployment of unskilled workers, the primary goal of these programs has been to create more "person days of employment," or more immediate employment as opposed to regular and sustainable employment.

6.4. ELEMENTS OF SUSTAINABLE LIVELIHOODS FOR WAGED AGRICULTURAL WORKERS

Some of the key elements of sustainable livelihoods for waged agricultural workers are included.

6.4.1. The Agricultural Labor Market and Its Changing Nature

Employment and income levels in rural areas are significantly influenced by labor markets. Although it is unrealistic to expect agriculture to employ every rural worker, it plays a crucial role in both the direct creation of jobs, such as wage employment and the indirect creation of jobs through increased economic diversification. Since hiring out their labor may be their only source of income, many rural poor people find that access to labor markets is especially crucial. For many waged agricultural workers, their labor is their only resource. Since this is the only practical means of increasing the productivity of their primary asset and, consequently, the livelihoods of the rural poor, it is crucial to improve the way rural labor markets operate.

6.4.1.1. Trends in Waged Employment, Wages, and Poverty

The three major trends in agricultural labor markets in the 1980s and early 1990s, reflected in data collected by the International Labor Organization (ILO), have been:

an increase in the share of agricultural waged employment in total rural economic activity; an increase in the share of women in agricultural waged employment; growing casualization of agricultural waged labor.

6.4.1.1.1. Waged Employment

Today, there are more people working in agriculture for pay than ever before. A key component of employment and income in rural areas, the percentage of waged employment in agriculture, including the number of smallholders who depend on wages, is steadily rising in almost every region. Put another way, waged workers become essential to agricultural production as the agricultural sector undergoes a process of ownership concentration that results in fewer and larger farms with a higher number of waged workers, as opposed to a sector based on smaller production units and self-employed farmers.

In 1996, the highest average shares of waged employment were found in Eastern and Central Europe and Central Asia (over 80%), followed by Latin America and the Caribbean (over 50%), Asia (over 40%) and sub-Saharan Africa (30%), Organization for Economic Cooperation and Development

(OECD) countries (35%) and Near East and North Africa (25%). Additionally, the percentage of waged employment appears to be inversely correlated with the percentage of the agricultural labor force; that is, the higher the percentage of waged employment, the smaller the agricultural labor force. The number of smallholders in agriculture declines as production units shift from small family holdings to large units depending on wage labor as the economy absorbs more labor in other sectors and the agricultural labor force declines.

6.4.1.1.2. Women and Waged Employment

Notably, women now make up 20–30% of all agricultural wage employment, and their percentage of waged agricultural workers has been increasing across all regions. The percentage is 40% in Latin America and the Caribbean. This percentage is probably higher in African nations, although accurate data is hard to come by due to the widespread use of informal labor arrangements. The growing feminization of the workforce is supported by a study conducted by Indian trade unions on the effects of globalization on agricultural workers. Male employees have demonstrated a stronger propensity to transition to non-agricultural occupations. They have been supplanted by women. Because they can pay them less and find them to be a submissive and reliant workforce, landowners also appear to favor female employees, according to the unions. Consequently, the disparity in pay between men and women has persisted. Food security and women's empowerment depend on expanding work opportunities and raising women's employment and income levels, not just in agriculture. Given that women are more likely than men to invest in their children's potential as human capital, a study conducted in the Latin American and Caribbean region shows that raising the proportion of women's income in the household significantly improves family and social welfare. It has also been demonstrated that raising women's incomes and their proportion of family income empowers them by giving them more negotiating leverage within the home. Women's incomes are particularly significant because more and more households are being run by women.

6.4.1.2. Casualization

In many regions of the world, the nature of agricultural employment is changing due to the global trend toward casualization of work, which has been largely fueled by deregulation, globalization, and competitive pressure to lower labor and production costs. The already weak protections for agricultural workers in terms of pay, job security, health, safety, and environmental standards, as well as social protection, are being undermined by global trade and economic pressures. Workers are increasingly being hired as temporary or casual labor on short-term, daily, or seasonal contracts with lower pay and working conditions, a trend known as "casualization," which is a shift away from permanent employment contracts. Additionally, there is a growing trend among agricultural employers to outsource work, which is the process of moving the internal production processes outside the company. The issue of labor employed by or through contractors is the most significant. This primarily affects temporary employees, particularly foreign nationals. Employers are increasingly using labor contractors with expertise in hiring, transporting, and managing waged agricultural workers because they need to be able to rely on a labor supply during times of high demand. The entire employment relationship is weakened by this middleman-based process, which also leads to a lack of respect for labor laws and a gray area surrounding the employer's obligations. Employment conditions are typically unstable, and labor contractors commonly misuse their power over employees by requesting commissions, charging excessive rates for housing, food, and transportation, withholding wages, and enforcing debt slavery.

The majority of Indian agricultural trade unions believe that the current global economic restructuring is a continuation of the agricultural commercialization that began in the 1960s. They believe that the biggest effect of globalization on agricultural workers is their loss of employment and means of subsistence. This has occurred as a result of crop patterns shifting to less labor-intensive crops and the adoption of new technologies, including labor-displacing mechanization.

6.4.1.3. Underemployment/Unemployment/Hidden Unemployment

Rural unemployment and underemployment are frequent and important aspects of the lives of waged agricultural workers and are also major contributors to poverty. With little money to support them in between seasons, waged agricultural workers usually find employment for 175 days and are available for work (i.e., unemployed) for 85 days, assuming full employment is 260 days per year (52 five-day weeks, excluding leave and holidays). Local wages are kept low by these sizable pools of underemployed or unemployed workers.

The high prevalence of hidden unemployment, or the underestimation of unemployment levels in labor statistics where only those who are "actively looking for work" are counted as unemployed, is another characteristic of rural labor markets. Transaction costs are high and information is faulty due to the generally unreliable nature of rural labor markets. Unemployed people often stop looking for work, and those who are employed often work fewer hours than they would like.

6.4.2. Sustainable Agriculture and Labor Markets

The effects of implementing sustainable farming methods on labor markets and job creation. There is currently very little evidence of either positive or negative impacts, so more information and research are required. According to a study conducted by Essex University in the UK, sustainable farming methods can significantly affect labor markets globally. As a result of improved rainfall harvesting, some sustainable agricultural practices led to the opening of entire new seasons for agricultural production, especially in dry land contexts, which increased the demand for labor on farms. Other sustainable agricultural practices actually decreased the demand for labor. When waged labor opportunities increased through sustainable practices (such as watershed improvements), when more productive agriculture resulted in higher wages and employment when there were higher returns to agriculture, and when village conditions, including infrastructure and services, generally improved, reverse migration from urban to rural areas could also take place.

Therefore, the true question is whether sustainable agricultural methods can increase labor productivity and correct labor supply and demand imbalances. It is crucial to emphasize that an analysis of the varying effects on men and women is necessary to assess the labor impact of sustainable practices, given the obvious gender division of labor in some agricultural tasks. This is because, depending on who ends up performing the tasks that are decreased, increased, or replaced with the adoption of a new sustainable agriculture practice, a technology that determines labor savings or increased use may have different effects on the workloads of men and women.

6.4.3. Wages

Many waged agricultural workers live below the poverty line because their labor is frequently poorly compensated. Additionally, industrial workers' average salaries differ significantly from those of their agricultural counterparts. The income determinants for waged workers are:

- For full-time workers: wage level, number of wage earners in household; and
- For casual workers: number of days worked, wage level, number of wage earners in household; share of income derived from plantation/farm wage work.

In addition to working fewer hours, full-time wage earners and casual workers may have very different incomes because they do not receive benefits like housing, healthcare, and education.

6.4.4. Factors Affecting Wages

Five major elements affecting wage levels in agriculture can be identified:

- Agricultural growth;
- Food prices and food security;
- Labor supply;
- Non-farm employment; and
- Minimum wages.

6.4.4.1. Agricultural Growth

According to the traditional "trickle-down" theory, as farm incomes rise in tandem with productivity gains, agricultural productivity growth can result in a long-term decrease in hunger and poverty. In rural areas, there is a greater demand for basic non-farm goods and services due to rising farmer incomes and wages for agricultural workers. Tools, carpentry, blacksmithing, clothing, and processed foods purchased from roadside stands are a few examples. It is frequently challenging to trade these goods and services across large distances. They have a great potential to reduce poverty and create jobs because they are typically produced and delivered locally using labor-intensive methods. According to FAO surveys conducted in four African nations, local goods and services accounted for one-third to two-thirds of rural areas' income increases. However, there hasn't been any automatic trickle-down effect of higher productivity on wages in the agricultural sector, per a recent study by Indian trade unions on the effects of globalization on agricultural workers. Instead, favorable wage rate changes appear to be linked to agricultural workers' ability to bargain or the availability of other employment options. The fact that consumption goods that were previously free (such as fuel, water, fodder, wild vegetables, and river fish) are now becoming marketable commodities has actually caused real wages to decline, despite the apparent gains.

6.4.4.2. Food Prices and Food Security

Food expenses frequently account for more than 70% of worker households' cash wages. A sizable portion of waged workers and their families may fall below or even below the poverty line as a result of rising food prices. For agricultural workers and their trade unions, increasing earning power and livelihoods and guaranteeing food security are intertwined concerns. Even though their nation has enormous supplies of food grains, Indian agricultural trade unions view "food security for all" as a pipe dream, although goals have been set at multiple World Food Summits to increase food security worldwide. Agricultural workers' low incomes prevent them from having the money to buy enough food from the market. Food insecurity has also gotten worse for many worker households as a result of shifts in crop patterns from staple food grains to more expensive cash crops. According to the unions, reliance on imported food will probably grow and eventually result in a loss of food sovereignty.

6.4.4.3. Labor Supply

Increased labor absorption in agriculture or land-augmenting agricultural practices based on higher cropping intensity, which were prompted by irrigation investment and occurred in Bangladesh and parts of India in the 1980s, are closely linked to real wage increases. Strong export-driven agricultural expansion can also be a significant driver of real wage growth.

6.4.4.4. Non-farm Employment

Several governments have over the years encouraged rural non-farm employment through emergency assistance or labor-intensive public works projects aimed at poverty. Given the advantages that the poor receive from public works initiatives, both directly through jobs and income created and indirectly through the employment effects brought about by the infrastructure built and the increased demand for locally produced goods, the ILO has long supported diversifying rural employment through such labor-intensive work programs. Rural areas may also be home to manufacturers of basic consumer goods, especially those that use regional raw materials. In certain places, agro-ecological tourism has grown in importance.

Non-farm operations in rural areas are typically labor-intensive, small-scale, and rely on local resources to meet demand. For that segment of the rural labor force that is severely underemployed during the lean agricultural season, such activities can offer additional employment. As has been shown, particularly in China, northern India, Pakistan, and some Latin American countries, the growth of non-farm activities can increase rural employment overall and lead to lower levels of poverty, even in countries where labor absorption in agriculture is declining.

6.4.4.5. Minimum Wages

To ensure that all workers receive a livable wage, national, sectoral, or negotiated minimum wages seek to safeguard the most vulnerable and low-paid workers. To prevent excessive exploitation of the lowest-paid workers, the ILO has long supported a minimum wage system. A national wage-fixing mechanism involving the government may be necessary to establish minimum wage levels because collective bargaining is frequently weak in the agricultural sector. These wage-fixing mechanisms vary depending on national systems and practices, but they are typically defined by a legal framework that includes a third-party arbitration process between employers and trade unions that is frequently enforceable, such as Wages Boards. Given the amount of excess labor and pervasive unemployment in rural areas, it is generally believed that enforcing minimum wages is challenging, if not impossible. Enforcing a minimum wage that is not determined by the market appears to be impossible in agriculture due to the labor contracts' primarily informal nature.

Regretfully, there are no thorough surveys or data sources available to support any evaluation of minimum wages in the agricultural sector. Government attempts to de-regulate wages and abolish wage-fixing mechanisms have been met with resistance from unions. For instance, in the UK, the Transport and General Workers Union's agricultural workers section has pushed for a minimum wage and vehemently opposed government attempts at deregulation against the UK Agricultural Wages Board.

6.4.5. Poverty among Waged Workers and Sustainability

In developing nations, one in four people currently live in extreme poverty, which is defined as making less than $1 per day. Further, 776 million people in developing nations—roughly one in six—remain undernourished, according to data from the FAO in 2002. In addition, undernourishment exacerbates other facets of poverty by impairing children's mental development and academic performance, as well as by decreasing their ability to work and fend off illness. In developing nations, about 70% of the impoverished reside in rural areas and make their living either directly or indirectly from agriculture. Since poverty estimates typically only differentiate between rural and urban areas, precise recent data are lacking, but waged agricultural workers and their families are included in the core rural poor. Agricultural wage workers' high rates of poverty are concerning because, in almost every region, their proportion of the rural labor force is growing. Poverty is a characteristic of commercial agriculture as well as subsistence farming. A large number of waged agricultural workers on plantations live in poverty. For instance, although making up less than 10% of the rural workforce, Indonesian agricultural workers had the highest rate of poverty in 1987, with 38% of all agricultural waged workers living below the poverty line. In many nations, waged agricultural workers have one of the highest rates of poverty among occupational groups, frequently surpassing 60%. By region, this may differ. According to ILO data, waged workers in Africa and Asia generally had higher rates of poverty than rural residents, but in Latin America—except Chile and Mexico—the opposite appears to be true. Wage levels and employment are key factors in eradicating poverty among waged agricultural workers and their families.

For these reasons, waged agricultural workers along with small farmers should be a key target group of the United Nations Millennium Development Goal: Eradicate extreme poverty and hunger:

- Halve, between 1990 and 2015, the proportion of people whose income is less than one dollar a day; and
- Halve, between 1990 and 2015, the proportion of people who suffer from hunger.

The perception of waged agricultural workers as a primary target group is unclear, though. Outside of the ILO, they are hardly ever mentioned in UN documents. How can their needs as members of the core rural poor be evaluated and programs to end their hunger and poverty created if they are never mentioned?

6.5. CONDITIONS OF EMPLOYMENT AND WORK

Concern for improved working conditions should at least be on par with concern for standards of product quality, given the growing proportion of waged labor in agriculture and the growing percentage of total output marketed either domestically or abroad. Enhancements in working conditions can boost economic efficiency by increasing labor productivity and are a prerequisite for modernization and market opening. Instead of being viewed as a social expense, such advancements ought to be seen as investments.

6.5.1. Labor Law and International Labor Standards

Many agricultural workers only have limited protection under national labor law when compared to those employed in other economic sectors. In certain instances, this is since the applicable laws—such as the Factories Act—only apply to industrial businesses and the people who work for them. The agricultural industry is expressly exempt from general labor laws in certain nations. When protective laws are in place, they might not be completely relevant to the agriculture industry, be outdated, or just not be used.

In rural areas, where employers and employees may not be as familiar with the specifics of the law, the application of general labor laws can present challenges. Compliance may be viewed as impracticable, especially for small-scale farms and businesses. Inspection and enforcement are often lacking, even in cases where specific laws have been passed to address the unique demands of agricultural labor, such as those about working hours, pay scales, and housing in isolated locations.

6.5.1.1. Promoting Rights at Work

The ILO Conventions and Recommendations, which collectively constitute the International Labor Code, comprise international labor standards. ILO standards cover the majority of topics pertinent to labor law and the social facets of development, so they offer member states direction for enhancing their domestic labor laws and social policies. ILO standards are universal in nature and sufficiently flexible to accommodate differences in conditions and practices across nations with varying degrees of development. All workers are covered by many ILO conventions. ILO core labor standards on freedom of association, the right to collective bargaining, nondiscrimination, equal pay for men and women workers, the abolition of forced labor, and the elimination of child labor are among them, but they are by no means the only ones. Human rights at work is a common term used to describe these fundamental labor standards. They act as enabling rights in addition to being significant in and of themselves. In other words, they establish circumstances that permit the exercise of other rights. One of the best examples of that function is freedom of association. The foundation for social discourse between employers and employees to regulate terms and conditions of employment through collective agreements is created by the freedom of both parties to form and join independent organizations of their choosing. A fundamental human right, freedom of association opens the door to better social and working conditions, such as through collective bargaining.

Even though the right to freedom of association is almost universally acknowledged, there are still legal barriers to agricultural workers' ability to organize in many nations where national laws either forbid the right to organize in agriculture or exclude the industry from the applicable legal protections. Some national governments might believe that their labor administrations cannot effectively enforce this right in rural areas. Others might argue that because of the unusual, seasonal, or casual nature of agricultural employment, the sector is less accessible to the right to organize. International labor standards provide member states with important guidance on what minimum standards should be applied in the workplace because they are developed and adopted by government representatives as well as employers' and workers' organizations. As employers and employees incorporate their values into collective bargaining agreements, ILO Conventions and Recommendations not only encourage advancements in national social and labor laws but also encourage best practices at the sectoral and enterprise levels.

6.5.2. Collective Bargaining and the Right to Organize

Employees' ability to negotiate collectively rather than individually with their employers has a significant impact on their rights as well as the terms and conditions of employment. Workers gain more power and are able to negotiate better terms and conditions when they organize into trade unions and bargain collectively. The process of negotiating terms and conditions of employment and developing regulations to control labor relations is known as collective bargaining. Employers and their associations and worker representatives, typically from trade unions, are the two parties involved in the process. On occasion, the government participates as a third party. Although it may go by other names, the process typically ends with the signing of an agreement known as a collective bargaining (labor) agreement (CBA). It is not considered collective bargaining when an employer and employees engage in individual negotiations.

There are two structural levels at which collective bargaining can occur: the enterprise level (farm or plantation) or the industrial or sectoral level. Wages, employment contracts, labor contracts, maternity rights, health benefits, work hours, leave, occupational health, safety, and environment, housing conditions, grievance procedures, worker transportation, the abolition of child labor, and even HIV/AIDS prevention measures are among the issues that can be collectively bargained. Understanding and addressing the gender implications of collective bargaining is especially crucial given the vulnerability of women and their predominance in the agricultural sector, particularly in developing nations. Collective bargaining can address issues like equal pay for equal work, maternity leave and benefits, child care concerns, reproductive health services, and equality of opportunity policies that are particularly important to women in practice. It is also critical to evaluate the gender implications of issues that appear to be neutral for collective bargaining, such as pay, leave, overtime, and bonus structures, since these frequently have different effects on men and women in practice.

6.5.2.1. Collective Bargaining an Important Element in Improving Working Conditions in Agriculture

Without a doubt, one of the main tools for controlling labor relations is collective bargaining. Collective bargaining, however, does not seem to be a prominent aspect of the agricultural sector in many nations, primarily due to a lack of appropriate institutions, a lack of government support for such discussions, and the generally weak state of trade unions and other rural workers' organizations. It is not a component of the industrial relations system. However, the significance of collective bargaining in the agricultural sector is growing as more and more workers enter the rural labor market in quest of wage employment.

CBAs are typically completed in industries and businesses with substantial full-time employment, and they typically represent the concerns of permanent employees. Even though they make up the majority of the workforce, other employees—whether daily, weekly, temporary, or seasonal—may not be covered by such agreements or may only be partially covered. In plantation agriculture, collective bargaining is more significant than in the industry overall. The wages and terms of service established through collective bargaining have typically been more favorable than those obtained in the rest of the agricultural sector because plantations are far more organized than the rest of the agricultural industry.

6.5.3. Contracts of Employment

In agriculture, verbal agreements are frequently used, and written contracts of employment are uncommon. These kinds of agreements are usually rather broad and conform to common or market conditions. Employment contracts that are common in agriculture show a lot of variation within and between nations. Alongside task or piece rate employment, there are seasonal, daily, and permanent employment contracts. The least common type of contracts are permanent ones, and in the majority of nations, their proportion of all agricultural employment has been falling.

The skill requirements of the production process can be connected to the nature of employment contracts. In low-skilled harvesting and processing operations where labor costs concerning skills are a major concern, casual, and seasonal migrant labor is frequently employed.

The difficulty lies in figuring out how to strike a balance between the fundamental standards and protections that ought to be included in every employment contract and the necessity for hiring workers with the flexibility to satisfy production and marketing requirements. This issue has not yet been resolved for the majority of temporary and casual employment types.

6.5.4. Contract Labor and Labor Contractors

Labor contracting is a widespread phenomenon in commercial agriculture, and the roles of labor contractors can differ significantly. Some are established companies that specialize in hiring, transporting, and managing wage laborers in agriculture, while others are less structured. One consistent feature of labor contracting arrangements is that the workers are not directly employed by the entity for which they are providing services. Generally, these contracted workers remain employees of the labor contractor. In some instances, the workers are merely supplied to the workplace by the contractor and do not have an official employment relationship with either the user enterprise or the contractor. This situation presents various operational and legal challenges. When an employee works for a user enterprise while being employed by a labor contractor, the user and the contractor share the role and responsibilities of an employer. In addition to providing transportation to and from work and, in certain situations, workplace supervision, the contractor may also pay the employees. In addition to establishing the general terms and conditions of the workplace, including with regard to occupational health and safety standards, the user enterprise frequently decides what needs to be done and how long it will take. This frequently indicates that the employee is unsure of their employer's identity and, consequently, how to assert their rights under the law. Because of these circumstances, the employment relationship is not legally clear.

The labor contracting system is occasionally abused. To hide the user's identity as the true employer, the labor provider may not be a legitimate business but rather a middleman of the purported user enterprise. This is a dishonest practice since it is an intentional attempt to evade the obligations of the employment relationship. Workers are especially at risk in these circumstances; numerous instances of mistreatment by so-called labor contractors are well documented, and include demands for commissions, exorbitant fees for housing, food, and transportation, withholding wages, and enforcing debt slavery.

Additionally, there is a growing trend of labor contracting across borders in many ways, some of which may be on the verge of being illegal or in the gray area between legal and illegal movement of people. An additional regulatory challenge for health and safety, particularly for migrant farm workers, is the growing use of farm labor contractors, according to a US study on agricultural injuries. Farm workers may be less inclined to report work-related conditions due to their heavy reliance on the contractor for housing, transportation, employment, and other amenities. Trade unions are increasingly calling for efficient regulation of labor contractors as well as the terms and conditions surrounding the hiring and provision of contract labor. As evidenced by the outcomes of multiple ILO Tripartite Sectoral Meetings and an ongoing conversation at the International Labor Conference, the ILO has long been concerned about the vulnerability of workers in labor contracting arrangements.

6.5.4.1. Hours of Work

When compared to other industries, waged agricultural workers typically put in more than 45 hours a week, and their work hours are still mostly unregulated. Many factors, including seasonal and climatic conditions, as well as peak times for sowing and harvesting, can cause hours to vary. Most often, it is still necessary to design, negotiate, and implement adequate working time arrangements that can both limit the total amount of time spent working on a daily, weekly, and annual basis and satisfy the unique requirements of agricultural work cycles. Article 20 of the ILO Safety and Health in Agriculture Convention, 2001, which addresses work hours, night work, and rest periods for agricultural workers, is a significant advancement in this area. It states that these hours must comply with national laws and regulations or collective agreements. For the first time, an ILO Convention has linked working hours, rest periods, night work, and workplace health and safety, reflecting current research and thinking on topics that were all too frequently overlooked in the past.

6.5.5. Housing and Living Conditions

The lives and occupations of many agricultural workers are inextricably linked because they live where they work. Housing, employee well-being, and productivity are closely related. Inadequate and crowded housing, lack of heating, poor ventilation, a lack of sanitary facilities, and nonpotable drinking water are characteristics of agricultural workers' housing that contribute to the spread of infectious diseases like influenza, TB, and upper respiratory tract infections. Many farms and plantations still have inhumane living conditions, with workers spending extended periods of time in tents, temporary plastic huts or shacks, or hostels. HIV/AIDS is also spread by substandard housing, such as families sharing cramped quarters or male employees living in hostels far from their families.

Some of the issues are demonstrated by housing examples on Kenyan plantations. Companies are required by Kenyan law to provide housing for their permanent employees. During an ILO visit, it was discovered that several plantations' housing facilities needed to be upgraded and repaired to raise worker living standards, and that some of the older homes needed to be completely replaced. While grass-thatched mud houses were still typical in some estates, workers were housed in rows of brick-built quarters with communal water and toilet facilities in others. Workers in certain estates complained about inadequate or nonexistent sanitation, which was exacerbated by the crowded living conditions. Many families shared a single pit latrine in the lower-grade housing estates, and most of the time there were no drainage systems in place. In certain cases, the lack of piped water forced workers to obtain their water from neighboring rivers and streams. Workers who lived in such circumstances frequently reported cases of waterborne illnesses.

Plantation housing is rarely offered to casual workers. Numerous migrants reside in shantytowns close to the estates, lacking even the most basic hygienic amenities. Frequent outbreaks of waterborne illnesses like cholera, typhoid, malaria, and dysentery are evidence of the poor hygiene conditions in these settlements. Historically, trade unions and employers on farms and plantations have engaged in negotiations over the provision of decent housing on these properties. In certain nations, employers are required by law to give their workers housing or a housing allowance. However, workers are frequently prevented from switching employers because they would lose their housing if the accommodation is included in the compensation package. When they retired, they would also lose their housing. Several countries have introduced laws that give farm workers more secure housing rights, including retirement provisions. However, many unions contend that wages ought to be high enough to enable workers to live wherever they choose and make purchases on the housing market, just like other citizens.

6.5.6. Transport of Workers

Agricultural laborers in many nations are regularly or occasionally transported over great distances from their homes to their workplaces. Too frequently, a lot of workers are crammed into open trucks and other vehicles that aren't meant to carry people. Safety regulations and weight restrictions are disregarded. There are numerous instances of severe traffic accidents involving this type of transportation. Whether such accidents are covered by employment injury insurance is frequently unclear.

6.5.7. Occupational Health, Safety, and Environment

One of the most important aspects of sustainable agriculture is raising the occupational health, safety, and environmental (OHSE) standards for small farmers and agricultural workers, including through regulation and enforcement. Additionally, it must be acknowledged that raising worker health, safety, and environmental standards can safeguard and raise consumer food safety, environmental protection, and public and environmental health standards—particularly when it comes to exposure to pesticides and other agrochemicals.

Together with mining and construction, agriculture is one of the top three most hazardous jobs. Among the many risks waged agricultural workers encounter on the job are hazardous machinery, hazardous electrical wiring and appliances, diseases spread by livestock, falls from heights, and exposure to harmful pesticides. According to ILO estimates, approximately 355,000 people die while working every year. According to estimates, half of them take place in agriculture, which employs half of the world's workforce. Of the estimated 270 million occupational accidents and 160 million occupational diseases that occur annually, waged farmers and agricultural workers also bear a disproportionate amount of the burden. On average, 6,000 people pass away every day from illnesses or accidents related to their jobs.

One of its most distinctive features is that, in contrast to factory or office workers, agricultural labor is performed in rural areas where living and working conditions are not clearly separated. Agricultural laborers and their families are consequently exposed to additional risks, including pesticide exposure. Additionally, poverty of small farmers and waged agricultural workers can raise their risk of work-related illnesses like malnutrition and poor diet. Women employees make up a large portion of those killed, wounded, or ill. They are particularly vulnerable because they frequently work part-time or on a casual basis, receive less training and education, perform repetitive tasks that can lead to musculoskeletal issues, and are exposed to pesticides, which can pose reproductive risks. Child laborers are also at risk; 22,000 children, many in agriculture, are killed on the job each year. Eliminating all dangerous work done by child laborers in agriculture is a top priority right now. Environmental, safety, and health practices can also be significantly impacted by the nature of the employment relationship. Subcontracting agreements on plantations, according to Kenyan trade unionists, have made it challenging to address some health and safety issues. Women weeders, for instance, were frequently exposed to pesticides at work, but the union was powerless to stop it because the subcontractors who hired the women were not subject to the collective bargaining agreement with the plantation company.

The economic losses resulting from subpar environmental, occupational health, and safety standards can be valued, but human suffering cannot. High rates of fatalities, illnesses, and accidents that cause missed workdays and higher medical costs have a significant detrimental effect on agricultural productivity and place a significant financial and social burden on businesses. According to ILO estimates, deaths, accidents, and illnesses related to the workplace cost the economy 4% of GDP. A successful and sustainable agriculture sector depends on having a workforce that is safe, healthy, and environmentally conscious.

For trade unions, upholding and enhancing workplace health, safety, and environmental standards depends on workplace organization and representation. The foundation of trade union organizations about workplace health and safety is made up of legally appointed and empowered worker health and safety representatives, or safety reps. They serve as trade unions' eyes and ears on issues related to occupational health and safety in the workplace and are essential in lowering workplace fatalities, accidents, and illnesses. Many also deal with environmental issues at work. These safety representatives contribute to the protection of the environment, the public, and employees.

However, because of the small size of these businesses, their dispersed location, and a general lack of funding and technical assistance, many farms lack safety representatives. Due to a lack of training and technical support, safety representatives may only have a limited ability to improve occupational health, safety, and environmental standards, even on larger farms and plantations where workers elect or choose them or where they are expressly permitted by law. Maintaining OHS and even environmental standards is also greatly aided by worker representatives on legally established, joint worker-management health and safety workplace committees (safety committees). Only larger workplaces—typically those with 50 or more employees—have such committees. Safety committees address persistent issues and consider longer-term fixes through collaborative problem-solving.

6.5.8. Labor Inspection

Effective national labor inspection systems are necessary to administer social and labor policy, supervise and enforce labor laws and standards, and offer advice to employers and employees. Labor inspection is a governmental duty and a public function that is best carried out as a system inside a larger state system. The ILO Labor Inspection (Agriculture) Convention, 1969, adds to the basic international standards established by the ILO Labor Inspection Convention, 1947, by taking into consideration the unique features of the agricultural sector. Laws on working conditions and protecting employees while they are at work, including those governing hours, pay, weekly breaks and holidays, safety, health, and welfare, the employment of women, children, and youth, and other related issues, are enforced by labor inspectors. Additionally, inspectors provide employers and employees with technical advice and information about the best ways to adhere to the law. Different nations have different labor inspection systems; some, for instance, have dedicated health and safety inspectorates. In many nations, labor inspectors also play a significant role in monitoring the protection of lawful workers' representatives (such as safety representatives), the observance of trade union rights, and the efficient operation of statutory bodies intended to promote social dialogue.

In many nations, labor inspectors are also tasked with educating worker representatives on all matters about labor protection. Inspectors will frequently take part in training events hosted by third parties if the inspectorate does not plan this training itself. However, labor inspection seems to be a low priority in many countries. It is impossible to overstate how difficult it is to arrange for a minimal number of labor inspectors to be present in vast areas with limited funding. There are not enough labor inspectors in many nations to carry out their responsibilities effectively across all economic sectors. In developing nations, labor inspectors in agriculture face unique challenges. First, there aren't enough inspectors in these nations to guarantee even a token presence at more than a small percentage of workplaces, even in cases where laws pertaining to agriculture are in place. The problem increases with the size of the plantation or the distance of the agricultural activity. Second, there is frequently insufficient training provided to agricultural inspectors in these nations. Third, a lot of inspectors lament the lack of resources, especially in the area of transportation, which prevents them from visiting farther-flung farms and plantations.

6.5.9. The HIV and AIDS Epidemic

A virus known as HIV (human immunodeficiency virus) destroys the immune system, resulting in AIDS (acquired immunodeficiency syndrome). In addition to affecting employees, HIV and AIDS are a workplace concern because workplace health, safety, and environmental training on the disease can be extremely important in preventing the epidemic's spread and consequences. About 37.8 million people had HIV in 2004, and 20.9 million people died from AIDS in 2003. 95% of people who have HIV and die from AIDS live in developing nations, according to the FAO. Women make up a disproportionate share of the rural poor, who make up the vast majority. According to the UNAIDS Report from 2004, the percentage of women living with HIV has been rising steadily. According to recent data, there are 13 infected women for every 10 infected men in Africa. Decades of social and economic advancement are being reversed by the epidemic, which is also leading to rural disintegration. For instance, HIV and AIDS are destroying the agricultural labor force in sub-Saharan Africa, reducing the number of farmers and food producers in the region for future generations. Around 7 million agricultural workers have died from AIDS since 1985 in the 25 most-affected African nations, and within 20 years, the most affected nations may lose up to 26% of their agricultural labor force. The majority of the 40 million individuals infected with HIV are in the prime of their working lives because the disease typically strikes the most productive age groups. Workers between the ages of 15 and 49 make up at least 25 million of the HIV-positive population. Workers and their families, businesses, and the national and local economies are all significantly impacted. Particularly hard hit have been the effects on food security, farming systems, livelihoods, and rural populations. The disease affects the world of work in many ways:

6.5.9.1. Households

- Growing burden on women workers as they have to earn a livelihood while at the same time caring for sick family members and even neighbors;
- In agricultural communities, an increasing number of households are headed by children, as the parents have died from the virus and extended family networks cannot always cope with the sheer number of orphans;
- Increasing pressure for AIDS orphans to be allowed to work in agriculture to cover the costs of their remaining on the farm or plantation and to pay school fees. However, there is a very real danger that these children will be exploited and their health put further at risk by exposure to occupational health and safety hazards; and

- Rural communities bear a higher burden of the cost of HIV and AIDS as many urban dwellers and migrant laborers return to their villages when they become sick.

6.5.9.2. Farms and Plantations

- Loss of valuable work skills and experience; and
- Falling productivity in industry and agriculture, and rising labor costs due to labor shortages.

6.5.9.3. Workforce

- Cutting the supply of labor and reducing income for many workers;
- Discrimination against people with HIV, often resulting in dismissal; and
- Other ways in which HIV and AIDS affect rural workers include changes in their ability to access and control productive assets and resources.

There is proof that the social, political, economic, and cultural environments all contribute to an individual's susceptibility to HIV and AIDS. Gender is one of the most significant variables influencing vulnerability. Poor housing and living conditions are major factors in the spread of HIV and AIDS, making them a workplace concern in agricultural estates as well, according to research. In some cases, spouses are not permitted to live together on the property, and a family or group of families may share a room. Due to seasonal labor migration, a large number of men without families reside in labor camps. The practice of extramarital sex, primarily by male spouses, was said to be facilitated by this. All of these actions, along with unfavorable working conditions and low pay, have sped up the disease's spread among employees. Individuals are unlikely to know how to prevent infection because they have limited access to health services and information. Many workers are now engaging in behaviors like commercial sex and sexual favors, which raise the risk of infection and disease transmission, as a result of poor working conditions and low pay.

6.6. AGRICULTURE, TECHNOLOGY, LIVELIHOODS, AND EMPLOYMENT

There is a persistent perception that India's agriculture is in crisis. Suicides by farmers, frequent price disputes, the high rates of poverty among peasants (particularly smallholders) and landless labor, relatively low productivity, and the erratic employment potential of agriculture all appear to reinforce the prevalent ideas of an agrarian crisis. However, since independence, total agricultural output has grown and stabilized, resulting in greater exports of food and non-food crops, decreased reliance on food imports, and national (but not household) food security. The changes in total factor productivity in agriculture can be partially explained by broad adoption and transformation of technology, substantial crop diversification, improved agricultural intensification, easier access to inputs, and increases in labor productivity. Such improvements, however, are mostly limited to areas with improved soil productivity, consistent rainfall, and easier access to irrigation. Feudal agrarian structures, subsistence peasant farming, high levels of climatic uncertainty, price, and market volatility, labor out-migration, labor shortage issues, and low factor productivity are the main characteristics of the country's semi-arid and arid regions.

From the extreme food shortages, productivity crisis, agricultural revolution, and extreme rural poverty of the 1950s and early 1960s, India's economy and society have advanced significantly. Economists, sociologists, and social anthropologists have all made substantial contributions to the subfield of agrarian studies, which has a long history in Indian social sciences. The agrarian question has remained hot in both the political and academic spheres due to the slow and gradual urban transition, the unfavorable terms of trade between rural and urban areas, the low level of public investment in agriculture and related infrastructure, and the obstinate and persistent rural poverty in light of agriculture's declining GDP share. The lack of an integrated approach to issues of livelihoods, employment, sustainability, and productivity has hindered the efforts of numerous and varied policy mechanisms to address the agrarian problem over the decades. These include:

- Changes in cropping pattern, i.e., more diversified cropping system;
- Agricultural intensification, i.e., bringing more land under cultivation and cultivation in more than one season;
- R&D, extension, and diffusion of new technologies for addressing productivity problems as well as problems of coping with drought, water scarcity, poor soil quality, nutrition, market demand, etc.;
- Addressing issues of factor endowment – irrigation, labor, credit, external inputs (fertilizers, seeds, pesticides) etc.;
- Increase in yields and productivity;
- Improvement in household asset base;
- Land and tenancy reforms;
- Increase in minimum wages for agricultural workers;
- Increased availability of employment throughout the year; and
- Reduction in indebtedness and easier availability of credit.

The agricultural sector continues to face many obstacles despite these efforts, including pervasive rural poverty, the depletion of natural resources, and achieving competitiveness in the increasingly globalized economy. The incomplete transformation of subsistence-oriented agriculture in marginal environments and the lack of technological advancement are major factors in this. In addition to endangering agriculture's sustainability and potential future growth sources, these issues could accelerate the process of marginalization in agro-economic zones that did not gain much from the green revolution unless new approaches are developed and put into place.

From the extreme food shortages, productivity crisis, agricultural revolution, and extreme rural poverty of the 1950s and 1960s, India's economy and society have advanced significantly. Nonetheless, the agricultural industry continues to face many difficulties, including pervasive rural poverty, the depletion of natural resources, and achieving competitiveness in the increasingly globalized market. To put the relationship between technology, livelihoods, and employment in perspective, the aforementioned challenges are briefly described.

Food insecurity and child malnutrition are still at unacceptable levels in South Asia despite the region's abundant grain reserves. The benefits of increased agricultural productivity were insufficient to reduce food insecurity and malnutrition because of the high rates of poverty and unequal access to productive assets. The incidence and severity of rural poverty are frequently higher in semi-arid and arid regions due to a lack of R&D investments and the challenging biophysical conditions that characterize dry land agriculture. Land productivity is low and market access is restricted in marginal areas; non-farm employment opportunities are also limited, though they have been on the rise for the past 10 or so years. Small farmers in arid and semi-arid regions have been facing low crop yields and high production costs, and the rate of agricultural productivity growth has been significantly slower than in irrigated regions. Some workers move to cities and high-production areas in quest of work as land becomes scarcer. However, the adoption of less labor-intensive technologies and increased production mechanization in green revolution areas restricts the influx of migrants from marginalized areas. In arid and semi-arid areas, poverty and marginalization are also linked to rising water scarcity, drought frequency, and the deterioration of the base of natural resources.

In the semi-arid tropics, biotic (the prevalence of pests and diseases) and abiotic (drought) constraints shaped agriculture and livelihoods. The two most limiting abiotic factors are low soil fertility and water scarcity. Agriculture is inherently risky in semi-arid regions due to the limited supply of fresh water, seasonal variations, and unpredictable rainfall. The ongoing threat of drought in rain-fed dryland agricultural systems makes livelihoods more vulnerable and exacerbates human insecurity. Low and erratic rainfall makes drought management a crucial tactic for agricultural development in these areas since water is essential for crop growth (Ryan and Spencer, 2002). According to future projections, population growth, aquifer depletion, and competition for non-agricultural water use linked to increased urbanization and industrial development are the main causes of the anticipated further decline in water availability in semi-arid regions (Seckler et al., 1998).

In addition to the growing water scarcity constraint, livelihoods and the sustainability of food production throughout India are at risk due to the degradation of soil resources brought on by salinization, waterlogging, soil erosion, and nutrient depletion. The remarkable increases in cereal production productivity that were attained in the regions that saw the green revolution are now beginning to wane or plateau. New empirical data indicates that it is challenging to maintain productivity over an extended period of time in intensive rice-wheat monocultures. Under the green revolution, lowland intensification has been linked to increased pest accumulation, depletion of groundwater supplies, soil nutrient imbalance, and the accumulation of salt in drier regions and water-logging in wetter regions (Pingali and Rosegrant, 2001). Widespread poverty, the SAT's water shortage and soil degradation, intensification-induced resource degradation issues, and the resulting productivity decline all call for a development strategy that is distinct from the green revolution's intensive monoculture systems, considers environmental externalities and aligns with the goals of more sustainable and equitable agricultural productivity growth. The demands of globalization adaptation make this task more difficult.

The efficiency and competitiveness of agricultural product production are becoming significant policy issues in the agricultural sector as a result of growing globalization efforts through domestic market reforms and liberalization of import and export markets (Gulati and Kelley, 1999). Historically, food security and self-sufficiency in major food products were the focus of macroeconomic policies and R&D investments in many developing nations. Since some food products may be less expensive to import than to produce domestically, national self-sufficiency may not be a feasible development strategy as the global economy becomes more open. However, given that millions of South Asians depend on agriculture for their livelihoods, increasing the sector's competitiveness by lowering average production costs is essential to many small-holders farmers' survival.

Production competitiveness depends on investments in small-scale irrigation to increase yields and lower production risk, extension services, and timely delivery of necessary inputs and credit facilities. Without them, there is a genuine chance that poverty and marginalization will increase as a result of globalization (World Bank, 2002b). Similarly, as agricultural markets become more liberalized and competitive, arid, and semi-arid regions that have historically received inadequate investment in productivity-enhancing technologies, basic infrastructure, and human resources may lose out even more. Therefore, these regions may become even more marginalized as a result of globalization and increased market liberalization, which could exacerbate poverty and environmental degradation. This process of marginalization in areas with limited resources is supported by historical empirical data on South Asian infrastructure investments and agricultural technology development. Fan and Hazell (1999) demonstrate that rain-fed districts consistently have lower rates of improved variety adoption, road density, market access (number of rural markets per 1000 km^{-2}), and fertilizer intensity than more favored irrigated districts. The relative competitiveness of smallholder crop-livestock production operations in these regions is impacted by the high transaction costs and low productivity of rain-fed dry land agriculture. It will also affect how farm households choose crops and technologies and how well they can manage risk from the market and the unfavorable biophysical environment.

6.6.1. Agricultural Technology and Impact

The agricultural benefits of the Green Revolution were significant, but they were limited to a few areas with favorable factors of production, had long-term negative ecological effects, disrupted society, and only benefited a small (upper) segment of the peasantry. Research revealed an uneven allocation of advantages from the spread of green revolution innovation, which stemmed from the characteristics of the factor endowments needed to profit from technology adoption as well as issues with scale neutrality. Labor-related ramifications included a decline in employment for some groups, particularly women, and an increase in employment prospects for men in specific regions of the nation. Questions like "Why is there a differential flow of benefits across farmer categories and to male and female-headed households from innovation?" and "Does inequality necessarily increase with the adoption of an innovation or does it do so under specific social conditions?" have been addressed and solutions sought through empirical studies conducted throughout the Indian subcontinent, particularly in the wake of the classic green revolution studies of the 1970s. Concerns about the intra-family distribution of technological benefits, specifically the security of food and nutrition for women and children, have also been brought up. It has been noted that gender-sensitive technologies help ensure that benefits are distributed fairly within the home (Kolli and Bantilan, 1997). Factors that promote innovation adoption at the household level might differ from those that promote impact. These, in turn, may be distinct from intervening conditions at the institutional and community levels that hasten adoption and effect. Additionally, R&D and technology diffusion strategies typically lack an understanding of the types of households based on asset ownership, access to resources and institutions, and other factors. Examining household typologies would determine whether the effects of poverty and labor differ depending on the kind of household or social category. It would also determine the implications of this for granting access to technology and establishing favorable conditions for reaping the rewards of adopting technology.

There is mounting evidence that the rural poor who depend on agriculture for their livelihoods have been left out of state policies, research, and development initiatives, market, and infrastructure development, and the delivery of other essential social and financial services. Policies and programs have had little impact because of subsistence-oriented livelihoods, low levels of resource availability and access, and a lack of welfare measures. Many technological interventions have overlooked important areas, such as market linkages, livelihood diversification, gender equity, and access to and management of productive water, land, and forest resources. As recent criticisms of globalization and the agrarian crisis demonstrate, the rural poor are even more vulnerable to marginalization and ongoing exclusion if market forces are allowed to rule their lives. This is mostly because of the limitations of a hierarchical agrarian structure, their increased susceptibility to drought conditions, ongoing displacement, and risks from other external shocks. The problem of the commons has not been given much thought in state strategies about agricultural technologies; issues of resource access and governance have received scant, hostile, or poorly thought-out attention. Similarly, R&D strategies as well as overall agricultural policies have in general failed to take into account the socio-economic characteristics of the rural poor, and the agro-ecological conditions in which they eke out their livelihoods; these become important if agricultural innovations are to be appropriate and relevant, enhancing the possibility of wider adoption and diffusion, and better impacts. The aforementioned analysis emphasizes the necessity of addressing long-standing issues and looking for new opportunities to address India's agrarian crisis. Farmers and agricultural workers are now forced to migrate to distant locations and urban centers in search of employment opportunities due to the sharp decline in income earning opportunities caused by the lack of technological advancement in agriculture. The importance of farming as a source of employment and income is declining due to soil degradation, declining water tables, and frequent droughts. The prospects for rain-fed farmers in India's arid and semi-arid regions seem bleak in the absence of calculated action. Up until now, neither crop production nor resource management technologies have been able to significantly affect rain-fed areas, at least not to the same degree as irrigated areas. Research must analyze and comprehend limiting factors (technological, policy, market, institutional, structural, etc.) and determine future development strategies in light of both established and new issues. Future sources of growth and development opportunities will probably differ among dry land agricultural and ecological system typologies because dry land agriculture is not a homogeneous system. To diagnose the barriers and opportunities for increasing productivity, creating jobs, and reducing poverty, a comprehensive and system-wide approach is necessary. Finding more sustainable options requires keeping an eye on changes in cropping patterns, the diversification of income-earning opportunities, poverty levels, livelihood strategies, investment opportunities (including incentives for productivity-enhancing and resource-saving investments), and the factors that drive these changes at various levels (household, community, district, etc.).

6.6.2. Technology, Agrarian Structure and Agricultural Transformation

Numerous studies have shown that post-independence India's technology-and productivity-focused agricultural strategies have had varying degrees of success in certain regions. They have had varying effects on livelihoods and employment. Overall, because of issues with agrarian power and social structure, improper R&D and extension, and disparities in factor endowments, such strategies have not been sustainable in terms of ecology and yield and have not shown the ability to be applied in other areas. Environmental critics have highlighted the serious repercussions of green revolution practices for ecosystem issues, genetic loss, soil degradation, water depletion and conflicts, and health effects. Women who worked in agriculture saw limited and unfavorable labor and employment gains. Scale neutrality issues widened the divide between farmers and peasants with varying land holdings. Numerous studies have provided valuable insights into the relationship between agrarian power and technology adoption, with the green revolution being the single largest source of the rapid technological transformation in agriculture. Although some groups of enterprising peasants widely embraced these technologies, early hopes for a significant technological revolution in Indian agriculture were dashed. This holds true whether one considers traditional technologies or those that are more suitable, sustainable, and advantageous for small-scale farmers in dryland agriculture. The social and political barriers to technology-driven labor, productivity, and agrarian transformation in rural India were highlighted in the seminal work on agrarian power and productivity by Desai, Rudolph, and Rudra. Similar to this, the "mode of production in agriculture" debate (Patnaik, 1990) brought up more significant concerns about forced commercialization, unfavorable terms of trade, and the circumstances surrounding the actual agrarian transformation. More recently, evidence of wealthy peasants choosing to take a crop holiday rather than increase agricultural wages has cast doubt on expectations of an MGNREGA-led change in labor market dynamics (Vakulabharanam and Prasad, 2011). It is evident that agrarian transformations are resisted by the current agrarian structure, which is entangled in exploitation and highly hierarchical caste and class disparities. Even unsustainable technologies have the potential to cause disruption, and the dominant rural class would only support change if it aligned with its own agenda. Lower castes are pushed to migrate by policy changes, which puts pressure on the labor market and results in a labor shortage in rural areas. Similarly, it is important to observe and provide an explanation for the significant variations in women's employment (Thomas, 2012). Although political shifts are forcing dependent lower castes to look for work outside of their areas, further straining the labor market and causing a labor shortage, they are powerless to compel more favorable working conditions for rural labor. Overall, these processes lead to a state of flux that necessitates more study and makes it more difficult to develop strategies and policies for creating jobs in agriculture. Furthermore, the feminization of agriculture and gender dynamics have an impact on the adoption of technology as well as the methods chosen to increase income and productivity. Some academics and activists who focus on gender issues in agriculture contend that biotechnology holds promise and potential for rural women in these kinds of circumstances. According to Omvedt and Kelkar (1995), Mitter (1995), and others, biotechnology can help women challenge male dominance in technology, support their entry into high-tech fields, and support sustainable agriculture with low external inputs. It is believed that new choices will strengthen the body of existing knowledge and improve technical expertise, thereby empowering women. Although social institutions are involved, technology design is also important. Additionally, it is crucial to concentrate on crops that are important to women in running their households rather than cultivating crops that merely increase market revenue, which male household members may steal.

In addition to portraying women as productive partners, food producers, traders, and family caregivers, studies on the role of women in agriculture have improved our understanding of the increasingly complex issues surrounding food production, farm structure, and rural development. Together with specific data on women's labor availability and potential, this type of information will aid in the better design and development of technologies that take into account the contribution of women to raising food production and raising the standard of living for the typical peasant household. Studies and summaries on the topic of agricultural extension have shown that public sector extension has tended to overlook issues of caste, gender, and other types of inequality, and has had a limited focus. The relationship between agricultural R&D and extension has historically been weak, despite major innovations and the testing of various strategies. This is especially true in dry land, and subsistence agriculture, where smallholders and marginalized peasants are most impacted (Parthasarathy and Chopde, 2000).

REFERENCES

1. Bantilan, M. C. S., & Parthasarathy, D. (1999). Efficiency and sustainability gains from adoption of short duration pigeon pea in non-legume based cropping systems. *Impact Series No. 6*, ICRISAT, Patancheru, Andhra Pradesh, India.
2. Parthasarathy, D., & Chopde, V. K. (2000). Building social capital: Collective action, adoption of agricultural innovations, and poverty reduction in the Indian semi-arid tropics. Paper presented in the panel on "Escaping Poverty" at the 2nd Global Development Network Conference, Tokyo.
3. Egger, P. (1995). Rural workers' organizations and employment and labor conditions in agriculture. *Labor Education, Special Issue: Rural Workers, 98-1995/1*, 6.
4. Fan, S., & Hazell, P. (1999). Are returns to public investment lower in less-favored rural areas? An empirical analysis of India. *EPTD Discussion Paper No. 43*, IFPRI, Washington, D.C.
5. International Union of Food, Agricultural, Hotel, Restaurant, Catering, Tobacco and Allied Workers' Associations (IUF). (2000). Good food, safe work: Trade unions and sustainable agriculture. Geneva, Switzerland.
6. Gulati, A., & Kelley, T. (1999). *Trade Liberalization and Indian Agriculture: Cropping Pattern Changes and Efficiency Gains in Semi-Arid Tropics*. Oxford University Press.
7. Hurst, P. (2002). Safer farms for safer food. In *Health and Safety at Work: A Trade Union Priority* (Labor Education 126, pp. 19–22). ILO, Geneva.
8. Kolli, R. D., & Bantilan, C. (1997). Gender-related impacts of improved agricultural technologies: Identification of indicators from a case study. *Gender, Technology and Development, 1*(3), 371–393.
9. Labor Bureau. (2010). *Report on Employment and Unemployment Survey (2009–10)*. GoI, Chandigarh.
10. International Labor Organization (ILO). (1973). *Minimum Age Convention, No. 138*; 143 ratifications as of 31 December 2005 and *Worst Forms of Child Labor Convention, No. 182*; 158 ratifications as of 31 December 2005.
11. Mitter, S. (1995). Does new technology bode well for working women? An evaluation and analysis (No. 12).
12. Mukesh, E., et al. (2009). Sectoral labor flows and agricultural wages in India, 1983–2004: Has growth trickled down? *Economic and Political Weekly, 44*(1), 46–55.

13. Omvedt, G., & Kelkar, G. (1995). *Gender and Technology: Emerging Visions from Asia.* Gender and Development Studies Center, School of Environment, Resources and Development, Asian Institute of Technology.
14. Papola, P. S. (2010). Livelihood in agriculture–Status, policies and prospects. In *State of India's Livelihoods Report 2010* (S. Datta & V. Sharma, Eds.). Sage-Access.
15. Parthasarathy, D. (2013). The poverty of (Marxist) theory: Peasant classes, provincial capital, and the critique of globalization in India. Paper presented at the Conference on "The Global E.P. Thompson: Reflections on the Making of the English Working Class after Fifty Years", Harvard University.
16. Patnaik, U. (1990). *Agrarian Relations and Accumulation.* Oxford University Press.
17. Pingali, P. L., & Rosegrant, M. (2001). Intensive food systems in Asia: Can the degradation problems be reversed? In D. R. Lee & C. B. Barrett (Eds.), *Tradeoffs or Synergies? Agricultural Intensification, Economic Development and the Environment* (pp. 203–225). CABI Publishing.
18. International Union of Food, Agricultural, Hotel, Restaurant, Catering, Tobacco and Allied Workers' Associations (IUF) & International Labor Organization (ILO). (2002). Promoting the role of agricultural workers and trade unions in sustainable agriculture and rural development. Leaflet for the World Summit on Sustainable Development, Johannesburg, South Africa.
19. International Labor Organization (ILO). (1949). *Right to Organize and Collective Bargaining Convention, No. 98*; 154 ratifications as of 31 December 2005.
20. Ryan, J. G., & Spencer, D. C. (2001). Future challenges and opportunities for agricultural R&S in the semi-arid tropics. *International Crops Research Institute (ICRISAT)*, Patancheru, India.
21. Seckler, D., Amerasinghe, U., Molden, D., de Silva, R., & Barker, R. (1998). World water demand and supply, 1990 to 2025: Scenarios and issues. *Research Report 19*, IWMI, Sri Lanka.
22. Thomas, J. J. (2012). India's labor market during the 2000s. *Economic & Political Weekly, 47*(51), 39–51.
23. Vakulabharanam, V., et al. (2011). Understanding the Andhra crop holiday movement. *Economic & Political Weekly, 46*(50), 13–17.
24. International Labor Organization (ILO). (1996). *Wage Workers in Agriculture–Conditions of Employment and Work.* Sectoral Activities Programmer, Report TMAWW/1996, Geneva.

25. World Bank. (2002). *Globalization, Growth, and Poverty: Building an Inclusive World Economy*. A World Bank Policy Research Report. The World Bank, Washington, DC.

CHAPTER 7

PROMOTING AGRICULTURE FOR SOCIAL PROTECTION OR SOCIAL PROTECTION FOR AGRICULTURE

Contents

INTRODUCTION

The development and poverty reduction agendas of donor organizations and developing nations place a high priority on social protection and agriculture. However, the development of social protection policies and agriculture have very different histories. Since structural adjustment and liberalization policies started to be widely promoted in the 1980s, funding commitments to agriculture, a once-dominant policy arena, have steadily decreased (with a few exceptions). Although there is renewed interest in advancing agricultural development, Ministries of Agriculture have not regained their previous resources and influence, and there is disagreement over how this should be accomplished. Conversely, social protection policy is more of a "new kid on the block." In fact, the need for social assistance and protection for groups whose welfare and means of subsistence had been harmed by structural adjustment and liberalization policies sparked donor interest in it. Since then, social protection policies' goals and scope have significantly broadened to include initiatives to assist the impoverished in managing stressors and shocks. This is done both directly to protect these individuals' livelihoods and indirectly to promote livelihood development and growth (by encouraging asset accumulation and lowering inefficiencies from risk-averse resource use).

There should be significant synergies between agricultural development policies and social protection programs if the former can assist impoverished rural residents in increasing their assets, making better use of them, and engaging in riskier but higher-rewarding activities than they otherwise would. If agricultural policies can improve people's livelihoods and food security, then reverse synergies should occur. Potential conflicts do exist, though, on two levels: in the effects of agricultural and social protection policies and in the funding of these policies. First, certain agricultural policies may make certain individuals more vulnerable (for instance, by raising food prices), and certain social protection programs may weaken incentives for investment in specific agricultural activities (for instance, food aid may impede the growth and production of the food market). Second, social protection and agricultural policies are likely to compete for scarce financial resources and influence if they are viewed as distinct policy domains and are carried out by different organizations.

7.1. STRESSES IN RURAL PEOPLE'S LIVELIHOODS

The main causes of stress in these livelihoods are discussed, along with how people respond to and are affected by these stresses in light of their dynamic livelihood strategies and how they relate to broader processes of structural change and economic growth. Even though we organize our discussion around:

(i) the causes of stress; (ii) how people react to these stresses; and (iii) the effects of these stresses and how people react to them, it is challenging to address these topics independently because they are so interconnected.

7.1.1. Sources and Impacts of Stress on Rural Livelihoods

One of the main characteristics of the livelihoods of impoverished rural people is how unfavorable changes threaten and impact them. The strains brought on by these shifts have a significant impact on the management and security of livelihoods, and consequently on welfare and growth. Designing and implementing policies that promote social protection and agricultural growth requires an understanding of the causes and effects of these changes and stresses. Therefore, it is useful to identify certain structural traits that are shared by the majority of impoverished rural residents' means of subsistence, as shown in Figure 7.1.

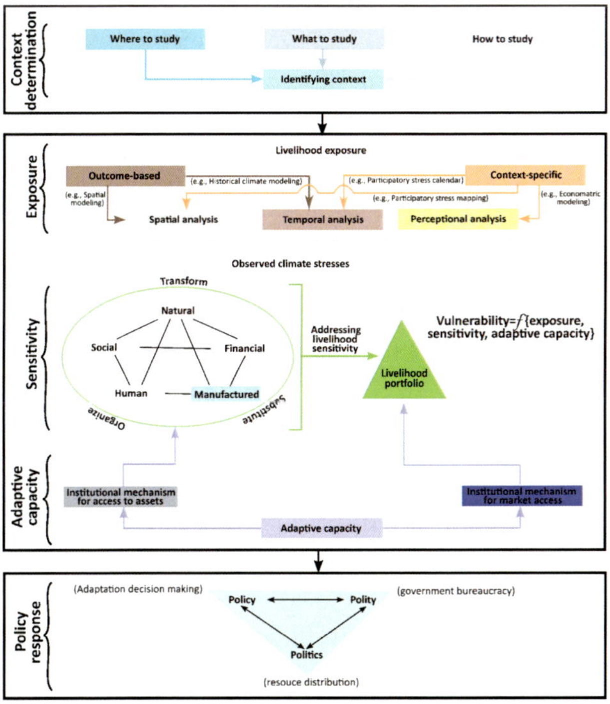

Figure 7.1. *Analyzing rural livelihoods.*

Source: Rahman HMT, Hickey GM. An Analytical Framework for Assessing Context-Specific Rural Livelihood Vulnerability. Sustainability. *2020; 12(14):5654. https://doi.org/10.3390/su12145654.*

Since shocks, trends, cycles (especially seasonal cycles), and "normal" random variation occur in many different aspects of the environment and interact to affect prices, resource availability, resource productivity, and livelihood opportunities, poor rural livelihoods are characterized by their exposure to a wide range of change types and processes. Similar to how some of the effects of change are predictable while others are not, some of these changes are predictable in their occurrence while others are not. Changes can be more idiosyncratic (affecting specific communities, households, or individuals) or covariate (affecting many communities and/or individuals at the macro and meso levels). Livelihoods also undergo change, albeit with varying dimensions, patterns, and predictabilities. While some are more (though rarely entirely) endogenous (for instance, the accumulation or loss of assets as a result of household members' actions; or births, marriages, and growing up and aging processes affecting household demographics, consumption needs, and labor resources), others are directly caused by changes in exogenous factors (such as those previously discussed). Relationships and power imbalances within families or communities, which both cause and worsen marginalization and exclusion, may also be linked to "endogenous" stress. As a result of both local and global processes and crises, such as market liberalization, the HIV/AIDS pandemic, civil conflict, the possible effects of climate change, and some elements of the globalization of agricultural trade, new forms and sources of change are also continuously emerging and having new effects on rural livelihoods.

The placement of impoverished rural livelihoods in narrow and imperfect markets is another characteristic that is emphasized. According to Dorward and Kydd (2004), poor rural areas are characterized by (among other things) low levels of economic activity, thin markets, and weak and expensive coordination in exchange. Ellis (1993) defines peasants as those who partially integrate into imperfect and frequently personalized markets. In terms of both transaction risks (buyers and sellers are exposed to risks of opportunism, coordination failure, and rent seeking in exchange) and broader market behavior (wilder price swings in response to local changes in production, for example), these institutional and market characteristics increase uncertainty and vulnerability.

When rural people are building livelihoods, imperfect (expensive and risky) markets are a major factor. They engage in a variety of activities to meet their various social and economic needs; for instance, subsistence production may be less expensive and risky than depending on markets to meet their food needs. The integration of "domestic" and "productive" activities is a particular aspect of this (this is also a feature in Ellis' definition of peasant livelihoods), and decisions in these areas are frequently "non-separable" (strongly interdependent, with changes in one causing changes in the other). When there are notable seasonal fluctuations in resource availability, prices, demands, and opportunities for productive resource use, diversification in livelihood activities is frequently a strategy for distributing risk and maximizing resource utilization.

Poor rural livelihoods tend to be intrinsically complex, meaning that exogenous and endogenous changes may have unpredictable effects due to the interplay of multiple objectives, multiple activities, and multiple dimensions of predictable and unpredictable change. Therefore, complexity both contributes to and is exacerbated by the previously discussed uncertainty in exogenous change. High vulnerability and risk are caused by complexity, exposure to uncertainty, and poverty (limited resources and opportunities for protection against unfavorable stresses).

The fact that the various domestic and productive activities that poor people engage in have a significant impact on their livelihoods is that their productive activities are susceptible to both predictable and unpredictable stresses that impact their domestic activities, and vice versa. Therefore, poor health (due to illness, accident, or aging, for example), food shortages, or significant expenses to fulfill social obligations or other household needs frequently impact the labor and financial resources available for productive activities, with detrimental short- and long-term effects.

The existence of different thresholds in rural livelihoods and the economies in which they are located leads to additional complexity, uncertainty, and vulnerability. These result in non-linear effects from change, making livelihoods especially vulnerable or sensitive to changes over specific ranges of some variables. To put it another way, changes that cross thresholds can have qualitatively and quantitatively different effects than changes that do not. In Figure 7.1, an asterisk next to "poverty," "imperfect and thin markets," and "low economic activity" denotes three different threshold types. First, "asset thresholds" occur in individual livelihoods, where specific asset sets (or amounts) are required to support specific welfare levels and/or engage in specific activities. Households lacking these minimum asset sets/amounts are at risk of becoming impoverished. Secondly, price thresholds can occur: (i) when certain activities become profitable (or unprofitable) for large numbers of people above (or below) a certain price; or (ii) across import (or export) parity, where prices become relatively constant below (or above) the parity price but highly variable above (or below) the parity price. Third, as market players and volumes increase, transaction risks (of opportunism and coordination failure) and transaction costs (though coordination costs may also increase in some situations) tend to decrease. These factors can create thresholds that determine whether an investment is profitable or not, which can result in low-level, under-investment traps for economies with low market volumes and few market players.(Rodenstein-Rodan, 1943).

Externalities and scale effects are a final source of complexity and uncertainty in the ways that change affects rural livelihoods: when many people behave in similar ways, it has an impact on the environments in which they operate. This is true for markets, where a large number of people purchasing (or selling) goods or services may cause price increases (or decreases), and the natural environment, where, for instance, a large number of people harvesting natural resources may cause their degradation.

The direct losses that result from stress actually happening and the indirect or behavioral effects that occur *ex ante* and in anticipation of stress are the two main ways that stresses affect rural livelihoods. People who invest in assets and strategies that lower the likelihood or severity of loss but also lower average incomes are responsible for these latter effects. Many of the direct costs and losses resulting from stressors are self-evident and require little explanation: crop losses may be caused by floods, droughts, or hail, while births, illnesses, accidents, disabilities, old age, or education may result in a variety of predictable and unpredictable increases in living expenses and decreases in output or profits. However, when impoverished rural people are compelled to adopt increasingly unsustainable practices to smooth their income or consumption, these sudden losses of assets and earnings may have far-reaching "knock-on" effects on other activities and asset holdings in peoples' livelihoods. These effects may not be as evident right away, but they may be more harmful. For instance, distress-driven hiring out of labor and distress sales may serve as "ratchets," locking people into poverty or low-level traps from which they are unable to escape without outside assistance due to irreversible losses of productive resources below asset thresholds. (Dercon, 2002). To lessen their susceptibility to stress, impoverished people usually have to focus on low-risk, diversified activities and forgo higher returns from specialization. This results in indirect impacts of risk and uncertainty. Because farmers with insecure land tenure do not invest in land improvement, families from ethnic minorities who have limited access to formal employment prefer to send their children to work rather than school, and entrepreneurs who lack access to microfinance or insurance will not engage in potentially profitable but risky ventures, this insecurity stifles investment and pro-poor growth. These risk-reactions may have a significant cost in the form of lower mean returns and, consequently, the continuation of chronic poverty (Townsend, 1995). There are significant potential synergies between social protection and agricultural development policies and interventions because of the numerous direct and indirect effects of stress on the livelihoods of rural residents. After talking about how individuals react to and handle stress in their daily lives and how social protection policies either support or enhance these reactions, we will look at these synergies.

7.1.2. Livelihood Responses and Strategies Against Stress

Depending on the resources at their disposal and the type of stress, people react differently to predictable and unpredictable stressors. Although it is impossible to precisely distinguish between the various response types due to the numerous interactions and overlaps between them, it is useful to recognize four general ways that people try—more or less successfully—to manage and cope with stress:

- **Removal:** Potential stresses are best handled by eliminating or at least partially eliminating them, if at all possible, either by lessening their frequency or intensity. Preventive healthcare (against diseases and pests of people, animals, or crops), irrigation (against drought), futures markets or bilateral fixed price agreements (against price fluctuations), and enhanced property rights (against transaction risks) are a few examples of stress reduction.
- **Resistance:** It may be possible to limit the impact of stresses if they cannot be eliminated, or if doing so would be extremely expensive. This could be done by diversifying so that losses in activities or assets that are susceptible to specific stresses only make up a small percentage of total livelihood income or wealth, or by making investments to lessen the susceptibility of specific activities or assets (e.g., building drainage systems that are resistant to floods, planting crop varieties that are resistant to drought or disease).
- **Recovery:** It might be possible to set up conditions that will enable a quicker recovery from the detrimental effects of stress if it is not possible to (partially) eliminate or resist stresses. Such recovery is aided by insurance systems that provide compensation based on losses, the development of social ties (in the event of idiosyncratic stresses), and the production of crops that are resilient to pest attacks.
- **Relief:** Finally, it might be possible to lessen the effects of stress to prevent total destitution if livelihoods are at risk of suffering significant losses as a result of the ineffective adoption of the aforementioned measures. To safeguard critical consumption levels, this usually entails the minimal application of one of the aforementioned strategies (for instance, by guaranteeing the maintenance of a very low level of production or by using insurance that offers very small payouts, insufficient for recovery but sufficient for minimal consumption).

The lines separating these four categories of stress responses are frequently blurred, although this classification can be useful in organizing thoughts about how people react to stress. Therefore, insurance can be viewed as a type of recovery in the most straightforward sense (payouts compensating losses), but it is accomplished by diversification across a far larger range of activities than individuals can. Additionally, insurance can be viewed as a type of "recovery" from losses as well as a stress-resistance strategy for preserving income when it compensates for the loss of productive assets and income generated by them.

People's broader livelihood strategies can include these particular stress-reactions. Dorward and associates make a distinction between the three main livelihood strategies that low-income households can use in advance:

- 'Hanging in,' where people undertake activities to maintain livelihood levels at a (sustainable?) 'survival' level;
- 'Stepping up,' where people make investments in existing activities to increase their returns; and
- 'Stepping out,' where people engage in existing activities to accumulate assets as a basis for investment in alternative, more remunerative livelihood activities.

People often use elements of two or more of these three strategies, so they are not mutually exclusive. However, "hanging in" activities are likely to be more common in areas where people are especially susceptible to stress. The focus on "hanging in" needs to give way to "stepping up" and "stepping out" in impoverished rural households. This is the typical course of economic growth. This typically entails a shift for small-scale producers from semi-subsistence to intensified crop production, more non-farm jobs, and eventually, for many, a profitable departure from agriculture. For people's livelihoods, this shift typically entails more specialization and integration into markets. It entails greater urbanization, a greater proportion of non-farm activities, and greater diversification of livelihoods for both rural and national economies. It is crucial to remember that these tactics produce comparable results when they are effective. However, giving up results in a fourth outcome that could be referred to as "falling out" to continue the terminology.

It should be evident that the livelihood responses and strategies discussed here have significant ramifications for and connections to agricultural development and social protection policies and interventions that are intended to assist individuals in "hanging in," "stepping up," and "stepping out."

7.2. A REVIEW OF SOCIAL PROTECTION POLICIES

The goal of social protection policy is to assist those who are at risk in managing their stress. This agenda has mostly resulted from an emphasis on various interventions created and carried out to assist various individuals with various vulnerabilities. However, more recent conceptualizations of social protection have also taken into account how social protection interventions either help or hinder individuals' and households' own coping mechanisms and stress management techniques. Although these expanded conceptualizations are crucial to social protection policy, they may be confusing if they fail to clearly differentiate between livelihood responses and the social protection tools (specific intervention types) that support them. An overview of the thriving social protection agenda is given below.

7.2.1. Social Protection Policies

Over the past 40 years or so, perceptions of the significance and character of "social protection" policies have drastically changed. A critical reaction to the "safety nets" debate of the late 1980s and early 1990s was the emergence of social protection policies. For example, in the 1990 "World Development Report," safety nets were envisioned as minimal social assistance in nations too impoverished and administratively weak to implement extensive social welfare programs. They were essentially the third pillar of the World Bank's three-pronged strategy for "attacking poverty." In the 1990s, however, as perspectives on livelihoods, risk, and vulnerability, and the multifaceted nature of poverty became more nuanced, safety nets came under growing criticism for being residuals and paternalistic. As its wider potential became apparent, more complex tools for attaining social protection were put forth, and at the same time, the objectives and reach of social protection policies and interventions were expanded. Larger claims are now being made about what social protection policies can and should aim to accomplish.

Earlier, limited ideas of social protection, which were frequently found in lower-income countries' labor or welfare ministries, still view social protection policies and interventions primarily as a safety net for the most vulnerable members of society, including the elderly, people with disabilities, orphans, and those living with HIV/AIDS. Nonetheless, there is compelling evidence in the literature today that social protection programs and policies cover a wide range of topics beyond welfare assistance. In contrast to social sectors like health and education and social welfare programs, this new agenda "represents a public commitment to reduce risk and vulnerability, as it is concerned at least in part with the interface between protective measures and engagement by the poor in productive, growth-oriented processes." The scope of social protection policies and interventions is far broader and more comprehensive than just a safety net package when the interconnections between the domestic and productive domains in rural livelihoods and the direct and indirect effects of stressors are acknowledged. This agenda's main theme is stress management, with an emphasis on the vulnerabilities of economic security (income and consumption) and how they positively relate to growth. The new social protection agenda's novelty lies in connecting the welfare and growth agendas, not in its well-known tools. Traditional safety nets were criticized for their exclusive welfare-focused approach, which may have encouraged dependency. The "graduation" from dependency to productive sector activities and growth is explicitly pushed in this updated version. The agenda's framing is where the novelty lies.

There are now several conceptualizations and terminologies used to describe and define social protection as a result of growing understandings of the concept and the corresponding expansion of its agenda. Here, we define social protection measures as actions that improve the management of stresses by individuals, households, and communities, taking into account both the earlier and more recent definitions of social protection as discussed above. Social protection measures can be exogenous, meaning they come from outside sources (like government policies and interventions, non-governmental organizations, and civil society), or endogenous, meaning they come from affected communities, households, and individuals themselves. The rest of this section will refer to exogenous social protection measures as social protection instruments. Based on their main purpose, which is seen as having an immediate effect on people's livelihoods, we differentiate between three major categories of instruments: resilience-building, insurance, and welfare instruments.

Social protection instruments according to the classification above are therefore defined as follows:

- Welfare tools offer respite and occasionally help people recover from hardship. Social assistance for the "chronically poor," particularly those who are unable to work and support themselves, is one of them. This is most comparable to "social welfare" in the mainstream. Targeted resource transfers, such as disability benefits, orphan-carer grants, free input provision, and "social pensions" for the elderly poor, are commonly included in social assistance programs. These programs are funded by the government, either through tax revenue, donor support, or NGO initiatives.
- Through the implementation of strong and easily accessible recovery mechanisms, insurance instruments aim to prevent deprivation in the face of stress. When trusted, these tools can help households transition from low-level subsistence strategies by balancing income and consumption. Another aspect of insurance is the release of liquidity through modifications in risky behavior. These tools include private, market-based insurance (like crop, weather, or livestock insurance), informal mechanisms (like savings clubs and funeral societies as well as familial, clan, and patronage relations between various members of society), and social insurance programs (formalized systems of pensions, health insurance, maternity benefit,

and unemployment benefits, frequently with tripartite financing between employers, employees, and the state). The conditions and nature of payouts, the costs and rules of access, the instruments' resilience to covariant stresses, and the institutional and financial demands of their operation on various stakeholders are all different.

- Through livelihood-enhancing initiatives that increase assets and foster stress resistance, resilience-building tools seek to improve actual incomes and capabilities. For instance, public works programs (which build physical infrastructure as well as financial and human capital), microfinance (which build financial assets and human, social, and physical capital), school feeding (which build human capital), and protecting minority ethnic groups' human rights (which build social capital by promoting social equity and attacking exclusion) are just a few examples of the many different tools that can be used to promote physical, financial, natural, human, and social assets. Sensitization campaigns to change public attitudes and behavior are examples of transformative interventions, as are modifications to the legal framework to safeguard "socially vulnerable groups" (such as individuals with disabilities or victims of domestic abuse) from abuse and discrimination.

Since all social protection instruments have direct components of other functions, this classification of exogenous social protection instruments by primary function permits overlapping types (similar to our classification of endogenous livelihood responses). For instance, school feeding programs give food to the underprivileged (welfare), promote education as a means of investing in human capital (building resilience), and offer some protection against the stresses of consumption in situations where the transfers are seen as steady and long-lasting.

7.2.2. Interactions between Social Protection Instruments and Livelihoods Responses and Impacts

People's (endogenous) reactions to stress, the effects of those stresses, and various forms of (exogenous) social protection tools interact in a variety of intricate ways. Before discussing some significant interactions that are not specifically covered by this framework, researchers first present a framework that characterizes the more general interactions. A straightforward sequence of events that occurs when stress impacts people's assets and/or income, which in turn impacts their welfare, is depicted by the horizontal axis of Figure 7.2 below in research. Therefore, by lowering stress, removal responses also lessen detrimental welfare effects; relief responses, on the other hand, only target the welfare effects of stress directly.

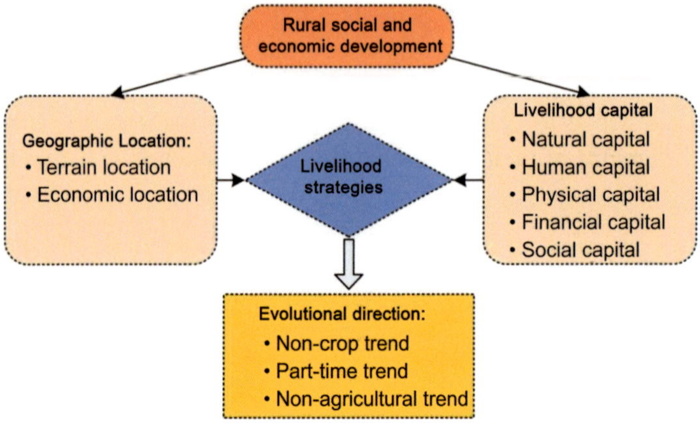

Figure 7.2. *Analyzing rural livelihoods.*

Source: Wu K, Yang X, Zhang J, Wang Z. Differential Evolution of Farmers' Livelihood Strategies since the 1980s on the Loess Plateau, China. Land. *2022; 11(2):157. https://doi.org/10.3390/land11020157.*

Additionally, Figure 7.2 illustrates the relationship between various exogenous social protection tools and livelihood stress responses and the effects of stress: resilience-building tools, for example, support people's resistance responses by increasing their assets; insurance tools support recovery responses more; and welfare tools support relief responses. This mapping is inaccurate, though, because it lacks clarity regarding the distinctions between the various livelihood response types and social protection instrument types. Furthermore, reliable and successful welfare programs will function somewhat like insurance programs, and reliable insurance programs will help people accumulate assets and take on greater risks, both of which promote resilience. Features of instruments can also influence people's behavior and stress reactions in many other ways. The behavior of recipients or potential recipients is therefore altered to increase their eligibility through certain conditional transfers, where receipt is contingent on recipient characteristics (e.g., being in a target group for "unconditional" cash transfers) or recipient behavior (e.g., attending school to receive school feeding). Other facets of peoples' lives may then be affected, either positively or negatively, by these behavioral shifts (attendance at school, for instance, has educational benefits but may withdraw labor from other activities). Likewise, receiving welfare transfers can have a variety of effects on productivity, including: (i) preventing the loss of productive assets; (ii) allowing otherwise unproductive people to enter the productive economy; (iii) undermining or strengthening incentives to engage in specific productive activities; and (iv) influencing the growth and welfare of others in the economy (either positively or negatively) through consumption or production linkages and multipliers (by affecting prices or other aspects of local or wider economic and social relations).

In terms of the size of individual transfers and the percentage of the population receiving them, threshold and scale effects may also have an impact on the effects of social protection tools. Because micro-level poverty traps exist, transfers that help people cross an asset threshold may have a significantly bigger impact on growth than transfers that don't. As a result, we cannot anticipate straightforward linear relationships between the size of transfers and the productivity gains for recipients; rather, benefits will vary depending on the recipients' distance from the threshold and will differ for each recipient in any given circumstance.

There may be thresholds, or at the very least, significant discontinuities, in transfers that bring individuals into the productive sector.

Similar to meso level traps (like the previously discussed under-investment trap), interventions that raise a supply chain above a certain market volume threshold or lower investment costs or risk to make investments profitable will have the biggest growth effects. Even in the absence of thresholds, the number of recipients and the size of the subsidy have a significant impact on price impacts (for instance, from input subsidies that boost local production and lower food prices). Due to the necessity of addressing several growth-limiting constraints, the growth effects of social protection interventions may also be highly context-dependent.

7.3. KEY ISSUES FOR SOCIAL PROTECTION AND AGRICULTURAL GROWTH POLICIES

This chapter address the main concerns that arise for social protection and agricultural development policies after taking into account the intricate and dynamic relationships that exist between people's livelihoods, economic growth, social protection, and these policies.

It starts by talking about two unique issues—food access and prices and HIV/AIDS—where the relationships between social protection and agriculture are especially significant. Next, it lists four general approaches to agricultural growth and social protection policies. This is followed by a brief discussion of potential interactions between the various tools that might be included in these social protection and agricultural growth strategies. The chapter is wrapped up by talking about design and implementation concerns that have a significant impact on the results of various interventions.

7.3.1. Special Issues for Agricultural and Social Protection Policies

We propose that food access and prices, as well as HIV/AIDS, are two particular issues where there are particularly strong and intricate connections between agriculture and social protection.

7.3.1.1. Food Access and Prices

Food price volatility is a crucial social protection issue because: (i) it frequently poses a serious threat to the food security of the poor, affecting both urban and rural poor people and contributing to their continued poverty; and (ii) it provides a strong deterrent to diversification into more lucrative cash crops and non-farm activities, tying up productive resources in low-productivity, often inefficient subsistence production. High food prices have a negative impact on the poor, those experiencing food insecurity, and others who are discouraged from diversifying out of food production due to price volatility. Low food prices, however, have a detrimental impact on surplus food producers and those who may decide to invest in more intensive food production. As a result, major population groups in impoverished areas are impacted by and have varying interests in the levels and volatility of food prices.

In the early phases of agricultural growth, the tensions between the advantages of high and low food prices are especially significant because high food prices are necessary to encourage investment in excess production and to initiate the growth process that will eventually drive down prices.

But as this growth continues, there are two tightropes that productivity advancements must balance: a) a price/total productivity tightrope, where the technical change must significantly lower the price of staples while increasing total factor productivity much more quickly to ensure that investment in the new technology remains profitable; and b) a wage rate/labor-land/productivity tightrope, where the output per labor-hour must increase significantly but output per hectare must increase much more, so that labor productivity and labor demand both rise and support real wage increases for the poor. The needs of those who are poor and food insecure must be balanced against these calls for higher prices to spur growth. During the early stages of growth, they face high prices until: (i) rising labor and land productivity boost supply and drive down prices; and (ii) broader processes of growth driven by changes in agricultural productivity raise wages and employment opportunities in both agricultural and non-agricultural activities. To protect their food security, impoverished urban and rural residents require focused welfare interventions before they can enjoy these advantages of agricultural growth.

The next challenge is to put in place efficient welfare programs to shield low-income households during this crucial transitional phase without compromising the financial incentives for wealthier smallholders to invest in more productive technology. The necessity of mechanisms to support price stabilization and the promotion of price levels that facilitate technical change and food security are hotly debated topics. Although these cannot be discussed here, it should be noted that they have an impact on food aid, targeted input subsidies, international, and regional food trade policies, and various institutional mechanisms that can be used to support the development of local, national, and regional grain markets as well as the coordinated provision of complementary services to emerging smallholder farmers with surplus grain. Here, it is crucial to stress that food availability and entitlements are both critical to ensuring secure access to food, and that in underdeveloped markets, food availability must be prioritized far more than in developed ones.

7.3.1.2. HIV/AIDS

Since groundbreaking studies on HIV/AIDS and household demographics were conducted in Uganda in the 1980s, attention has focused on the effects of the disease on household demographics and livelihoods. However, very little research has been done to look at how local labor and food markets are affected by the high rates of HIV/AIDS morbidity and mortality, as well as how this affects rural livelihoods. Impacts at the individual household level have been generalized into conceptual frameworks that follow the probable connections between HIV infection and economic, demographic, and health outcomes at the household level. According to Devereux and Sabates-Wheeler, one of the main findings of this study is that HIV/AIDS increases household vulnerability in many ways and to an unprecedented degree.

At least from the standpoint of livelihood, its detrimental effect on household labor capacity is arguably the most important. HIV/AIDS turns "workers" into "dependents" and "producers" into "consumers" of household resources by weakening and displacing working adults. This is particularly problematic in impoverished households where labor is one of the few resources. Additionally, Slater and Wiggins observe that "the HIV/AIDS epidemic raises acute questions for social protection and has multiple and uneven implications for rural livelihoods." HIV/AIDS-related labor shortages suggest that work-based interventions may be challenging, crop diversity and production are decreased, increasing the risk of shortages, and subsistence crops are substituted for high-yield cash crops. Savings and productive assets to pay for care are being depleted at the same time that labor becomes incapacitated. The ability to manage stress informally is diminished.

Numerous social protection initiatives have been implemented that specifically target HIV/AIDS. When HIV/AIDS patients' nutritional status is improved to the point where they can engage in the economy, this could have a major impact on productivity. As an alternative, focused initiatives can incorporate nutrition, education, and occasionally productivity improvements.

This approach seems to ignore the fact that the pandemic is systematically undermining entire livelihood systems, even though food or cash transfers, education bursaries, and other forms of support for HIV/AIDS patients are essential and frequently crucial. For instance, removing many adults from the workforce necessitates changing agricultural practices (such as growing less labor-intensive crops) or leaving farming entirely and pursuing alternative livelihoods in rural farming communities where livelihoods rely heavily on labor power. However, there is little strategic thinking, few policy documents, and even fewer projects that address this reality. However, Dorward and Mwale (2005) contend that widespread HIV/AIDS morbidity and mortality may paradoxically loosen labor markets and depress wages. If "labor saving" techniques are adopted more widely under such circumstances, they could harm the livelihoods of both PLWHA and impoverished individuals who are not directly affected by HIV/AIDS. The promotion of "labor saving" crops and techniques is frequently proposed as a way to alleviate labor shortages for PLWHA.

The application of targeting in agricultural social protection programs has not advanced much. Stricter targeting will result in more cost-effective social protection for PLWHA, but specifically focusing on PLWHA presents at least two issues. First, in many (but not all) social contexts, it runs the risk of stigmatizing the targeted groups. This can be addressed through community targeting in an area where HIV prevalence is known to be high (assuming that the targeted group will include the HIV positive) or through close but expensive community engagement and awareness-raising. Although the latter strategy has been employed, questions still surround whether it is feasible to scale up programs that are specifically targeted without increasing exclusion errors. The second issue is that people living with HIV/AIDS might not be able to fulfill the requirements of certain social protection programs, particularly the physical labor that many PWPs require. Reducing workloads, hiring households rather than individuals to cover for sick family members (though this may limit the scope of intra-household care), and taking up low-intensity jobs like horticulture (though this tends to create private rather than public assets) are some possible solutions to this.

Some policymakers have viewed the transfer of resources to HIV/AIDS-affected groups as an ineffective investment. But as highlighted, these transfers can increase overall productivity in many ways. When HIV/AIDS patients' nutritional status is improved to the point where they can engage in the economy, there may be notable productivity effects. The beneficial effects of nutrition on ARVs compound these.

7.3.2. Broad Policy Strategies in Social Protection and Agricultural Growth

Over time, expectations for agricultural growth and social protection policies have evolved, as has the relative attention given to them. Four general strategic approaches to agricultural growth and social protection are useful to take into account.

- Social protection (inter alia) from agriculture and agricultural growth;
- Social protection independent of agricultural growth;
- Social protection for (inter alia) agricultural growth; and
- Social protection through (inter alia) agriculture.

We can classify historical shifts in social protection and agricultural growth policies as follows: agricultural liberalization and early social protection policies were largely independent (strategy B), state-led agricultural development policies fell under strategy A (social protection from agriculture and agricultural growth), and then social protection policy was extended beyond welfare instruments to include insurance and resilience building (strategy C), with a focus on agricultural growth. Social protection through agriculture, or Strategy D, is a slightly unique example that combines aspects of A and C. It involves the use of specific tools to support agriculture to achieve specific and immediate social protection impacts (targeted agricultural inputs programs, for instance), but these are not incorporated into broader agricultural development strategies.

There are overlaps and ambiguities in this broad strategy classification. Nonetheless, it is helpful because it (i) outlines general historical approaches to agricultural growth and social protection policies; and (ii) outlines the strategic options available to policymakers. It also offers a straightforward yet efficient framework for evaluating the contributions of various policy tools to agriculture and social protection.

7.3.3. Policies and Instruments for Social Protection and Agricultural Growth

Over the past four decades, policy approaches to agricultural growth and social protection have undergone significant change. International "fashions" that take a "one size fits all" approach have frequently been the driving force behind changes in specific nations. Clear policy principles should be developed whenever feasible, but it's also critical that policies work well for addressing various issues in various situations.

- Conditions, constraints, and opportunities in agricultural and non-agricultural sectors;
- The structure of people's livelihoods, and the opportunities, constraints, and stresses that they face;
- The state of market and supply chain development;
- Existing formal and informal social protection mechanisms;
- Capacity and resources for implementation of agricultural and social protection interventions; and
- Particular characteristics, strengths, and weaknesses of different agricultural and social protection interventions in achieving particular outcomes.

7.3.3.1. Social Protection (Inter Alia) From Agriculture and Agricultural Growth

The general strategic approach to agricultural development in (mainly) state-led agricultural development policies after independence is referred to as "social protection from agricultural growth." As they sought to encourage agricultural growth, harvest agricultural surpluses to support industrial development, and offer some social protection primarily aimed at promoting food security, these policies frequently had several and occasionally conflicting goals. Here, the main policies were market and output price interventions, input delivery and subsidy systems, agricultural credit provision, infrastructure development (market, irrigation, and transportation facilities), technical change (supported by all of these policies, as well as research and extension), and (less commonly) land reform. Policies were put into effect both nationally (for instance, through tariffs, market regulations, and pricing policies) and locally (through projects and programs targeted at specific regions). In certain nations, these policies were synchronized through "integrated rural development" initiatives that sought to integrate transportation, various agricultural service delivery and investment activities, and health and educational service investments and provision.

7.3.3.1.1. Output Price And Market Interventions

Reduced food price volatility benefits everyone, despite diverse and competing interests in low food prices (important for impoverished consumers) and high food prices (important for promoting intensification and growth, with immediate benefits for producers and longer-term indirect benefits for others). Historically, these worries have offered a compelling case for food market interventions aimed at stabilizing prices and providing producers with guaranteed minimum prices and consumers with guaranteed maximum prices. These systems, however, can be expensive and challenging to administer since they necessitate the ability to step in and purchase and hold produce when prices are dropping and to sell it when supplies are limited. They are also highly politicized and can pay dishonest politicians or officials high rents because of the volume of trade. Therefore, political or patronage abuses often subject them to and render them ineffective. Though it can be used in both the early stages of growth through non-market mechanisms (by directly promoting food production in remote areas with food deficits) and later using market-based mechanisms, food price management can, when it is effective, provide significant social protection welfare and insurance mechanisms for both producers and impoverished consumers. (Byerlee et al., 2006).

There are significant questions regarding the appropriate level and methods for stabilizing food prices. Regarding the first question, it is possible that their indirect impact on rural wages may significantly offset higher food prices, especially if increased incentives for intensification effects across specific price thresholds are taken into consideration. In the short term, deficit households will surely suffer from high prices. More study is needed to better understand (i) rural labor markets and how they function across different regions, and (ii) how sensitive food price increases are to the range of prices they fall within (especially the nature and impact of price thresholds) and other policy measures. Any interventions supporting producer prices must be accompanied by appropriate social protection interventions to smooth consumption for those affected in the short term and to provide long-term assistance to those unable to take advantage of increased labor demand, as deficit households suffer short-term and often long-term negative effects from higher food prices.

There is continuous discussion about the best ways to stabilize food prices. A range of market and non-market mechanisms can be employed, and their relative efficacy and efficiency differs from nation to nation based on factors such as local market development, production opportunities and constraints, connections to regional and global markets, political will and the ability of the government to directly intervene in markets, and the institutional, financial, and human capacity to use advanced market instruments to hedge risks. As long as private investors are guaranteed that governments won't arbitrarily interfere in the food markets, proponents of market-based instruments typically contend that they can function effectively and efficiently on a large scale to stabilize markets.

Others, however, contend that while such arguments may appear to provide technical solutions to market stabilization problems, in fact, they ignore the following:

- The political realities that cause governments and politicians to intervene in food markets;
- The way that high cost and time requirements for food importation into some land locked countries lead to very wide import-export parity price bands with the potential for damaging price peaks following delayed imports;
- The significant benefits and modest costs that may be involved in promoting at least some in-country storage. and
- The role of output markets and food prices in the development of complementary services for small farmers in poor rural areas.

7.3.3.1.2. Input Subsidies and Delivery Systems

Asian Green Revolutions have made extensive use of input subsidies to expand access to and adoption of contemporary inputs. It is argued that by reducing investment risks and assisting in the removal of significant price and transaction cost distortions in impoverished rural areas, such subsidies can help jump-start pro-poor growth in the early phases of agricultural transformations. In the context of failing credit markets, input subsidies may be a significant way to encourage the development of supplementary services required for agro-chemical-based intensification, as high transportation costs frequently put inputs beyond the reach of the majority of smallholders, particularly in landlocked nations and remote areas. However, because input subsidies are hard to control and vulnerable to political and patronage interests, they may also promote inefficient use of inputs. They can quickly become a significant financial burden because their expenses also have a tendency to increase quickly and they are politically challenging to remove.

Although successful agricultural intensifications and, consequently, growth and national food security have historically benefited greatly from input subsidies and delivery systems, they may not have a significant effect on the direct access of the poor to inputs unless the subsidies are extremely high (with extremely high fiscal costs) or specifically targeted at the poor. As a result, they have contributed significantly to the production of staples required for national food security but have had a comparatively minor impact on direct social protection outside of these targeted input programs.

7.3.3.1.3. Credit Subsidies & Delivery Systems

In agricultural growth policies, credit subsidies and delivery systems have been largely supported to (i) speed up the adoption of modern technology, (ii) overcome obstacles to smallholder access to seasonal finance due to a lack of collateral or information, and (iii) overcome a critical constraint to growth from a lack of cash to finance farm investments and variable inputs. With notable exceptions, large-scale agricultural credit programs that were put in place as part of state-led development policies were typically unsustainable, had low repayment rates, high operating costs, and did not encourage the poor to access financial services. As a result, their direct social protection function was quite limited. Even though many of these programs were ineffective and unsustainable, they were linked to the success of the green revolution, and there is some empirical evidence that they significantly boosted agricultural growth in India.

Credit availability is frequently viewed as enabling consumption smoothing, which makes it a type of insurance that lowers vulnerability. Ironically, though, credit linked to the acquisition and use of agricultural inputs may make people more vulnerable and at risk because loans typically account for a sizable amount of the assets and yearly income of poorer farmers, which in turn causes high gearing. High-risk exposure results from this as well as the inherent unpredictability in the production of rain-fed crops, livestock, and agricultural markets. Naturally, there are trade-offs between greater growth and productivity and greater vulnerability; the merits of these trade-offs will vary depending on how they affect particular regions and populations as well as the state's capacity to offer social safety nets to counteract vulnerability increases.

7.3.3.1.4. Infrastructure Development

Although they can provide some protection to local producers, poor roads result in low farm gate prices for agricultural produce, high transportation and communications costs, and high input purchase and service delivery costs. Therefore, the development of infrastructure is a fundamental prerequisite for agricultural and overall economic growth. Almost everyone agrees that poor transport infrastructure is a major obstacle to growth in Sub-Saharan Africa, even though donor allocations have decreased as a result of high implementation and maintenance costs. It is less frequently acknowledged that inadequate infrastructure contributes to high transaction costs, such as those associated with measurement, search, and contract enforcement, which in turn lead to transaction failures and thin markets. Irrigation infrastructure investments are essential for reducing risks and boosting incentives for agrochemical-based intensification investments. Therefore, infrastructure investments greatly reduce the investment thresholds needed to spur growth, though additional coordination interventions are probably needed in areas with sparse or nonexistent agricultural markets.

Infrastructure expenditures may produce positive externalities for social protection even though they are typically justified in terms of their growth-promoting effects. Better transportation infrastructure tends to enhance temporal and spatial arbitrage (e.g., storage infrastructure), which could lessen seasonal price volatility and enhance the availability of food in areas with shortages. Lower food prices for consumers and national food security can both be facilitated by increases in productivity brought about by improvements in irrigation infrastructure or higher farm gate prices. Along with improving the spatial distribution of these multipliers and increasing the labor and consumption links resulting from agricultural growth, improved infrastructure can also reduce the costs associated with implementing social protection programs. Even though these effects might lessen the strain on social protection programs, they don't always benefit all rural poor people equally, particularly when they avoid less favored areas and aren't accompanied by other interventions meant to break the cycle of micro-level poverty and meso-level underinvestment. On the other hand, as will be covered later, public works projects (PWPs) are unlikely to have a major and long-lasting effect on growth unless they are large enough and of a high enough caliber to lower transaction and transportation costs below critical levels. Recognizing these kinds of challenges, state-led policies for a while implemented "integrated development programs" to make sure that various growth and social protection constraints were addressed in a complementary and coordinated manner. However, because these initiatives—and growth-oriented infrastructure in general—tend to concentrate on high-potential regions, they might have had little impact on social protection issues in LFAs. PWPs under social protection programs, on the other hand, are probably going to concentrate on LFAs and offer less growth-oriented benefits.

Although state-led development policies frequently placed a high priority on infrastructure development, most African nations had very little infrastructure when they gained their independence, and low agricultural population densities meant that infrastructure investments, especially in the transportation sector, were very expensive per unit of capital. Additionally, state-led systems have very weak incentives for encouraging maintenance, and rapid infrastructure investments have outpaced the financial and other resources required for maintenance. Consequently, there has been a problem with the rapid deterioration of large-scale irrigation systems and roads (although there are growing efforts to improve local ownership and management of infrastructure, which may be related to PWPs).

7.3.3.1.5. Technical Change

For agriculture to grow, technical advancement that boosts labor and/or land productivity is a prerequisite. Modern seed varieties, fertilizers, crop protection chemicals, irrigation when feasible, and, in land-rich areas, expanding access to farm mechanization have been the main focuses of technical change in impoverished rural areas. (i) the challenges of developing and adapting appropriate technology to diverse, resource-poor conditions; (ii) the challenges of promoting the development of and ensuring coordinated access to the complementary input, output, financial, and technical services necessary for successful and lower risk adoption of these technologies; (iii) the capital, labor-saving bias of some forms of technical change; (iv) the potential uptake biases towards large, commercial farms; (v) the bias in research and extension towards the needs of the commercial sector; (vi) the challenges of developing efficient, two-way research and extension systems. These are the key obstacles to technical change as a driver of pro-poor growth.

Because of these difficulties, technical change is frequently skewed in favor of the commercial sector or, at the very least, larger smallholder farms concerning its capital intensity, technical requirements, and ease of adoption. Due to their easier access to credit and the generally large size of seed and fertilizer packages, larger farms are more likely to be integrated into markets, even with scale-neutral technology like hybrid seed varieties. The agricultural growth that has resulted from labor-intensive technical change focused on staple food crops (such as "green revolution" technologies) has raised the demand for wage labor and, in many cases, raised labor costs and decreased food prices, which benefits the impoverished in rural areas. Both direct and indirect "social protection benefits" have been substantial as a result. By breeding crop varieties that are more tolerant or resistant to drought, pests, and diseases, as well as by making advancements in crop protection, water harvesting, and animal health, agricultural research has also been crucial in the development of lower-risk technologies.

More drastic measures, like competitive public contracts, are needed to make poverty oriented research by private companies profitable and thus appealing. Applied biotechnology research, which may offer new opportunities for increasing yields, especially in environments with limited resources, is primarily privately owned, exploited, and motivated.

Providing similarly varied packages that are adapted to local conditions is a significant challenge for agricultural research and extension serving diverse, complex, and resource-poor farming systems. A more differentiated approach to research and extension systems is advocated by Berdegué and Escobar (2002). These systems include (i) a private, market-driven system that only indirectly benefits the poor, (ii) a market-oriented, asset-constrained system that targets smallholders who are "better-off" in terms of their asset base and production environment, and (iii) a context and asset-constrained system in areas where agricultural development potential is limited. This strategy involves a broader policy emphasis on expanding producers' asset base and/or overcoming environmental constraints first, and it partially overlaps with asset-building social protection strategies.

A frequently mentioned outcome of technical changes related to state-led agricultural development policies after independence is a decrease in the variety of livelihoods during the early phases of development. Although there are many facets to this and significant regional variation, the overall result could be an increase in risk and vulnerability. Both a decreased range of farm activities (crops and/or livestock, such as switching from millet and/or root crops to maize) and a decreased range of crop varieties and livestock breeds being raised or maintained (and thus less genetic diversity) are examples of reductions in diversity. Diversity may be diminished both within and between livelihoods. Similar to the previously discussed use of purchased inputs and input credit, there should be some trade-off between increased productivity and growth and increased vulnerability in each of these situations. The desirability of these trade-offs will differ depending on the region and the individual, and it will rely on the state's ability to provide social protection tools to counteract increases in vulnerability.

7.3.3.1.6. Land Reform

It is commonly acknowledged that to encourage investment and sustainable resource management, clearly defined and protected land rights are essential. However, many regions of Sub-Saharan Africa still have high levels of land inequality, and as the continent's population grows, land reform is probably going to become more important on the political agenda. These two factors, along with the well-known transaction cost advantage of small family farms in terms of using and managing farm labor, continue to make a compelling case for ongoing land reform initiatives.

With a few notable exceptions, land reform has received less attention in rural development policy over the past 20 years or so, although land reforms carried out up until the 1970s were frequently more successful than is acknowledged and played a significant role in creating the conditions required for agricultural transformation. "New-wave," negotiated approaches that rely on market-based incentives and place more emphasis on the empowering aspects of land reform have replaced traditional, administrative, and redistributive approaches. More recent strategies concentrate on enhancing the efficiency and transparency of the rental and land sale markets as well as supporting market-based reform by encouraging profitable projects. In addition to being more cost-effective and efficient than centrally managed procedures, decentralized implementation of negotiated reform processes allows for greater participation from beneficiaries in the negotiation process.

To succeed, these new strategies must be more cost-effective and efficient than the previous ones. They should also try to integrate effective aspects of the pre-reform system, such as the cost advantages of large farm transactions in credit markets, input and output marketing, and their economies of scale in post-harvest and transportation operations. This necessitates the creation of new institutional structures that lower transaction costs both within and between holdings while combining the relative benefits of large and small holdings. Lipton contends that secure access to even very small land parcels (for example, only enough for a house) can provide important risk management and resilience building benefits, providing reservation income and strengthening the bargaining position of rural labor. Therefore, greater attention to the empowering aspects of land reform within new approaches is also relevant to our discussion, as secure land tenure is likely to be particularly important for socially excluded groups.

Therefore, in addition to the more obvious agricultural benefits, land policies may have significant social protection and growth benefits. In discussions concerning the consequences of policies that support land markets, social protection viewpoints are also crucial. Although these are frequently advertised as having the ability to permit the use of land as collateral to obtain credit, doing so is frequently fraught with unofficial institutional challenges. Reforms that enable the poor to sell land may also allow them to "step out" of agriculture with the money raised from the sale, but on the other hand, they may harm the interests of the poor by forcing them to sell their land in distress when land prices are extremely low, making it impossible for them to rely on subsistence farming as a social safety net.

7.3.3.1.7. Livestock Services

Some rural residents depend heavily on livestock for their livelihoods; occasionally, they are used for production and revenue, and other times, they serve as assets for insurance, buffering, and accumulation. Livestock development services tend to prioritize the revenue-generating potential of livestock over low-risk, low-cost livestock keeping, which serves more "social protection" purposes like insurance, buffering, and accumulation. Nonetheless, there have been notable achievements in promoting small-scale dairying, which gives the impoverished access to both protein and income. Both production and social protection from livestock keeping have benefited from better animal health services. For pastoralists and impoverished keepers of small stock, like poultry, the value of property rights, market mechanisms, and technological advancements supporting social protection through livestock keeping is becoming more widely acknowledged. However, it's also critical to acknowledge that as rural economies expand and microfinance systems become more widely available, the value of raising livestock for insurance and savings is probably going to decline.

7.3.3.1.8. Complementary Coordination

The way the state provided (or attempted to provide) complementary coordination between (i) the various services that were individually necessary but insufficient for agricultural growth; and (ii) some social protection through food pricing and marketing interventions to reduce price volatility for producers and consumers has been a key component of state-led "social protection from agriculture" policies. Regulation and intervention in the national produce, input, and agricultural finance markets, as well as investments in infrastructure, research, and extension services—sometimes preceded by land reform—were the main sources of this complementary coordination. Special efforts were made to create coordinated service delivery systems in designated areas as part of integrated rural development projects (IRDPs).

7.3.3.2. Social Protection Independent of Agricultural Growth

The broad strategic approach that we refer to as "social protection independent of agricultural growth" describes social protection policies that were primarily promoted in the 1990s, when welfare instruments were heavily emphasized in the early stages of the development of social protection policies. This was linked to market liberalization policies that, like those in other industries, placed a high priority on growth promotion while paying little attention to the ways that agricultural development could directly provide social protection. As a result, we focus a lot of attention on social protection welfare instruments here, with a brief mention of agricultural policies.

7.3.3.2.1. The Need for Social Protection Instruments in Liberalized Agricultures

Market liberalization is one of the main agricultural development strategies that have been implemented separately from social protection measures. This entailed eliminating price controls (often by switching from fixed prices to price bands first), dismantling or privatizing parastatals, and removing tariffs and regulations safeguarding state monopolies. As previously mentioned, there is a lot of discussion regarding the political economy challenges associated with putting these policies into practice, as well as concerns regarding their effectiveness in various contexts and their inability to produce growth and price stability, especially for food staples. Ethiopia and several southern African nations have experienced extreme food price volatility in recent years; however, this is frequently not attributable to liberalization policies in and of themselves, but rather to the inconsistency of liberalization with poorly executed interventions. It is commonly acknowledged that there have been shortcomings in the development of systems for delivering inputs, and in particular fertilizer, to smallholder farmers. However, while some analysts contend that these shortcomings are inherent in liberalized systems, others counter that they are the result of partial liberalization and a lack of complementary investments in public goods, particularly roads, irrigation, and agricultural research.

Lack of access to seasonal finance for food production is a major issue in liberalized markets. Many different institutional models and financial products are trying to fill the void left by the demise of formal lending programs in Sub-Saharan Africa by meeting the needs of seasonal financing. Few of these, nevertheless, are found in rural areas with lower densities or in places with little agricultural and non-agricultural activity. None seem to be offering seasonal financing for the production of food crops or functioning under the circumstances that most impoverished farmers in sub-Saharan Africa do. While the high risks and low returns of agricultural investment in resource-poor areas tend to stifle demand, high costs and risks make such services generally unprofitable to provide in impoverished rural areas. Better-off smallholders might be able to obtain loans from microfinance institutions in regions where agriculture has already expanded and where the non-farm sector is expanding. In impoverished, undiversified rural economies, the absence of credit markets and high lending risks may be addressed institutionally by interlocking arrangements. Even though this might give buyers an unfair share of the output margin, such agreements usually only work for certain cash crops and in situations where output markets are concentrated among one or two buyers.

Increasing focus on market development and expanding smallholder farmers' access to markets has been a significant response to these challenges. This has evolved alongside a school of thought known as "Making Markets Work for the Poor." A variety of models, including innovations like the creation of fertilizer supply systems through small agro-dealers, have been developed to increase smallholders' access to services. However, the growth of farmer organizations has received the most attention.

Farmer organizations face many challenges, have a very mixed history, and members and outside supporters have varying and sometimes conflicting goals and expectations. However, the majority of prosperous farmer organizations prioritize enhancing farmers' access to supply chains for higher-value cash crops and tend to steer clear of staple food markets, where atomistic markets and low prices deter farmers from making investments in suitable coordination arrangements. By their very nature, these organizations exclude poorer producers with marginal or irregular surpluses, even though they are essential for smallholder market access and agricultural growth. Although the scale of informal transfers within the community will be constrained by the bottom line of operating a competitive business, farmer organizations may nevertheless contribute once they are established. However, farmer organizations also run the risk of having their business justification compromised by outside groups, especially government and non-governmental organizations, who might see them as easy ways to deliver social protection interventions.

As a result, market liberalization-driven agricultural policies have not specifically prioritized a social protection agenda. Nonetheless, a few insurance products have been released to encourage the expansion of agriculture. Crop insurance and ongoing (albeit sparse and sporadic) investments in land reform, infrastructure development (particularly transportation infrastructure), and technical change (through extension and research) are a few of these. These are acknowledged as being crucial for the provision of public goods required for smallholder agriculture to function in liberalized markets; however, as previously mentioned, a major contributing factor to the stagnation of African agriculture over the past 20 years is frequently attributed to a lack of investment in infrastructure, research, and extension. This can be explained by the fact that the reduction in the extent of state responsibilities, which was a key component of liberalization policies, was accompanied by a general decline in the state's ability to carry out all of its obligations, as well as a reduction in the resources at its disposal to engage in activities deemed to be within the purview of legitimate state responsibilities. Concurrently, there has been an effort to get the private sector more involved in providing research and extension services. As covered in greater detail under "social protection from agriculture" above, land reform, infrastructure development, and investment in technical change were key components of early state-led agricultural development.

7.3.3.2.2. Unconditional Cash Transfers (UCTs)

UCTs are "unconditional transfers of cash made by government or non-governmental organizations to individuals or households identified as highly vulnerable, to alleviate poverty, providing social protection, or reducing economic vulnerability" and include social pensions and child support grants. The principal connections to agriculture reside in 1) the ability of assisted households to participate in the rural economy as a) consumers and b) producers, and 2) the uses to which the transfers are put either a) directly, or b) indirectly, through intra- or interhousehold fungibility. We will deal with the multiplier and incentive effects of transferring cash (as opposed to food or inputs) later. Here, we point out that (i) UCTs are cheap; (ii) UCTs' poverty impact is highly sensitive to design; and (iii) UCTs can have behavioral effects.

The low administrative costs of UCTs in comparison to other programs are a major benefit. Cash transfers are generally less expensive than in-kind inputs; however, as with other tools, costs rise significantly with the target population's distance and the strictness of the targeting. Because UCTs lack a complementary behavioral intervention, they are less expensive than conditional cash transfers. To lower the risk of inflationary effects, some supplementary work might be required, which could raise expenses.

The size and mode of transfer determine how UCTs affect poverty. For a fixed budget, there are undoubtedly significant trade-offs between the effects on the extent of poverty (maximized by poverty-line targeting and low-value transfers to many) and the severity of poverty (maximized by strict targeting and high-value transfers to a few). The ability to target, political economy concerns, the traits of various target groups, and the presence and type of threshold effects in and on local markets and individual livelihoods all play a significant role in resolving these trade-offs. For example, because many elderly people in Southern Africa live with and care for their grandchildren, old age pensions are occasionally thought to be effective in lowering the vulnerability of all age groups. To evaluate the broader effects of pensions on, say, HIV/AIDS orphans, a thorough examination of poverty demographics is necessary, as the majority of adults and children are likely to live in households without elderly members.

7.3.3.2.3. Food Aid

Food assistance is essential during humanitarian crises, especially when it comes to preventing consumption cuts when households would rather not sell assets, lowering household vulnerability, and easing temporary hunger. It is particularly crucial for populations impacted by HIV/AIDS and can also have major positive health effects. There is no question about the potential for food aid to have a significant positive impact.

Barrett and Maxwell (2005) note five potential positive effects of food aid in stimulating local livelihoods:

- Relief of short-term borrowing constraints that prevent farmers from buying agricultural inputs.
- Safety net provision, allowing producers to take on more risk.
- Reduction of food import, freeing up foreign exchange for import of other inputs.
- Prevention of irreversible health problems, contributing to a healthy labor force.
- Increased transport capacity in response to demand.

All interventions that routinely provide vulnerable groups with food or money (where food is available) experience effects 1 and 2. Although effects 3, 4, and 5 may be more specific to food aid, their importance is unclear, and the overall impact on factor and product markets may be unclear. Therefore, careful planning and resources are needed to create food aid interventions that are effective without destroying livelihoods. For populations affected by HIV/AIDS, food assistance can be especially crucial. Food security and HIV/AIDS interact in intricate ways. Food insecurity and malnutrition are made worse by HIV/AIDS, and they can also make people more vulnerable to AIDS and HIV. Although there is conflicting evidence, HIV has been identified as a primary cause of the food crisis in Southern Africa. People living with HIV/AIDS may also have high dependency ratios and struggle to produce enough food. Additionally, PLWHA requires more nutrients to fend off illness and boost the efficiency of antiretroviral medications. Although targeting PLWHA entails significant resource demands, Kadiyala and Gillespie (2003) contend that HIV/AIDS indicators ought to be included in food aid targeting (as in the WFP's VAM) (De Waal and Whiteside, 2003).

The impact of food aid on agricultural production is significantly influenced by its source. Like any food-based intervention, local production may be impacted in both positive and negative ways. Food aid is often associated with external sourcing and harm to local factors and output markets because it is typically thought of as being meant for areas experiencing food shortages. This poses a serious risk. Additionally, food aid may alter consumer preferences (for example, from white maize grown in Africa to yellow maize grown in Western industrialized nations), which could jeopardize the sustainability of food production. These factors make a strong case for triangular sourcing, which involves buying food from a neighbor or another region of the destination nation. This necessitates surplus food supplies in the sourcing areas as well as inadequate market integration between the source and destination markets.

There are three reasons to be wary of triangular sourcing of food aid, especially when it comes to agricultural livelihoods, though this may be the case more frequently than is generally acknowledged.

A dependence on rain-fed agriculture is extremely risky, in terms of availability, capacity, and quality. It is important to have alternative back-up sources which include

- Poor integration of markets may reflect infrastructural or institutional problems that;
- complicate the management of food aid; and
- Buying large quantities of food affects not only local sellers (positively) but also local buyers (negatively), and it is important that attempts to address a food crisis of one vulnerable group do not damage other vulnerable people.

One strategy to guarantee a prompt response to food aid and to stabilize grain prices to lower the risk to producers and consumers is to maintain national food stocks. Nevertheless, there has been conflicting practical experience, and there are worries that they: (i) displace private storage; (ii) have major management issues; and (iii) fail to meet price targets, and iv) are expensive.

However, Barrett and Maxwell (2005) contend that the 2002–03 food crisis was exacerbated by Malawi's Strategic Grain Reserve's depletion, while Poulton et al (2006) contend that if appropriate management can be guaranteed, local storage may offer additional noteworthy benefits and that fiscal costs for landlocked nations may be lower than those of imports.

7.3.3.2.4. Public Works

To (i) increase employment; and (ii) create an asset, PWPs are defined as "activities which entail the payment of a wage in return for the provision of labor, with the overall objective of promoting social protection." The main goal of previous PWPs was to increase employment (through Employment Based Safety Nets, or EBSN). There is currently a policy conundrum between these two goals as useful asset creation—through Labor Based Infrastructure Programs, or LBIPs—becomes more and more important. The perceived trade-off between growth and welfare impacts is a major topic in PWP literature.

7.3.3.2.5. Conditional Cash Transfers (CCTs)

Although conditional cash transfers aim to alter the structure of household incentives to encourage specific behavior that aligns with broader social goals, they are likely to have multiplier effects that are comparable to those of unconditional cash transfers. The most common areas of intervention are agriculture, children's nutrition, healthcare, and education. The livelihoods of farmers may be significantly impacted by these interventions. Where there are direct links to (conditionality of transfers on) agricultural activities (affecting, for instance, distribution of access to and the scale of transfers) and where induced behavioral changes have significant effects on agricultural activities, there may be connections between CCTs and agriculture that go beyond those found with UCTs. However, other limitations that agricultural households face, as well as their access to markets and other services, frequently also affect behavioral changes and their outcomes. However, little is currently known about the nature and scope of these links to agriculture, and what is known is mostly centered on program experience in Latin America. Generalizing from Latin American experiences to less developed African nations with less developed markets, institutions, and services at an earlier stage of agricultural transformation calls for caution.

However, it is unclear how well CCTs have performed in reducing poverty, even in Latin America. The components of CCTs that deal with behavioral change and poverty reduction most likely have to be traded off. When the transfer is not dispersed evenly, targeting also becomes a crucial problem. It's possible that CCTs are less cost-effective than UCTs due to their higher administrative costs if the intervention's main goal is to reduce poverty.

7.3.3.2.6. Food For Education (FFE)

FFE programs comprise school feeding interventions, where children at school are provided with food, and food for schooling, where children are given rations to take home. The intended impacts are primarily a) behavioral (concerning educational attendance and gender relations), and b) nutritional, but there are also effects on c) agricultural production.

The effects on behavior are widely known. It has been discovered that school feeding and food for schooling initiatives both improve enrollment and attendance, lower repeater and dropout rates, and close gender gaps. Although there isn't much evidence to support this or better nutrition under normal conditions, some studies have also found enhanced cognitive development and learning capacities. However, FFE interventions are most effective for highly vulnerable groups, such as children who are stunted, wasted, or undernourished, or groups that are experiencing periods of hunger. The primary determinant of the impact on agricultural output is the food source. If food can be bought locally or complements local produce, there might be major effects on market growth and agricultural outputs (Grantham-MacGregor et al.,1991).

7.3.3.3. Social Protection for (Inter Alia) Agricultural Growth

There are synergies even though there are occasionally trade-offs between agricultural growth and welfare outcomes in "welfare" social protection interventions. However, other forms of social protection interventions may not have a clear welfare focus and are much more directly focused on agricultural growth.

7.3.3.3.1. Insurance

Producers can take chances and engage in riskier and more lucrative endeavors thanks to insurance. Therefore, farmers can use more productive technology and participate in more specialized and intensive production through social protection interventions that support risk management through insurance. Crop and livestock insurance (against production or price risks) as well as savings and credit services are examples of explicit insurance instruments that are pertinent to this discussion.

Microfinance and crop insurance both face significant obstacles in the context of resource-poor, diverse, and complex agriculture. The diversity of production conditions and output in smallholder agriculture in Sub-Saharan Africa, even across small areas, presents significant challenges to such a system, even though area-based index insurance, such as regional rainfall insurance, may be able to address many of the traditional issues with crop insurance in smallholder agriculture (such as moral hazard, adverse selection, and high administration costs). Though this experience typically does not apply to staple crops in developing agricultural economies, there is increasing experience with systems that use index insurance and/or risk layering to facilitate risk transfer. Therefore, for many rural poor people, self-insurance through official or informal saving methods frequently offers the only explicit risk management tool available.

However, there are other ways to lower risk. Because producers are aware that they will be eligible for welfare transfers if their crop fails and their incomes decline, the established and consistent availability of "safety-net" transfers for the impoverished, such as food assistance or unconditional cash transfers, can serve as insurance. The extent to which insurance influences risk-taking behavior beyond the safety-net income level, however, and the magnitude of these effects are not well understood. Policies that encourage different livelihood strategies—such as crop diversification, seasonal migration, and the accumulation of small assets—can also lower risk and boost output. Policies that support institutional stability and macroeconomic stability also lower risk and boost incentives for production.

7.3.3.3.2. Building Resilience

Resilience-building tools directly increase rural productive capacity and unleash otherwise latent or constrained productivity (e.g., through asset creation, social inclusion, and improved access). Transfers could be made in exchange for capital creation through public works projects or food for education, or they could be made to sustain human productivity over the long term by supplying nutrition at key developmental stages or encouraging health care seeking. The ability of resilience-building transfers to help recipients break free from micro-level poverty traps is one of their main purposes. Therefore, transfers must be substantial enough to raise people or households above important asset thresholds. To gain a deeper comprehension of poverty traps and the type and magnitude of transfers required to help recipients cross crucial thresholds, more research is needed.

The goal of "transformative" social protection is to lessen exclusion, which could lead to more equitable access and cooperative measures, both of which have been linked to positive growth effects in recent studies. In the areas of land reform and producer collective action, there are evident connections between agricultural growth policies and "transformative" social protection goals. Additional research on these and other links would be beneficial. It's also important to acknowledge that social protection policies may have detrimental effects on growth, primarily through crowding out.

7.3.3.3.3. Public Works

If the resources produced by PWPs have substantial and pro-poor output effects, they can significantly contribute to the reduction of poverty. There is compelling evidence that labor-based methods are more cost-effective when it comes to building physical infrastructure. However, as Devereux points out, while it is frequently argued that successful PWP can lessen future reliance on social protection, this has rarely been realized in reality. The creation and upkeep of rural infrastructure in Maharashtra, as well as improvements to irrigation, pro-poor assets, roads, and water conservation in Bangladesh, are examples of the evidence that PWPs can be successful in producing agricultural assets, according to Farrington et al. However, success is not a given. It is not that PWP cannot be used to create productive rural assets; rather, PWP a) reduces welfare impacts due to the high cost of non-wage items like training or materials, b) necessitates effective implementation, which is expensive, and c) has issues with unskilled labor. A distinction between short-term and long-term horizons is one that policymakers may make to resolve this conflict. Accepting a smaller short-term wage transfer could result from a stronger emphasis on longer-term opportunities for asset creation and maintenance. To manage this focus, political commitment and robust institutions are probably essential.

Analysts have observed a pro-rich bias in asset creation, even in the more successful instances of PWP asset creation, like the Maharastran EGS. The nature of agricultural growth that results from the creation of specific assets is significantly impacted by the fact that technical change is not distribution neutral.

Participants in PWPs face significant costs, both in terms of opportunity and direct costs. Lipton (1988) calculated opportunity costs of 20–30% of wages in South Asia, while Maxwell (1993) estimated direct participation costs of up to 1000 calories per day. The timing of the intervention can have a significant impact on agricultural output. However, labor-constrained households—which frequently consist of the most vulnerable members of communities—also face difficulties. This problem is especially important in areas where a high prevalence of HIV/AIDS leads to high dependency ratios and makes hard physical labor unfeasible for many households. There are serious concerns regarding PWPs' potential as useful social protection tools for sizable HIV/AIDS-affected populations, even though some creative, small-scale, and resource-intensive programs have been successful in tackling this issue. PWPs have an indirect impact on local labor markets through multipliers and a direct impact on labor supply. Although poorly timed PWPs can have a direct negative impact, well-designed PWPs can raise wages by decreasing the labor supply and increasing workers' options and bargaining power. The project design mediates this upward pressure: Labor Based Infrastructure Programs (LBIPs) have low employment coverage, and very low wages in Employment Based Safety Nets (EBSNs) generate very little upward pressure. However, Devereux notes that because of consistent work and comparatively high incomes, participants in an LBIP project in Zambia bought seeds and fertilizer and employed agricultural labor. In addition to 1) creating a sizable employment multiplier, 2) relieving women participants of labor-intensive agricultural duties, and 3) most likely increasing agricultural output. PWPs frequently draw roadside vendors on paydays, which results in the generation of particular multipliers.

Lastly, the cost of EBSNs to the poor is higher than that of cash transfers; in Malawi, PWP costs 13.9 per unit, while cash transfers cost 1.73. This appears to necessitate choosing between PWP for assets and PWP for welfare, given the low capacity of implementing agencies, the high costs for participants, and the dubious value of the assets generated. However, as was already mentioned, effective LBIPs can significantly reduce the poverty of low-income participants while also generating productive assets and other beneficial multiplier effects on local economies (especially if the wages are paid in cash) that benefit nonparticipants. These factors point to a possible synergy between growth and poverty alleviation, albeit at the price of a clear emphasis on social protection.

7.3.3.3.4. Inputs for Work Programs

Inputs for works programs are one type of public works program that has especially strong ties to agriculture. These have many of the same characteristics as PWPs as previously mentioned, but instead of giving participants money or food, they give them agricultural production inputs, frequently in the form of vouchers. Inputs for work programs may be criticized for limiting participants' options for how to spend their earnings when compared to cash-for-work programs. However, because they are a coerced method of saving money, trial inputs for work programs in Malawi have been very well-liked by participants. These programs may also expand networks of input suppliers, depending on how they are designed. The benefits for participants are postponed, contingent on land access, and vulnerable to the typical risks associated with agricultural production in the subsequent season unless they sell inputs or vouchers. Delivery of the input at the right time is also crucial.

7.3.3.4. Social Protection Through (Inter Alia) Agriculture

Although this strategy is closely related to the aforementioned categories A and C, it differs from them in that it focuses primarily on providing social protection welfare (instead of insurance or resilience) in the short term through poor people's contributions to their own production. It can also be differentiated from strategy A (social protection from agriculture) by the lack of a longer-term, more comprehensive framework that addresses complementary coordination issues. However, this distinction might not always be obvious. Targeted input programs are the main tool in this strategy, but we could also argue that some aspects of land redistribution, work program inputs, and even cash transfers fall under this category (cash transfers, for instance, when they alleviate significant cash flow and consumption constraints to enable people to cultivate their own land instead of seeking off-farm employment).

7.3.3.4.1. Targeted Inputs Programs (TIPs)

TIPs are intended to lessen the susceptibility of populations to food shortages by supplying agricultural production inputs. As a result, they directly connect the promotion of agricultural livelihoods and social protection. These interventions may have a significant impact. A well-known and extensive example is the Starter Pack in Malawi, which is said to have raised participants' output by 100–150 kg/pack. However, there are many cross-cutting issues surrounding TIPs, and the effectiveness of TIPs is currently not entirely clear.

First and foremost, it is critical to comprehend how the TIP will accomplish the program's goal and how these factors interact with the program's design, sustainability, and dedication. For example, the goal of Malawi's recent Starter Pack program was to (i) increase national aggregate food security by achieving self-sufficiency in maize production; and (ii) give impoverished households a crucial boost in food access and production. Since the crucial contribution of starter packs to food security was not increased production by target households but rather (i) lower food prices; and (ii) higher labor demand and wages due to the widespread cultivation of starter packs by both poor and less poor households, the 2000 scale-down of starter packs from universal distribution to small-holders to reduce costs and increase cost-effectiveness may have contributed to the 2002 food crisis. Targeting the food insecure was a major challenge for the reduced, targeted Starter Pack as well.

Whether TIPs displace private sources of agricultural inputs is the second concern surrounding them. The record is not entirely consistent. Nyirongo (2005) contends that this was only partially true for seeds and not true for fertilizer with the Malawian Starter Pack. According to Sperling et al. there is mounting evidence that repeated seed assistance under conditions of chronic stress distorts farmers' own procurement strategies in Malawi and Kenya, impairs the operation of local seed and grain markets in Burundi, and jeopardizes the growth of more commercial seed supply systems in Zimbabwe. The crucial question for design is whether poverty or a lack of seeds causes seed insecurity. Sperling et al. contend that since poverty is typically the issue, local seed markets ought to be encouraged. They contend that rather than direct seed assistance, which tends to weaken local markets, seed vouchers or fairs are more suitable in this situation. If there are local sources, a misdiagnosis of the issue that results in a default assumption of seed unavailability can be very harmful.

Third, it is critical to understand that technical change is subject to change. The distributional effects of input provision are significant for growth patterns and future inequality in addition to social protection goals. Because benefits are obtained at harvest, unless inputs are sold on, input provision for TIPs does not immediately and directly benefit the cash or land-constrained poor. However, if the intervention is implemented on a large scale, the majority of impoverished households should benefit from lower food prices as well as increased wages and/or opportunities for earning wages. Patronage and political systems also have an impact on input access. For households that cannot directly benefit from additional inputs, alternative forms of social protection are therefore required, and policymakers must be extremely cognizant of distributional, social, and political issues. In this context, TIP design—in particular, targeting—is essential.

7.4. DESIGN AND IMPLEMENTATION ISSUES

The selection of suitable social protection and agricultural development tools is contingent upon the design and execution of each tool.

7.4.1. Type of Transfer – Cash, Food, Inputs, Vouchers

Numerous instruments covered above necessitate selecting between different transfer types. For example, PWP accepts cash, food, or inputs as forms of payment. To some degree, food aid and UCTs can be viewed as a trade-off between cash and food transfers. The context in which the intervention occurs greatly influences the suitability of various transfer forms. Cash transfers, for example, will have a negative (inflationary) effect when the supply of food is completely inelastic. However, supply elasticity depends on the size of the demand (larger scale transfers will both exert greater inflationary pressures and increase incentives for market players to invest in supply to meet increased demand) and the existence and efficacy of markets to respond to increased demand for goods and services (which in turn depends upon infrastructure, market institutions, trader access to credit, etc.).

The characteristics of each transfer type's relative fungibility—cash being the most fungible—can also be used to differentiate between them. These factors consist of:

- Multiplier effects on local economies (food and inputs may have some multiplier effects, especially on the areas from which the food or inputs come, but cash is generally thought to have stronger multipliers).
- Crowding out (or in) of informal intra- or inter-household transfers (the evidence here is mixed and further research is needed).
- Effects on incentives (for example food aid may disincentivize food production whereas inputs encourage it).
- Corruptibility (cash is more easily looted and more desirable to the non-poor, leading to delivery problems, but food can also be misappropriated).
- Fungibility and liquidity (these are normally considered as desirable characteristics, but as sometimes the poor may prefer less liquid and fungible transfers to assist them with forced savings).
- Specific program objectives (food may achieve better nutritional outcomes than cash, while inputs may have stronger effects on agricultural productivity).
- Delivery costs (cash is usually assumed to be the cheapest to deliver, but risks of diversion and costs of providing security must be taken into account).
- Ease of targeting (cash is thought to be less easy to target because it is attractive to the nonpoor, where certain types of food or quantities and types of input may not be; where the transfer is earned, as in public works, the choice of transfer can be used as a targeting mechanism).
- Gender effects (women tend to control food provision and men cash).

The specific structures, opportunities, and limitations in various people's livelihoods; the overall level of economic activity in the local area; the operation of various markets; and the intervention's goals all influence the relative benefits and drawbacks of various transfer types.

7.4.2. Timing

For seasonal agriculture, which faces vulnerabilities and production opportunities that change over time, the promptness of social protection interventions is especially crucial. However, incentive and crowding-out effects also vary over time because labor market and goods market contexts also change seasonally. Different instruments also have different seasonal effects.

Food assistance can be scheduled to 1) provide a macro-level counter-cyclical transfer at times when recipients are most vulnerable, such as during emergencies or before harvest. There are several issues that arise are:

- Lags between commitment and delivery due to complex logistics. The median lag for even urgent shipments is 139 days. As a result, the cost of effective early warning is greatly increased.
- Food aid volumes co-vary negatively with international prices, so food aid volumes are more volatile than food production or trade.
- The provision of food aid suffers from high inertia.
- In addition to having a major negative impact on markets, mistimed food assistance may not be able to stop asset sales or malnutrition. When food aid is delivered after the supply is no longer inelastic, it can have a significant negative impact on prices by lowering agricultural output, producer incomes, supply chain development, and production incentives while raising producer risks.

According to Lipton (1998), the impact of seasonal public works on poverty can be "doubled or destroyed" by private labor market reactions. PWP must be planned to align with the slack season in rural areas due to the real worry that it may interfere with agricultural operations and force workers off the fields if it occurs at the wrong time. However, it can also allow participants to hire others in their fields, and food and cash needs are frequently highest during periods of high demand for agricultural labor.

Thc provision of inputs at the right time is also crucial. Inputs obviously need to arrive before planting season. Selecting the right season may also be crucial in multi-season farming. It was determined that limited financial resources should be focused on a single season because winter starter packs in Malawi only gave households an additional half-months' worth of maize at most.

7.4.3. Scale

The scale of the intervention becomes crucial in terms of the size of the transfer and the number of beneficiaries, especially in light of the above recommendations regarding possible thresholds in agricultural growth at the livelihood and local economy/market levels. Although more research is needed in this area, social protection interventions that prevent falls below or raise households above these critical levels could be designed if we could better understand the critical characteristics of households on either side of these thresholds. In a similar vein, comprehending the elasticities of labor and food supply and demand as well as the production and consumption traits of various household types is essential to comprehending scale, multiplier, and targeting effects.

7.4.4. Conditionality

Positive or negative behavioral effects are a part of every social protection transfer. As was previously discussed, conditional transfers can produce the intended behavioral effects. But it's important to keep in mind that conditionality frequently has unforeseen consequences and may compromise other goals.

7.4.5. Stability/Reliability of Payment over Time

The ability of welfare programs to provide insurance benefits is significantly impacted by their stability. People can only transition from low-risk, low-return activities to higher-risk, high-return activities if they can rely on welfare programs to help them in the event of a crisis. The transfer's stability and people's confidence in their capacity to access it when needed are crucial concerns. This significant potential benefit of welfare programs has been seriously undermined in many African nations by their inconsistent and frequently delayed responses to food insecurity.

7.4.6. Targeting Social Protection in Agriculture

One strategy to lower administrative costs in rural communities is to target social protection to specific groups, but there are several drawbacks to this approach. Although the majority of these are not unique to agriculture, they nevertheless have a significant impact on how social protection tools and agriculture interact:

- Bias against remote areas. In many communities, more isolated groups are more in need of social protection because they are more vulnerable. Targeting and distributing it to them, however, might be more difficult, particularly in places with low population densities. People who must travel to pick up transfers may incur significant opportunity and direct costs.
- Bias against individuals not clearly part of any household, such as orphans or the homeless.
- Stigmatization and exclusion of target groups as, for example, use of low wages to encourage self-selection by the very poor in PWPs may contribute to the exclusion, marginalization, and maintenance in poverty of those employed.
- Resentment among untargeted groups who are denied transfer benefits.

- Exclusion and inclusion errors as, for example, geographic targeting of food aid can exclude vulnerable households outside drought prone areas. (Clay et al., 1999).
- Inappropriate definition of group membership, leading to targeting of the 'wrong group.'
- Political and cultural factors that (i) make it difficult to match 'community' perceptions of vulnerability to 'objective' indicators more amenable to policymakers; and (ii) lead to political patronage biasing the distribution of social protection.

Large-scale targeting is especially challenging because of these issues. Targeting challenges specific to agricultural-related social protection include potential targeting-related administrative delays, issues when a sizable portion of rural populations live in poverty, and the significance of large-scale interventions in providing multiplier, food, and labor market benefits.

Prioritizing resources in agricultural growth policies is another facet of targeting. With complicated short- and long-term trade-offs, public spending on agricultural research, infrastructure investments, and subsidies for services, inputs, and other technologies must be prioritized to maximize results and meet a variety of political or patronage agendas. Although short-term political agendas may also push the interests of poorer smallholders up the agenda, such decisions will frequently be biased towards the higher investment and patronage returns offered by high-potential areas and organized commercial interests (Clay et al., 1999).

REFERENCES

1. Byerlee, D., Jayne, T. S., & Myers, R. J. (2006). Managing food price risks and instability in a liberalizing market environment: Overview and policy options. *Food Policy, 31*(4), 275–287.
2. Clay, D., Molla, D., & Habtewold, D. (1999). Food aid targeting in Ethiopia: A study of who needs it and who gets it. *Food Policy, 24*(4), 391–409.
3. De Waal, A., & Whiteside, A. (2003). New variant famine: AIDS and the food crisis in Southern Africa. *The Lancet, 362*, 1234–1237.
4. Dercon, S. (2002). Income risk, coping strategies and safety nets. *The World Bank Research Observer, 17*(2), 141–166.
5. Dorward, A. R., Kydd, J. G., Morrison, J. A., & Urey, I. (2004). A policy agenda for pro-poor agricultural growth. *World Development, 32*(1), 73–89.
6. Dorward, A. R., Kydd, J. G., & Poulton, C. D. (2005). Beyond liberalization: "Developmental coordination" policies for African smallholder agriculture. *IDS Bulletin, 36*(2), 80–85.
7. Grantham-McGregor, S., Powell, C., Walker, S., & Himes, J. H. (1991). Nutritional supplementation, psychosocial stimulation, and mental development of stunted children: The Jamaican study. *Lancet, 338*(8758), 1–5.
8. Hardaker, J. B., Pandey, S., & Patten, L. H. (1991). Farm planning under uncertainty: A review of alternative programming models. *Review of Marketing and Agricultural Economics, 59*, 9–22.
9. Jayne, T. S., Govereh, J., Mwanaumo, A., Chapoto, A., & Nyoro, J. K. (2002). False promise of false premise? The experience of food and input market reform in Eastern and Southern Africa. *World Development, 30*(11), 1967–1985.
10. Jayne, T. S., & Jones, S. (1997). Food marketing and pricing policy in Eastern and Southern Africa: A survey. *World Development, 25*(9), 1505–1527.
11. Jayne, T. S., Zulu, B., & Nijhoff, J. (2006). Stabilizing food markets in eastern and southern Africa. *Food Policy, 31*(4), 328–341.
12. Joshi, A., & Moore, M. (2000). Enabling environments: Do anti-poverty programs mobilize the poor? *Journal of Development Studies, 37*(1), 25–56.
13. Lipton, M. (1993). Land reform as commenced business: The evidence against stopping. *World Development, 21*(4), 641–657.

14. Maxwell, S. (1993). Can a cloudless sky have a silver lining? The scope for an employment-based safety net in Ethiopia. In H. Thimm, H. Hahn, I. Reinhard, & D. Schulz (Eds.), *Regional Food Security and Rural Infrastructure* (pp. 101–119). Verlag Munster.
15. Poulton, C. D., Kydd, J. G., Wiggins, S., & Dorward, A. R. (2006). State intervention for food price stabilization in Africa: Can it work? *Food Policy, 31*(4), 342–356.
16. Rodenstein-Rodan, P. (1943). Problems of industrialization of Eastern and Southeastern Europe. *Economic Journal, 53*(210, 211), 202–211.
17. Salama, P., et al. (2001). Malnutrition, measles, mortality, and the humanitarian response during a famine in Ethiopia. *Journal of the American Medical Association, 286*(5), 563–571.
18. Townsend, R. M. (1995). Consumption insurance: An evaluation of risk-bearing systems in low-income countries. *Journal of Economic Perspectives, 9*(3), 83–102.

CHAPTER 8

PROMOTING AGRICULTURE FOR SOCIAL PROTECTION OR SOCIAL PROTECTION FOR AGRICULTURE

Contents

INTRODUCTION

In the agriculture industry, labor unions and worker organizations are essential because they fight for the welfare, safety, and rights of agricultural workers. Low pay, hazardous working conditions, a lack of benefits, and job security are just a few of the issues that these organizations aim to address. Their roles are examined in detail below:

1. **Advocacy for Workers' Rights:** Labor unions represent agricultural workers collectively and fight for better working conditions and equitable pay. They negotiate contracts that set higher wages, job security, health benefits, and safety standards by engaging in collective bargaining with employers. For example, the United Farm Workers (UFW) have successfully negotiated contracts that protect 1000s of farmworkers by using boycotts and strikes to fight for labor rights.
2. **Efforts to Organize:** Unions strive to unite agricultural laborers into unified teams capable of tackling workplace problems as a group. This organization is essential because, because farms and rural areas are so dispersed, many agricultural workers are frequently isolated. Workers' demands for improved treatment from employers can be strengthened by unions when they band together behind a common goal. An example of how organizing efforts can result in notable changes to labor relations in agriculture is the UFW's founding in the 1960s.
3. **Legal Protections and Representation:** Unions offer alternative forms of support, even though federal laws such as the National Labor Relations Act (NLRA) frequently deny protections to agricultural workers. When possible, they promote state-level protections and assist in navigating the legal complexities surrounding employment rights. California, for instance, has passed legislation providing farmworkers with certain protections not found at the federal level.
4. **Resolving Health and Safety Issues:** Because agricultural labor is notoriously dangerous, unions place a high priority on health and safety laws to safeguard their members. They push for advancements in workplace safety procedures and support adherence to Occupational Safety and Health Administration (OSHA) standards. By establishing employment guidelines targeted at protecting migrant workers, the Migrant and Seasonal Agricultural Worker Protection Act (MSPA) also contributes to this.
5. **Assistance for Undocumented Workers:** Many agricultural laborers are undocumented immigrants who encounter particular

difficulties with regard to their rights at work. Despite federal law restricting their ability to formally organize, unions have started to advocate on their behalf as they have come to understand the significance of this group within the workforce. By combating wage theft and hazardous working conditions, groups such as WeCount aim to assist low-wage immigrant workers.

6. **Political Advocacy:** To influence local, state, and federal laws that impact the rights of agricultural workers, unions also participate in political advocacy. They advocate for changes that would enhance working conditions or broaden farmworker protections, like the Farm Workforce Modernization Act, which attempts to address labor shortages and expedite immigration procedures.
7. **Community Engagement:** In addition to having direct talks with employers, unions frequently run campaigns that highlight workers' fights against unfair treatment or exploitation to increase public awareness of agricultural labor issues. By involving the community, labor initiatives gain public support.

The rights of agricultural workers, the improvement of working conditions through collective bargaining, the provision of legal representation when feasible, the maintenance of health and safety regulations, the support of the rights of undocumented workers, political engagement on behalf of their members, and the development of public awareness of these important issues are all made possible by labor unions and worker organizations.

8.1. HISTORY OF LABOR UNIONS IN THE AGRICULTURE SECTOR

Agriculture was one of the last sectors to resist unionization because of social and legal barriers until the United Farm Workers (UFW) achieved success in the 1960s. Unfair legal protection, discrimination, dependence on foreign, exploitable labor, and resistance from state and federal representatives who either worked for agribusiness or were large agricultural landowners themselves were all challenges faced by workers and organizers. U.S. laborers, agricultural, and otherwise, continue to face these challenges today. Despite union victories of the 1800s and 1900s, such as child labor laws, the eight-hour workday, and the five-day workweek—i.e., the creation of the weekend—there are still movements aimed at undermining workers' rights in the United States. Despite being frequently disregarded, agricultural labor's history should be studied, particularly in light of COVID-19's revelation of historical problems that persist in agriculture and other facets of society (Figure 8.1).

Figure 8.1. *"Not all essential labor is equally celebrated in our society. Have you seen celebrations for those who provide less glamorous essential services, like farm work?"*

Source: https://americanhistory.si.edu/explore/stories/essential-and-expendable-rise-agricultural-labor-and-united-farm-workers.

Because they frequently come from underprivileged backgrounds, agricultural workers are extremely susceptible to unreported abuse and exploitation, which has hindered unionization. Workers lived and worked at the whim of growers and agribusinesses because they were marginalized and their jobs were remote and rural. If there were labor laws, they were frequently not enforced, and there were no watchdog groups concerned with the treatment of agricultural workers. In the South, large-scale organization was impossible due to sharecropping and the structural and racial legacy of slavery. To the detriment of Black and non-white citizens, lynching, segregation, and other forms of racial terror and policing practices preserved the status quo. Deportation was used as a threat for both legal and illegal workers, as many agricultural laborers in the West were immigrants. Agricultural workers were not included in the legal protections afforded to unions and collective bargaining, even when these rights were established in 1936 with the passage of the National Labor Relations Act.

Agricultural organizations before the UFW were irregular and met with extreme hostility. Between 1931 and 1941, there were roughly 30 attempted strikes in the San Joaquin Valley of California, but growers and local law enforcement brutally put an end to them. With the assistance of expert organizer Luisa Moreno, 20-year-old Tejana Emma Tenayuca successfully led a pecan shellers' strike in San Antonio, Texas, in 1938. Moreno collaborated with several unions before being compelled to leave the US in 1950 due to threats of deportation. The Latina labor activist was criticized for posing a communist subversive threat to the nation. Nevertheless, subsequent generations of activists and organizers were influenced and motivated by these movements and their leaders.

Effective unionization was also thwarted by the Bracero Program. Between 1942 and 1964, the United States and Mexico entered into agreements that brought in roughly 200,000 braceros, or temporary Mexican agricultural laborers, each year. Growers mistreated and underpaid braceros, arbitrarily withheld their pay, and threatened to deport them for protesting, despite legal promises. The Bracero Program ended, in part, because U.S. leadership was forced to act on the reality that the presence of the exploited braceros depressed the earnings of U.S. farm labor to the sole benefit of growers. In response, growers made an unsuccessful attempt to replace braceros with mechanization. The fact that unionization expanded throughout the agricultural sector in the 10 years following the conclusion of the Bracero Program is no accident (Figure 8.2).

Figure 8.2. *These photos show how braceros were physically inspected, en masse, in degrading conditions and forced to pay for cramped, shabby living quarters which provided limited comfort after hours of intense physical labor in the hot Texan or Californian sun.*

Source: https://americanhistory.si.edu/explore/stories/essential-and-expendable-rise-agricultural-labor-and-united-farm-workers.

Agricultural unions like the Agricultural Workers Organizing Committee (AWOC), the Agricultural Workers Association, the Filipino Farm Labor Union, and the National Farm Workers Association (NFWA) emerged in the late 1950s and early 1960s. In 1966, the NFWA merged with other unions to form the United Farm Workers. They called for better wages and working conditions. There was widespread child labor. The housing that the underpaid laborers were forced to live in at exorbitant prices frequently lacked cooking and plumbing facilities, and growers frequently neglected to provide restrooms for their employees. There were serious health risks due to overwork and a lack of safety. In stark contrast to the 67-year national average, the average life expectancy of farmworkers in the 1960s was 49 years.

The boycott and strike of grape growers in Delano, California, from 1965–1970 was the UFW's first significant strike. Larry Itliong started the AWOC strike with more than 1,000 Filipino farmworkers. In an attempt to set up newly hired Mexican laborers against Filipino laborers, grape growers turned to Itliong and the NFWA for assistance. The nonviolent demonstrations led by Martin Luther King Jr. and union leaders like Chavez, Itliong, and Dolores Huerta drew inspiration from Mahatma Gandhi for their tactics and actions. In addition to organizing grassroots and community boycotts of grapes and grape products across the United States, the UFW led nonviolent picketing of the grape fields, persuaded strikebreaking workers to join the strike, and marched 300 miles to Sacramento. The first-ever agreements between growers and an agricultural union were the outcome of these extremely successful efforts. These agreements improved working conditions and increased wages by 40% over the standard Bracero wage, which is still in use today. Some of the benefits included paid time off and insurance.

The UFW still exists and defends the rights of farmworkers, even though the museum's collection focuses on Cesar Chavez and the organization's activities in the 1960s and 1970s. The UFW has fought against sexual harassment, wage theft, and other issues, and its membership has doubled since 2000. COVID-19 has recently exposed the lack of comprehensive protections for farmworkers: many agribusiness companies only provide masks to supervisors and do not enforce safety measures, despite studies showing that farmworkers are among the most vulnerable to COVID; many cases of farmworkers being fired for protesting after worker COVID outbreaks; laborer families earn less than $20,000 annually on average; and farmworkers frequently do not receive sick leave, with 65% of workers lacking health insurance.

Due to the nature of their work, farmworkers are isolated and out of the spotlight, unlike employees in many other vital fields. Nearly all of the 72% of foreign-born people in this highly racially segregated field are from Latin America. This investigation into the origins of agricultural unions is significant because it shows the extent of historical injustices and legacies, as well as how challenges and problems from "back then" continue to influence our world today. How is worker value established if their labor is necessary but they are not? COVID-19 has made these problems worse, necessitating much more thought and discussion.

8.2. WORKERS AND THEIR TRADE UNIONS IN AGRICULTURE

To safeguard and advance their livelihoods and labor rights in the workplace, workers founded and managed trade unions, also known as labor unions. Patronage has no place in a free trade union. It is not a workers' organization that was openly or covertly founded by an employer to regulate the demands or actions of the workforce, nor is it an organization that the ruling class or elite supports for the same ends. Trade unions strive to improve the pay and working conditions for their members as well as to guarantee their job security through collective bargaining with employers. To uphold and enhance labor rights, standards, and policies, unions also collaborate with governments and lobby them.

Unions are funded by the dues paid by their members. Delegates chosen by the members democratically decide union policies at conferences and congresses. Additionally, unions give employees a voice in broader political, social, economic, and ethical matters. The following are examples of agricultural trade unions, federations, and confederations.

8.2.1. National Level

Only agricultural workers on commercial farms and plantations are members of national agricultural trade unions. In developing nations where agriculture continues to be the main economic activity, this kind of union is typically found. This type of union is normally found in developing countries where agriculture is still the dominant economic activity. Such unions may organize workers across all agricultural sectors or only have membership in specific crop sectors, e.g., tea, sugar, tobacco. Large unions are frequently associated with plantation agriculture. The Kenya Plantation and Agricultural Workers Union in Kenya and the National Union of Plantation and Agricultural Workers in Uganda are two examples.

In nations with comparatively small agricultural labor forces, national general workers' unions unite members from various commercial and industrial sector industries. Within these sizable general unions, waged agricultural workers were organized into sections or trade groups. Developed nations frequently have these kinds of unions. Examples include the Transport and General Workers in the UK, the Agricultural Workers' sections of Kommunal in Sweden, and SiD in Denmark.

Additionally, there are hybrid organizations known as agricultural trade unions that represent both independent farmers and wage earners. For instance, there are about 9 million members of CONTAG in Brazil (3 million waged workers and 6 million smallholders). Typically, these unions have distinct political and organizational structures to handle the two kinds of members. The two forms of membership in the same organization can occasionally have a difficult relationship, particularly when small farmer members also work for pay. Whether they are general, agriculture-specific, or hybrid, national trade unions typically affiliate with or are part of a national trade union center. Unions that organize workers in various commercial and industrial sectors are represented by a national center. A nation may have more than one national center. Certain agricultural unions, like the one in India, are set up on a federal-state basis. Labor laws in certain nations restrict unions from organizing within a single company. The Guatemalan banana union SITRABI serves as an illustration of this.

8.2.2. International and Regional Levels

The International Union of Food, Agricultural, Hotel, Restaurant, Catering, Tobacco, and Allied Workers' Associations (IUF) is one example of a global union (labor) federation to which national unions may affiliate on a sectoral basis. Workers on farms and plantations, in businesses that manufacture food and beverages, in lodging facilities, dining establishments, and in the tourism and catering industries worldwide are organized by the IUF, a network of 336 national affiliated unions. Bananas, cocoa, coffee, sugar, tea, and (cut) flowers are the six main crops that the IUF has singled out to build ties between unions along the food chain and unite workers at every level of production.

The International Confederation of Free Trade Unions (ICFTU), which is composed of national trade union centers, collaborates with global union federations. 148 million workers are members of the ICFTU, which has 233 affiliated organizations spread across 154 nations and territories on all five continents. Its three main regional organizations are ORIT for the Americas, AFRO for Africa, and APRO for Asia and the Pacific. Additionally, it keeps tight ties with the Global Union Federations and the European Trade Union Confederation (ETUC). There are also two other national center-based international trade union organizations. With more than 26 million members, the World Confederation of Labour (WCL) is an international trade union confederation that brings together 144 democratic and autonomous trade unions from 116 different countries. Brussels, Belgium, is home to its headquarters. To establish a new international trade union organization, the ICFTU and WCL are currently in merger negotiations. home to the World Federation of Trade Unions (WFTU).

8.2.3. Union Organization in Transnational Corporations in Agriculture and Along the Food Chain

Global union federations, like the IUF, represent and organize workers in large multinational corporations (abbreviated TNCs) in the food chain, including food and drink processing companies, restaurants, hotels, and tourist facilities, as well as in retail and wholesale distribution. Through their international federations, national, and regional trade unions can play a significant role in collaborating with these multinational enterprises (MNEs) and tracking their advancements toward sustainable agriculture, particularly concerning labor standards.

8.2.4. Levels of Trade Union Organization in Agriculture

Generally speaking, most countries have low levels of trade union representation among agricultural workers, especially among women agricultural laborers and temporary workers. There is a dearth of complete data on agricultural trade union membership. Agricultural workers continue to be poorly organized for a variety of reasons. Some of these difficulties are practical or financial, e.g., difficulties of organizing over large geographical areas, lack of transport for organizers, low membership dues resulting in only basic union services, and so on. However, workers who want to exercise their fundamental human right to freedom of association by joining the union of their choice are frequently prevented from doing so by administrative and legal obstacles. For instance, Turkey's labor laws were updated to include agricultural workers only in 2003. However, the public notary must approve each new application for union membership, and there is a significant fee associated with this process. This requirement hinders freedom of association and makes it expensive for unions, particularly low-income agricultural ones, to recruit members. For indigenous workers, who are frequently denied the legal ability to form or join such organizations, the situation is even more dire.

However, the ILO Committee on Freedom of Association has examined cases that demonstrate the prevalence of violations of freedom of association among agricultural workers. VI These infractions include everything from practical obstacles and regulatory limitations to physical attacks and the killing of trade union officials and leaders in rural regions. The Committee of Experts has regularly reviewed cases of forced and bonded labor involving agricultural workers in various nations and regions. Additionally, the ICFTU releases an annual report on global trade union rights abuses. Additionally, it can be challenging for waged workers employed by labor contractors or subcontractors to register as union members. The growing number of workers employed by these labor-supplying intermediaries is raising concerns about this.

8.3. ROLE OF UNIONS IN ADVOCATING FOR AGRICULTURAL WORKERS' RIGHTS

The rights, pay, and working conditions of agricultural workers—who frequently make up a sizable percentage of the labor force in farming and food production—have historically been fraught with difficulties. Because they give these workers a collective voice that can bargain with employers for better terms and conditions, unions play a critical role in their advocacy.

8.3.1. Contribution of Trade Unions to Sustainable Agricultural Production, Food Security and Sustainable Development

Trade unions can increase their contributions to the attainment of sustainable agricultural production, food security, and sustainable development by utilizing their workplace representation and structures, organizational skills, collective bargaining experience, technical know-how, education and training programs, and political commitment. They can accomplish this in many ways like:

- Starting union campaigns on a local, national, regional, international, and workplace level;
- Making greater contributions to issues of sustainable development, food security, and sustainable agricultural production through the use of union structures, institutional capacities, and networks;
- Employing committees and organizations within union branches to increase employee commitment to and knowledge of sustainable development, food security, and sustainable agriculture;
- Discussing "sustainability clauses" with employers on farms and plantations as part of collective bargaining agreements. Employees have a stake in the long-term viability and profitability of the companies they work for;
- Encouraging joint worker-management health and safety workplace committees to address environmental, sustainable agricultural production, food security, and sustainable development issues; making sure that worker safety representatives handle environmental issues as well;
- Modifying union education and training programs to include instruction on sustainable development, food security, and sustainable agriculture for shop stewards, safety representatives, safety committee members, and union officials. Currently, the great majority of grassroots agricultural workers and even leaders do not understand these "buzzwords."

- Advancing these causes through national tripartite processes and structures that involve the government, businesses, and labor unions. Additionally, using the legislative and parliamentary processes to guarantee, for instance, standards and policies on these matters.
- Encouraging inter-regional cooperation between unions, and promoting these issues in appropriate regional fora and processes;
- Encouraging international cooperation between unions, and promoting these issues in appropriate international fora and processes;
- Strengthening links between agricultural workers and their families and local communities on these issues;
- Networking and alliance building with non-governmental organizations and civil society organizations; and
- Building links and alliances with agricultural producer organizations.

8.3.2. Examples of the Contributions of Workers and Their Trade Unions to SARD, Food Security, and Sustainable Development

Future contributions from waged agricultural workers and their trade unions to food security, sustainable development, and sustainable agricultural production could be substantial and largely unrealized. Examples of the contributions waged agricultural workers are already making to sustainable development, global food security and agricultural production are given in this section, both as employees and as citizens. It explains how their trade unions are assisting them in their endeavors and aiming to boost their civil society contributions in this field. Through communication, collaboration, and occasionally partnership formation with employers, governments, agricultural, and rural development agencies, intergovernmental organizations, science, and research institutions, agricultural banks and credit institutions, and civil society organizations and groups, the examples show how workers and trade unions are addressing the new challenges.

8.3.2.1. Framework Agreements between Trade Unions and Multinational Enterprises

A framework agreement is a comprehensive agreement on labor rights, standards, and employment that is directly negotiated between a business and a trade union for all aspects of the business's operations. For instance, a framework agreement was signed in 2001 between the IUF, the Coordination of Latin America Banana Workers' Unions (COLSIBA) and the multinational banana company Chiquita.

The agreement states that Chiquita will uphold fundamental labor standards and demand that its joint venture partners, suppliers, and contract growers do the same. To make sure that labor standards are being fulfilled, there is also a provision for collaborative assessment and monitoring.

8.3.2.2. Supply Chains in Agriculture, Codes of Conduct, Ethical Trading Initiatives, and Fair-trade Labeling

Trade union work on supply chains in agriculture, codes of conduct, fair trade labeling and ethical trading initiatives includes:

8.3.2.2.1. Codes of Conduct

The work on corporate social responsibility has grown significantly over the last 10 or so years. Civil society organizations (CSOs) that were worried about environmental and child labor issues initially sparked this interest. This has recently broadened to include a greater interest in the social and labor conditions in producing nations. Many companies responded to these concerns by issuing unilateral codes of conduct which often had little basis in accepted international norms, e.g., International Labour Standards, and which were little more than publicity stunts.

In response, the global trade union movement released a model code of conduct that was firmly grounded in international labor standards. This code could be used by trade unions to negotiate more authentic codes and to challenge companies. Nonetheless, a lot of international trade unions, like the IUF, would rather negotiate directly with the businesses using the framework agreements mentioned above.

8.3.2.2.2. Agricultural Examples - IUF and Cut Flowers

The cut flower industry is known for using a lot of (young) women on seasonal contracts, using a lot of pesticides, having trouble disposing of waste, and using a lot of water. Concerns regarding flower production, particularly but not solely the environmental costs, were immediately voiced by non-governmental organizations (NGOs) in the North. In response, the flower industry released numerous unilateral codes of conduct.

The IUF created a model International Code of Conduct for the Production of Cut Flowers to address the issue of unilateral codes. This was done in collaboration with affiliates and many non-governmental organizations under the auspices of the International Flower Coordination. The International Labor Organization's standards serve as a solid foundation for this code. To persuade them to accept the International Code of Conduct, importers were specifically targeted, particularly in Germany and the foundation that establishes the environmental standards for the flower auction held in the Netherlands. Trade unions in East Africa attended workshops on the International Code of Conduct and its application to worker organization and better working conditions. To teach shop stewards how to use the Code, a training manual has been created. Following more discussions with the flower growers, Union Fleurs, an industry association, launched the Fair Trade in Flowers and Plants program.

IUF affiliates have numerous issues with codes of conduct, even those that are multilateral and based on ILO standards, as the work surrounding the promotion of the International Code of Conduct for the Production of Cut Flowers has been brought to light. According to available data, trade unions in producing nations find it extremely difficult to implement codes of conduct to enhance working conditions. There are currently no examples of a code that would result in the creation of a new union, even if freedom of association were its guiding principle. There are some instances of better living and working conditions, particularly when a union has been able to add the code to its collective bargaining agreement. However, this is still rather unusual, as some unions have been able to use the code to open new branches.

8.3.2.2.3. Ethical Trading Initiative

To find and encourage best practices in the application of codes of conduct, the Ethical Trading Initiative (ETI), a membership organization of businesses, trade unions, and non-governmental organizations with headquarters in the UK, was founded in 1998. All members of the ETI negotiated its own base code, which is based on international labor standards. The ETI base code was created as a tool to bring about constructive change on the ground, according to all stakeholders.

By promoting business practices that embrace social, environmental, and financial responsibility, the ETI seeks to increase the private sector's contribution to sustainable development. In developing nations, managing the supply chain ethically is essential to doing business responsibly. The IUF is participating in the ETI at the board level and has also been involved in pilots in the agricultural sector, e.g., horticultural products and bananas.

8.3.2.2.4. Fair-Trade

Fair-trade is a trading partnership that aims to increase equity in global trade and is founded on communication, openness, and respect. By providing improved trading conditions, such as protecting the rights of marginalized producers and workers, particularly in the South, it promotes sustainable development. Agreements between producers in developing nations and commercial buyers who want to buy and market goods based on stable, "just" or "fair" prices and production criteria that respect labor and environmental standards are thus commonly referred to as "fair-trade." Fair trade seeks to improve producers' incomes, expand their market reach, and guarantee that their output is grounded in sustainable development principles.

For instance, the Fair-trade Labelling Organizations International (FLO) establishes uniform standards for fair trade bananas, tea, coffee, cocoa, honey, and orange juice. FLO primarily labels products from small farmers, but plantations are also involved in the tea and banana industries. The IUF has been interested in how fair trade affects employed workers and how it can support their ability to organize trade unions and improve their living conditions without compromising collective bargaining.

8.3.2.3. Workers Promote Integrated Production and Pest Management

When combined with other crop husbandry techniques, Integrated Production and Pest Management (IPPM) maximizes the control of pests by their natural enemies, which include parasites, diseases, and pests. By keeping pest populations below economically harmful levels and limiting the use of pesticides to economically justified amounts, this management strategy seeks to lower risks to the environment and human health. The four key principles of IPPM are:

- Grow a healthy crop, and conserve healthy soil;
- Conserve natural enemies - pests, parasites, and pathogens;
- Observe the crop on a regular basis; and
- Farmers and agricultural workers are the experts in pest control.

Farmers frequently remark, "We are aware that chemical pesticides are harmful to our health as well as the health of our families and communities. What are the alternatives? How can we stop using these poisons? Making sure employees are taught IPPM techniques is one way to do this. IPPM training is typically only given to farmers, particularly through an educational approach known as "Farmer Field Schools" (FFS). Through farmer field schools in its national programs around the world, the FAO has been encouraging the use of such techniques. The FAO's integrated pest management program (IPM) has been especially effective in Asia. To introduce its effective IPM approach to interested policymakers from other regions, the FAO intercountry program on IPM rice in Asia hosted a global IPM meeting in 1993. Therefore, the Global Integrated Pest Management Facility was established in 1995 by the United Nations Development Programme, the World Bank, the FAO, and the United Nations Environmental Programme. The FAO is home to this collaborative program, which serves as the primary global organization advancing IPPM.

The Global IPM Facility and the IUF are currently collaborating to teach agricultural workers IPPM techniques through the FFS method. Similar to farmers, field schools involve workers going into the field to observe how a crop grows, learn how to spot weeds, diseases, and harmful insects, and discover how to preserve and promote beneficial insects. The employees then create their own agro-ecology plan for that specific crop and field, outlining non-chemical methods for growing a healthy crop and defending it against weed competition, pest, and disease attacks, and other threats.

With this newfound understanding, employees can then bargain with employers to include provisions in collective bargaining agreements mandating the use of IPPM programs. The purpose of IPPM training is to equip agricultural workers with the knowledge and abilities to argue that IPPM methods offer a safer alternative to toxic pesticides when directed to do so by a manager or employer. This is safer for the crop (which could subsequently fetch a higher price), the community, the environment, the managers, the supervisors, and themselves. In 2001, Tanzanian (TPAWU) and Ugandan (NUPAW and NUCMAW) agricultural trade unions participated in the first-ever pilot IPPM courses for waged agricultural workers. Additionally, a few NGOs and organizations representing organic farmers were invited to participate in the courses by the concerned unions. Professional IPPM trainers supplied by the facility conducted the training. To expand it to other unions and nations, the pilot training is still in progress.

8.3.2.4. Workers Promote Improved Health, Safety, and Environmental Standards for Pesticides

In 1998, the IUF launched a Global Health, Safety, and Environment Project to enhance workplace occupational health, safety, and environmental standards, with a particular focus on deaths, poisonings, illnesses, and pollution from heavy pesticide use. To address occupational hazards and advance integrated production, pest management, and sustainable agriculture, the Project seeks to strengthen the ability of affiliated national unions as well as the regional and global networks of the IUF. The IUF and ILO have created a series of trade union education manuals for agricultural workers on health, safety, and the environment. These manuals are intended for use by nongovernmental organizations and small farmers alike.

Training is intended for branch and national union officials and committee members, as well as grassroots members, worker health and safety representatives, and worker representatives on joint worker-management health and safety workplace committees. A core group of trainers has been trained by each of the seven participating African agricultural unions in Ghana,

Malawi, Tanzania, and Uganda thus far. For the target groups employed in the production of cocoa, coffee, flowers, horticulture, sugar, tea, and tobacco, these trainers conduct frequent training sessions. Improvements in workplace OHSE standards, the addition of new members, and the unions' increased national influence over OHSE laws and policies are all outcomes of the training.

Additionally, there is close collaboration with the Pesticides Action Network (PAN) on a local, regional, and global scale. In Latin America, IUF and PAN have a cooperation agreement in place and are collaborating on projects like the prohibition of toxic pesticides. Tanzanian and Ugandan agricultural affiliate unions of the IUF are assisting in the expansion of the PAN network in other anglophone African nations.

The 1985 FAO Conference adopted the FAO International Code of Conduct on the Distribution and Use of Pesticides, which established voluntary standards of conduct for all parties involved and outlined the shared responsibilities of governments, industry, users, and international and non-governmental organizations in all areas of pesticide distribution and use. Because they use pesticides in their farming operations, farmers, and agricultural workers in developing nations are particularly affected by the Code. The Code was amended in 1989 to include the Prior Informed Consent (PIC) procedure, which states that a chemical that is prohibited or subject to severe restrictions (for the protection of human health or the environment) should not be shipped internationally without consent, if one exists, or against the ruling of the appropriate authority in the importing nation. The IUF actively participated in the Code's 2000 revision. The UN's Rotterdam Convention on the PIC Procedure for Certain Hazardous Chemicals and Pesticides in International Trade now incorporates the PIC principle. The FAO administers the PIC procedure for pesticides, while the UNEP administers the PIC procedure for industrial chemicals. The IUF continues to take part in the meetings of the convention's governing body, advocate for its ratification and implementation at the national level, and represent trade unions globally in the negotiations that led to the development of the Rotterdam Convention. Both the FAO and the ILO actively participate in the work of the Intergovernmental Forum on Chemical Safety (IFCS), which the IUF has been actively involved in helping to establish. The Forum is a gathering place for governments, trade unions, industry, scientific associations, intergovernmental organizations, and public interest NGOs that will:

- Develop and coordinate policies on the environmentally sound management of toxic chemicals; and
- Provide advice and make recommendations to governments, intergovernmental organizations, industry, trade unions, scientific bodies, and public-interest NGOs.

Currently, the IUF serves as the international trade unions' representative on the IFCS Steering Committee. To create National Profiles to Evaluate the National Infrastructure for Chemical Management, the IUF has also been actively involved in another international initiative linked to Agenda 21. The working group that created the profiles format included the IUF. Ghana, Tanzania, and Uganda's IUF agricultural affiliate unions have contributed to the creation of their nations' national profiles. To raise national pesticide health and safety standards for the benefit of farmers, agricultural workers, local communities, the environment, and consumers, they are currently assisting in the implementation of the action plans that were produced.

8.3.2.5. Worker and Pesticides Industry: Joint Training on Pesticides

To jointly train agricultural workers and even management on pesticide hazards and risk management prevention and control measures, the IUF's Global HSE Project is collaborating with Crop Life International (CLI), the international trade association of the transnational pesticide manufacturing companies. For each course, one to two trainers are provided by the IUF and CLI, who then work together to train management and employees. Two Ugandan trade unions, NUPAW and NUCMAW, started offering pilot courses in 2000, and they have been offered regularly ever since.

An ILO ACTRAV-sponsored study of the pesticide industry's voluntary "Safe Use" training program in Guatemala in 2000 was a major factor in the development of this collaboration. The industry project's failure to offer training to waged agricultural workers or their trade unions was particularly criticized by the IUF, which also insisted that the training be grounded in sound educational and training practices and be less presentational in nature (Hurst, 1999).

In 2001, the IUF presented the results at the Annual Conference of the CLI. The collaborative work mentioned above was the result of additional discussions between CLI and IUF.

8.3.2.6. Tripartite Cooperation to Develop Schemes for Roving Safety Representatives in Agriculture

Given the strong correlation between productivity and environmental, occupational health, and safety standards, employers, and trade unions may be able to work together to ensure safe and healthy work practices. There is a lot to gain in terms of employee happiness and business productivity.

8.3.2.6.1. South Africa

The project to create a roving safety representatives program in South African agriculture, modeled after the legally functioning system in Sweden, is an illustration of tripartite cooperation between the government, employers, and trade unions. The project's objectives are to increase the availability of safety rep programs on larger farms and plantations and to create programs for small to medium-sized agricultural operations. Roving safety representatives can visit farms where they are not employed, but otherwise they have the same legal authority as regular safety representatives. To prevent and lessen workplace fatalities, accidents, and illnesses, roving safety representatives assist in increasing awareness, disseminating information, and offering advice on issues based on shared objectives between employers, trade unions, and employees.

A tripartite delegation from South Africa, consisting of six members, was sponsored by the IUF and the Swedish trade union Kommunal in June 2002 to examine the Swedish roving safety representatives program. The purpose of the government, agricultural trade union, and agricultural employers' representatives' visit was to: (i) observe the roving safety representatives program in operation on Swedish farms; (ii) foster knowledge of the legal rights and responsibilities of the union-nominated roving safety representatives, including the role of each stakeholder in the system; (iii) learn about the training programs for roving safety representatives, the types of backup support that are required, who can provide it, and how the system is funded; and (iv) generate ideas about how the program, or portions of it, could be implemented in South Africa. Kommunal and the IUF are continuing to support the tripartite talks that are currently taking place to establish pilot RSR schemes in three South African regions (IUF, 2003).

8.3.2.6.2. United Kingdom

In 2001, the Health and Safety Executive, a government agency in the United Kingdom, made the following public statement: "There is compelling evidence that workplace-based safety representatives contribute positively to raising health and safety standards. We must ascertain whether comparable knowledge can enhance employee consultation and health and safety in establishments that do not currently benefit from safety representatives. To compare the efficacy of professional health and safety advisers with roving safety representatives appointed by trade unions and trained primarily by the trade union, an 18-month study was funded by the Health and Safety Executive. The project was overseen by the Agricultural Development Advisory Service (ADAS) and started by the tripartite Agricultural Industry Advisory Committee (Health and Safety Executive).

The project was fully supported by the Transport and General Workers Union's (TGWU) agricultural workers' section. In collaboration with ADAS, they chose and assisted in training the eight volunteer worker roving safety representatives. The National Farmers Union, the employers' group, also contributed to the project. To create a baseline, an initial assessment of the health and safety conditions was carried out on about 100 volunteer farms. After that, the farms were split up into three groups.

The eight worker roving safety representatives visited a total of about 40 farms in one group, with each representative visiting a certain number of farms in their respective regions four times. The professional advisers (ADAS staff) visited the second set of 40 farms. The 20 farms that remained served as a control group and were not visited. A final status review of the farms in each group was conducted at the conclusion of the process. The project discovered that roving safety representatives worked well. Securing access to workers, a crucial requirement for promoting increased employee involvement in raising health and safety standards, proved challenging in certain cases, though.

8.3.2.7. Environmentally Sustainable Development

Trade unions worldwide have been more active on environmental issues, as well as on networking and forming alliances with other organizations, particularly since the 1992 UN Conference on Environment and Development. By tying workplace problems to the environment and establishing a link between the environment and health, the Labour Environmental Alliance Society (LEAS) in Vancouver, Canada, for instance, actively engages employees in environmental issues. It also aims to increase environmentalists' comprehension of the labor movement. More than 50,000 employees, environmentalists, and social justice advocates have united under LEAS's broad network.

The Society's primary goal is to rid the community and workplace of toxins. The effects of pesticides in food and their effects on consumers and workers, particularly children, are of particular concern. Together with the Canadian Labour Congress and the British Columbia Federation of Labour, it conducts educational workshops and collaborates with consumer, environmental, and citizen organizations. Farmfolk/Cityfolk, one of its member groups, is a national authority on food-related issues. Additionally, the Society works with local and coastal fishermen on issues related to water quality protection, food distribution, sustainable fisheries, and fish habitat.

In collaboration with the Spanish government, SUSTAINLABOUR, an international labor foundation that advances trade union interests in sustainable development, was founded in October 2004. The ILO, UNEP, and significant national and international trade union organizations that actively address sustainable development issues were also involved. The foundation, which is based in Spain, aims to investigate the main issues, opportunities, and challenges that trade unions face in this area and to create regional networking, capacity-building, and training initiatives. Through research and measurement, national campaign design, and corporate assessments, it will create country profiles from a labor perspective.

It will create methods for workplace evaluations and training on clean development mechanisms at the workplace level. It will specifically carry on and strengthen the work initiated by the International Trade Union Network on Sustainable Development, which resulted in the strategic attendance of more than 400 trade union representatives at the 2002 World Summit on Sustainable Development. This work will focus on topics that affect workers globally, including clean water, sanitation, human settlements, and SARD good practices. SUSTAINLABOUR will specifically investigate the social aspects of globalization and sustainable development. Effectively contributing to and keeping up with the ongoing discussions and preparations for the Commission on Sustainable Development, which meets yearly at the United Nations in New York, is one of the main goals of the Trade Union Foundation.

8.3.2.8. Food Safety and Food Security

Based on its historical and legal commitment to "actively promote the organization of the world's food resources for the common good of the population as a whole," the IUF seeks to advance food security and safety.

8.3.2.8.1. Food Safety

The Transport and General Workers Union's (TGWU) involvement in the Bovine Spongiform Encephalopathy (BSE) crisis, also known as the "mad cow" disease crisis, in the UK livestock industry during the 1990s is an example of trade union action on food safety that sought to protect both workers and consumers. The TGWU made significant efforts to prevent job losses during the crisis and counseled its members on the health and safety implications, especially with regard to Creutzfeldt-Jakob disease.

The union also demanded that the Ministry of Agriculture, Fisheries, and Food transfer oversight of food safety and hygiene to the Department of Health and the Health and Safety Commission, and that a public investigation be conducted into the causes of BSE and its connection to Creutzfeldt-Jakob disease, and that independent studies be conducted to assess the risks of catching the disease for those employed in the food and farming sectors. On these matters, the union also presented arguments to the European Commission and the UK government.

8.3.2.8.2. FAO Food Safety & Quality Standards

Maintaining and enhancing food safety and quality at the national, regional, and worldwide levels is the focus of the FAO's Food Quality and Standards Programme. It encourages national regulatory frameworks to be established and run following international standards, especially those set forth by the Codex Alimentarius Commission.

To guarantee food quality and safety along the entire food chain, it also offers technical guidance and support for the development of food control programs and systems at the national and local levels. It offers countries and the Codex Alimentarius Commission scientific evaluations of the risks to food safety along with relevant recommendations. This covers the evaluation of food additives, veterinary medication residues, naturally occurring toxicants, chemical, and microbiological contaminants, and foods made using contemporary biotechnology.

Policy advice on specific issues, institutional development and/or strengthening, food legislation review and updating, harmonizing food regulations and standards with Codex and other international regulatory instruments, training technical and managerial staff in various food safety-related disciplines, and studies and applied research on specific food-related subjects are all examples of the FAO's capacity-building efforts in the field of food safety. Organizing national and regional workshops and seminars on food safety-related topics, as well as creating and disseminating manuals, guidelines, training materials, and other resources required to support food control and food safety development programs, are additional aspects of capacity-building (FAO www.fao.org/es/ESN). Although the FAO is undoubtedly a major player in advancing food safety and quality globally, international trade union participation in this effort has been minimal thus far. Future research could focus on strengthening FAO-trade union collaboration on food safety concerns and how they affect the food trade.

8.3.2.8.3. Food Security

Many waged agricultural workers are among the most vulnerable to food insecurity, despite making up 40% of the agricultural labor force and contributing significantly to food production. In addition to being a major area for future FAO-ILO-IUF collaboration and joint work, promoting greater food security for waged agricultural workers, their families, and communities would be a significant step towards more sustainable agriculture.

According to the WFS Plan of Action (1996) of the FAO, food security is achieved when everyone, everywhere, has physical and financial access to enough wholesome food to satisfy their dietary requirements and food preferences for an active and healthy life. Since it helps explain how people become food insecure, vulnerability to becoming food insecure is a crucial part of the food security equation. It includes the entire spectrum of factors that put people at risk of food insecurity. Individuals, households, or groups of people's susceptibility to food insecurity is based on their exposure to risk factors and their capacity to handle or endure stressful circumstances.

8.3.2.8.4. The FAO and Food Security

For waged agricultural workers and their trade unions, the FAO's specific work on promoting greater food security is especially pertinent. For instance, to better understand who the food-insecure and vulnerable are, where they live, why they are vulnerable, and what can be done to address their needs and concerns at various levels, the Food Security and Agricultural Project Analysis Unit (ESAF) of the Food and Agriculture Organization (FAO) has conducted many studies on vulnerable groups. The research was carried out in Vietnam, Nepal, and Guatemala (ESAF. FAO, 2004). Similar research is being conducted in the Indian states of Himachal Pradesh and Orissa.

To determine and comprehend why specific groups of people with the same livelihood, residing in a relatively large geographic area (beyond the household or community level), are food insecure or vulnerable to food insecurity, this study used a livelihoods-based approach to assess, monitor, and map vulnerability to food insecurity in Nepal. The SLA examines how individuals use both tangible and intangible assets to achieve their livelihood goals and how the institutional and policy environments in which they are embedded, as well as the shocks and trends they are exposed to, affect their ability to achieve these goals. The SLA's multi-sectoral nature enables us to consider the various elements that impact susceptibility to food insecurity, enabling analyses to expand beyond the conventional emphasis on food availability to encompass aspects of stability, utilization, and access. In addition, the SLA offers a framework for examining macro-, meso-, and micro-linkages, which helps to explain how household well-being is influenced by both macro- and meso-level elements, including national economic development. This aids in determining the best kinds of interventions for food security. According to the SLA's participatory principles, the analysis takes into account the opinions of all parties involved, including those whose vulnerability we are worried about (Lovendal, Knowles, and Horii, 2004).

The study's summary in Nepal identifies agricultural laborers as one of the groups experiencing food insecurity and demonstrates that coping mechanisms for increased susceptibility to food insecurity include changing livelihood strategies, with special reference to rural employment patterns, as labor market participation was a common coping mechanism.

Other coping strategies included shifting spending and consumption habits, taking out loans, selling assets, using resources from common property more frequently, moving in pursuit of wage work, shifting livelihoods or looking for local wage labor, and altering social identity. To augment their traditional source of income, many households from the seven vulnerable livelihood groups profiled rely on seasonal and long-term migration. The majority of the time, the household's working men leave for other regions of India or Nepal in search of wage labor. Porters, agricultural laborers, marginal farmers, and rural service providers all stated that seasonal and/or long-term migration is critical to their survival.

When there is little work available in their traditional occupation, groups that are at risk of food insecurity are forced to change their means of subsistence or look for local wage labor jobs, sometimes permanently. Households with inadequate agricultural output or income to make ends meet seek out local wage opportunities in addition to relocating to other regions. After fulfilling the requirements of their regular contract, some marginal farm households attempt to rent more land for cultivation as sharecroppers. Many rural service providers and porters look for work as daily laborers during times when there is little or no demand for their services. To deal with the decline in demand for their services, some rural service providers have shifted to wage labor or agriculture as their primary source of income.

8.3.2.9. The Sustainable Agriculture and Rural Development Initiative

The FAO's Director-General, governments, and major civil society groups launched the SARD Initiative in 2002 during the World Summit on Sustainable Development (WSSD). The SARD Initiative is a multi-stakeholder framework that aims to support the shift to people-centered sustainable agriculture and rural development and to increase participation in the creation of programs and policies. The Director-General of the ILO also conveyed the support of his organization for the Initiative.

FAO, in its capacity as SARD's Task Manager, facilitates the SARD initiative, which is spearheaded by civil society and backed by governments and intergovernmental organizations. With the hope that FAO could serve as a suitable and impartial convener, more than 65 governments and civil society organizations from 90 countries showed their support for this initiative at the WSSD.

The Agenda 21 program, which was adopted by governments and interested parties at the 1992 UN Conference on Environment and Development, established the Major Groups. These include the scientific and technological community, farmers, indigenous peoples, women, children, youth, business, and industry, nongovernmental organizations, local government, and agricultural workers and their trade unions. The SARD Initiative was created to do more than just compile a list of successful case studies; it now aims to increase the number of successes by promoting communication between major groups at all levels and the pertinent government, non-governmental, and international organizations.

The SARD Initiative's goals are to decrease rural residents' vulnerability, promote their self-reliance, and aid in the eradication of poverty:

- Build the capacities of rural communities, organizations, and networks to better access resources, promote fairer conditions of employment in agriculture, and adopt innovative practices for SARD; and
- Strengthen the cooperative relationships and work among Major Groups.

For agricultural workers and their trade unions, the IUF is serving as the SARD Initiative's Major Group focal point. Specifically, it is spearheading the development of concepts and initiatives to advance more equitable working conditions in agriculture, one of the Initiative's three main thematic areas. This report contributes to this program in part. To promote SARD and develop SARD implementation initiatives related to the following areas: Fair Conditions for Employment in Agriculture; SARD Good Practices, Principles, and Case Studies; and Access to Resources, FAO organized an Electronic Forum on Sustainable Agriculture and Rural Development (the SARD E-FORUM) from June 24 to August 17, 2002, in advance of the World Summit on Sustainable Development.

Among these, the discussion of fair conditions for agricultural employment was closely related to the contribution of agricultural workers to SARD. Its objectives were to gather diverse viewpoints and insights regarding fair employment in agriculture and the role of waged agricultural workers in sustainable agriculture; establish a database of successful cases about the enhancement of waged agricultural workers' conditions and their contributions to sustainable agriculture; and determine areas of agreement regarding the attainment of fair employment conditions for agricultural workers. Documentation of achievements about agricultural workers, sustainable agriculture, and rural development, as well as important action areas for the future concerning fair employment in agriculture, were among the outputs. There was also agreement on the main issues surrounding fair employment for agricultural workers.

The problem of child labor and young workers in agriculture, international standards, particularly freedom of association, the right to collective bargaining, non-discrimination in the workplace, the abolition of forced labor, and the elimination of child labor, as well as specific ILO standards on agriculture that cover safety and health, labor inspection, rural workers' organizations, and plantations, as well as health, safety, and environmental standards in agriculture, community-oriented employment models, and fair employment in agriculture and the WSSD (particularly through promotion of policies that recognize employment as central to poverty reduction, and initiatives to increase labor productivity, the level of decency of work, and income-generating capacity).

Through the improvement of health, safety, and environmental standards for farmers and agricultural workers, as well as more equitable employment conditions for women in agriculture, the SARD campaign on fair employment focuses on the efforts of the Major Group (in collaboration with other stakeholders) to strengthen economic and social standards/conditions of SARD concerning promoting fair conditions of employment in agriculture for wage-dependent smallholders, self-employed farmers working as contract laborers, self-employed workers employed as contract laborers, and sharecropping.

8.3.2.10. The World Food Summit and follow-up

Food security initiatives have attracted the attention of trade unions that represent waged agricultural workers. The 13th FAO/International Trade Union Consultation, held in the lead-up to the 1996 World Food Summit (WFS), concentrated on food security and the ways in which trade unions could support the Summit and the execution of the WFS Plan of Action.

To encourage trade union participation in the food security policy dialogue during the WFS preparations, the FAO prepared the Partners for Food Security document in collaboration with trade unions and other organizations. This document, which aims to support the WFS process, compiles the trade unions' mandates, approaches, and experiences in advancing food security as well as their stances and opinions on the main concerns covered by the Plan of Action and the WFS. It outlines potential areas of collaboration for carrying out the WFS Plan of Action.

The trade unions stated their willingness to inform national, regional, and global policymakers of the opinions and stances of their members on matters pertaining to food security. For instance, they cited the involvement of farmers' organizations and trade unions in the Latin American and Caribbean Movement for Food Security and in the formulation of the 1995 Managua Declaration, wherein governments, people's organizations, and non-governmental organizations committed to collaborating on methods to attain food security in the region. Several trade unions expressed their ability to conduct research on food security-related topics and serve as information and research dissemination channels. The trade unions also stated that to address the needs of the food insecure, collaboration between governments, cooperatives, trade unions, and other civil society sectors is essential at both the local and national levels. The provision of safety nets and social welfare programs, as well as integrated initiatives for rural development and sustainable agriculture, are areas where collaboration is needed.

The IUF attended the World Food Summits in 1996 and 2002. In the International Non-governmental Organizations/Civil Society Organizations Planning Committee on Food Sovereignty and Food Security, it has also served as the main point of contact for trade unions. NGO/CSO follow-up to the World Food Summit: Five Years Later, 2002, is being coordinated and promoted by this body. As one of the primary outcomes of the summit, the NGOs and CSOs proposed an International Code of Conduct on the Right to Adequate Food, on which the FAO has now established a working group. The IUF wrote a section on workers' rights for this code.

8.3.2.11. HIV/AIDS Prevention

The scope of trade union efforts to prevent HIV/AIDS is growing quickly. For instance, with assistance from an Irish non-governmental organization, the Tanzanian Plantation and Agricultural Workers Union (TPAWU) has drafted an HIV/AIDS Code of Practice for plantation workplaces.

The ILO Code of Practice on HIV/AIDS, which offers governments, employers, and employees worldwide guidelines based on international labor standards for addressing HIV/AIDS and its impact at the national, community, and enterprise levels where the majority of infections occur, is another initiative supported by agricultural unions. The ICFTU has been mapping the efforts being made by trade union organizations worldwide in the fight against HIV/AIDS in anticipation of an international union campaign on the disease.

8.3.2.12. Land Reform

The degree to which waged agricultural workers are regarded as stakeholders in land reform is unclear. Even though they labor on the land and contribute to its increased productivity, waged workers frequently lose out on the benefits of land reform. In fact, if land reform disregards this sizable occupational group, workers may lose their jobs, housing, and means of subsistence, as well as their ability to work as migrant laborers in a nation, further impoverishing and starving them.

When it comes to land reform or redistribution, agricultural workers frequently lack clear policies or guidelines at the national level. Due to many limitations, including a lack of organizational, political, financial, and technical capacity to effectively represent and advance the interests of their members, agricultural workers' unions have so far played a very limited role in addressing workers' issues during the land reform process.

In nations like Zimbabwe, Namibia, and South Africa, the issue is quite severe, as major shifts in land possession have led to exclusion and substantial declines with harmful repercussions for agricultural laborers. Regarding legislation and policies pertaining to land redistribution, relocation, or remuneration for displaced farm workers, there is a notable scarcity. The South African White Paper on Land Reform of 1996 represents an exception, as it addressed matters of land tenure stability and land redistribution. In South Africa, farm workers even have the right to profit from land reform and are granted land to work on. Nevertheless, the ongoing Zimbabwean land reform serves as a reminder of the importance of addressing the situation of farm workers during land reform initiatives.

The International Land Coalition (ILC) and the IUF are currently conducting research and developing guidelines to help improve the status of waged agricultural workers as stakeholders in land reform processes:

- Help ensure that policymakers consider waged workers and their trade unions as stakeholders in land reform and as potential beneficiaries; and
- Help policy makers ensure that "just transition" measures are put in place for workers displaced by land reform (i.e., where for whatever reasons they will not benefit from land reform by becoming small farmers), so as to guarantee some form of sustainable livelihood.

8.3.2.12.1. FAO Study on Waged Agricultural Workers and Land Reform in South Africa

The FAO conducted a study in 2004 to look into the part that unions for agricultural workers played in South Africa's land reform movement. As a largely underutilized group of actors who are thought to be able to contribute to and create a more profound and sustainable agriculture and broader rural development, the FAO's mandate includes the creation of an enhanced set of support arrangements and initiatives for agricultural workers and unions. The goal is to understand the situation and organization of South African agricultural workers and unions, the legislative environment that supports or influences their actions, their actual or potential participation in policy-making and implementation, especially in the country's land reform program, the limitations on their engagement with the agricultural labor community, farmers, and government, and the goal of developing specific policy recommendations. This is naturally desirable in the context of significant structural changes in rural and agricultural economies. Large-scale labor shedding in response to rapid liberalization is a defining feature of recent employment and labor changes in South African agriculture. Unlike many other agricultural economies, the proportion of skilled and semi-skilled labor is increasing as reliance on "casual" or informal workers rapidly declines, even though the number of commercial farms employing skilled and semi-skilled labor is declining overall. The three main goals of South Africa's Land Reform Programme are land redistribution, tenure security for those currently farming, and restitution for those who were forcibly removed from their lands. These goals are hotly debated. In the quest for more sustainable agriculture and rural development, this final component is thought to be the most suitable for research and for access by agricultural workers, via their unions. Their trade unions have not been particularly involved in the process at the institutional level, and agricultural workers have very little knowledge of the program itself. Although there is no operational program or funding for such union services, individual trade union officers occasionally offer assistance and counsel to their members on land reform matters. Although there are cases of widespread union support for land reform, there is also harsh criticism of how schemes are run, and there is little public sector involvement with agricultural workers to encourage them to access land reform options.

Despite their obvious lack of training and policy and program knowledge, the majority of union organizers and unions want to take on more active roles in support of land reform initiatives on behalf of workers. The report makes the following recommendations in this area: Many land reform beneficiaries to date come from backgrounds that are very different from those of workers. Land reform and the potential advantages of productive agriculture have essentially excluded agricultural workers. To encourage members' participation in the land reform process and as beneficiaries, it is advised that agricultural workers' unions receive support. The assistance could be provided in the ways listed below:

- Fostering a better understanding of the laws, tools, agreements, and choices related to land reform and redistribution for union organizers by implementing capacity-building initiatives in a few chosen unions. Such capacity-building should begin as a program and then gradually transition to a project-based approach. Knowledge-based and sharing-based capacity-building should be the initial focus.
- Expanding the range of services provided by unions to foster a "development agent" or "broker" role between the program and potential land redistribution program beneficiaries. Additionally, it would allow the union to efficiently provide its members with a range of options in the event that they are evicted, laid off, or their place of employment is expropriated. It goes without saying that this kind

of capacity-building should be planned to support the ministries' efforts to introduce agricultural workers to the possibilities of land reform. Complementarities between the public and labor sectors can result in significant developmental synergies, but they are contingent on political viability before any capacity building initiatives are designed.

- To expand the scope of discussions from labor issues to topics of shared concern, such as land redistribution, unions should be encouraged to more formally address the nature of communication with actors in the agricultural sector. Additionally, the unions should be viewed as social partners in the process of land redistribution.

According to the study, despite worker-targeting land redistribution policies, union participation has been low, primarily because of a lack of knowledge and resources as well as political obstacles. Land reform also calls into question the mission and membership of workers' unions because it makes workers self-employed and possibly employers.

8.3.2.13. Building Links between Trade Unions and Agricultural Producers' Organizations

How to improve ties between waged worker trade unions and small farmer organizations is one of the main issues that agricultural trade unions have been addressing since the 1990s, particularly in light of the effects of globalization and structural adjustment policies. The interactions and relationships between waged agricultural workers and small farmers can present trade unions with challenges, although many issues could bring the two groups together. What happens, for instance, when small farmers are also employers? Similarly, during the land reform process, disputes between farmers and workers may arise, particularly if there is no arbitration mechanism in place. Workers, frequently migrant laborers who are not allowed to remain in the country, lose their jobs when farmers seize control of the land.

From 1998–2005, the IUF operated a Land and Freedom Project to foster relationships and partnerships between trade unions and organizations representing small farmers. [IUF Geneva, 2002] For instance, the Project sought to counter the concentration of power of multinational corporations in the agricultural sector and to guarantee the collective defense of legal rights, better access to land, and increased influence on local, national, regional, and international policies. It created models to facilitate closer collaboration between trade unions and small farmers, as well as to modify trade union rules, structures, and training initiatives to enable unions to enlist small farmers as members.

The FAO conducted a study on Ghana's General Agricultural Workers Union (GAWU) in 1994 as part of its partnership with international trade unions to support the work of rural workers' organizations in sustainable rural development. The organization of independent contractors in rural areas and the function of agricultural unions in rural development were the main topics of the report. As part of the broader goal of boosting workers' involvement in rural development, the union placed particular emphasis on fostering collaborations with other rural organizations and encouraging women's trade union participation.

One of the first agricultural unions to tackle the issue of rural union membership declines was GAWU, which also suggested forming alliances with other locally based community-based organizations. To expand the function of trade unions beyond the worker-employer relationship and pursue more general objectives like agrarian reform and rural development, GAWU also sought to open its membership to all rural workers, including wage earners and independent contractors. Through a variety of initiatives, including self-employed support services, group enterprise promotion, the transfer of advanced technology and techniques, economic support services, and social and community development initiatives, GAWU has contributed to rural development. Nonetheless, the primary obstacles that GAWU faced during its strategy were during the stages of program operationalization. To ensure that small farmers and waged agricultural workers, as well as the trade unions and producers' organizations that represent them, participate in national decision-making processes and to achieve sustainable agriculture and rural development, it is essential to establish networks and alliances between these groups and organizations. Improving working conditions and opening up job opportunities can be achieved by boosting the involvement of workers and small farmers in these processes and fortifying the organizations that represent them.

8.3.2.14. Understanding Changing Employment Patterns in Agriculture

An analytical report on understanding shifting agricultural work patterns in the Ugandan sugar industry has been jointly produced by the ILC and the IUF. It addresses (i) full-time waged workers; and (ii) temporary and/or seasonal waged workers, including self-employed farmers hired as wage laborers. To assist the IUF and ILC in better focusing their respective work programs, allocating their resources, and providing more precise indications of possible partner organizations, the research sought to increase knowledge of the quickly evolving patterns of production, employment, and labor in agriculture (Castro, 2003).

The report discovered the following:

- Ongoing downsizing of the permanent waged workforce on sugar company nucleus (i.e., directly managed) plantations;
- An increase in the number of waged workers on short-term contracts of employment on the nucleus plantation;
- Increased use of casual waged workers on nucleus plantations;
- Increased hiring of casual waged workers by self-employed farmers, producing sugar under contract as “outgrowers” to the sugar plantation companies;
- Outgrower associations acting as labor contractors, hiring casual waged labor to work on the farms of its outgrower farmer members; and
- Increasing casualization of employment.

Increased job insecurity, decreased wages, worse working conditions, rising food insecurity, and rising poverty are all consequences of these changes for waged workers.

8.3.2.15. Developing Alliances between Indigenous and Tribal Peoples and Trade Unions

According to an ILO discussion document on initiatives between trade unions and indigenous and tribal peoples, “The labor movement and indigenous and tribal peoples have ignored each other until recently.” Their relationship, if any, has been marked by mutual mistrust and a lack of interest (Tomei, 1995).

The report continues by examining potential avenues for collaboration between the two groups. It notes: “What emerges clearly is that, in the current context of labour market deregulation and growing concern for degradation of the environment, there are grounds for joint ventures between the indigenous and labour movements.”

Some examples of such cooperation include:

- Numerous banana plantations in Panama are situated in areas inhabited by indigenous people. The indigenous people’s community is the source of leadership for SITRACHILCO, Panama’s main banana union; and
- Strengthening the capacities of indigenous and tribal peoples, assisting them in creating and carrying out their own development plans and initiatives through their own organizations, and ensuring the preservation of traditional values and culture are the goals of the ILO program to Support Self-reliance of Indigenous and Tribal Peoples Through Cooperatives and Other Self-help Organizations

(INDISCO). In India, for instance, INDISCO has established national-level consultative committees with participation from trade unions.

8.3.2.16. Rural and Migrant Workers

The Argentinean government passed legislation in 1999 to establish the National Register for Rural Workers and Employers (RENATRE in Spanish) to enhance the security and working conditions for rural workers. This Register was created in collaboration with four employers' organizations, UATRE, and the agricultural trade union; these groups also assist in its administration. A rural worker receives a Rural Worker's Card upon completing the mandatory registration process. In addition to providing workers and their families with access to retirement, health, education, and social security benefits, the Card can serve as documentation of the maintenance of positive labor relations. Approximately 70,000 rural workers have signed up for the Register thus far.

A proposal from the third world conference of the IUF Agricultural Workers Trade Group to start addressing the massive issue of migrant labor in agriculture was approved by the IUF World Congress in 2002. Among other things, the proposal called for creating a charter of rights for migrant workers in the industry that unions could use as a platform. In June 2003, the IUF and IG BAU, a German agricultural trade union, organized an international union workshop that resulted in a draft charter of rights for migrant workers in agriculture.

The National Farmers Union (employers) and the Transport and General Workers Union are involved in the UK's Seasonal Agricultural Workers Scheme, which is run by the government agency DEFRA. The union works to guarantee that these employees receive the same benefits and rights as other employees in the sector. Approximately 25,000 students, primarily from Eastern Europe, are enrolled in the program at this time.

8.3.2.17. Labor Contractors

Similar to employment agencies, "gangmasters" in the UK hire seasonal workers for farms, horticultural businesses, pack houses, and, more recently, other industries. Numerous gangsters mistreat their employees by threatening them, paying them too little, and taking unlawful deductions from their paychecks. They deceive the tax authorities as well. For many years, the Transport and General Workers Union (TGWU) in the UK has advocated for gangmasters to be registered. Despite the government's reluctance to enact legislation to address this issue, the National Farmers Union, the association of farm employers, recently agreed to support the campaign.

8.3.2.18. Water Security, Pollution, and Management

Up to 70% of all renewable water resources that are diverted for human use worldwide are used in agriculture. In developing nations, the percentage can reach 80–90%. Water availability is essential for agricultural laborers. The average person in the 40 poorest nations—of which more than half are in Africa—must meet all of their water and sanitation needs with 30 liters or less per day, which is significantly less than the 50 liters per day per capita that the UN states is the absolute minimum for water needs. People are compelled to survive on an average of less than 10 liters per day in the nine poorest African nations, which is astounding.

In India, the Self-Employed Women's Association (SEWA), which also represents waged workers, organizes a water campaign to expand cultivatable land and enhance access to safe and dependable drinking water sources. Many women have been freed from laborious and time-consuming water collection thanks to improvements in the villages' water supply, which has also improved their families' health.

Agricultural trade unions from the Nile basin and the IUF met for two days in 2002 at the Egyptian General Trade Union of Agricultural and Irrigation Workers. The conference examined ways to protect this important source of water for many East and North African nations in the future as well as ways to lessen pollution, particularly from pesticides. The adoption of a statement called for information sharing between trade unions in the Nile basin as well as water-related education and awareness-raising.

8.3.2.19. Organic Agriculture

In Northern Uruguay, 150 producers of vegetables and sugar cane make up the Association of Agroecological Producers of Bella Unión (APABU). These producers have been collaborating with the IUF's Latin American Regional Office (Rel-UITA) since 1999, offering guidance on ecological practices and assistance in establishing their organization.

A pilot "union farm" for employees of orange-growing farms in the Department of Salto has been established in collaboration with the Uruguayan agricultural workers union, SUDORA, and Rel-UITA. These workers do not have full-time jobs growing oranges. Therefore, during times when they are unemployed, the union farm enables these workers to grow small plots of land to augment their income and standard of living. In the initial phase, the workers and their families have been planting orchards, realizing a long-held dream of "land for food production." The US Greengrants Foundation is providing financial support for the project.

8.3.2.20. Literacy Campaigns

Many rural workers suffer greatly from illiteracy, and trade unions collaborate with NGOs, governments, and other organizations. to address this issue. For instance, in collaboration with the Ministry of Education, the Argentine agricultural trade union UATRE conducts literacy initiatives for its members.

8.3.2.21. Savings and Credit Cooperatives

Trade union-founded savings and credit cooperatives give low-income employees a simple way to save money and get small loans at affordable interest rates. For instance, two African agricultural trade unions, TPAWU in Tanzania and NUPAW in Uganda, have established 10 savings and credit cooperatives since 1998 through the ILO's COOPNET Program, which benefits 30,000 members in total.

REFERENCES

1. Castro, M. (2003). *Changing Patterns of Agricultural Production, Employment, and Working Conditions in the Ugandan Sugar Industry*. IUF/ILC.
2. ESAF, FAO. (2004). *Food Security and Vulnerability in Nepal: Profiles of Seven Vulnerable Groups* (ESA Working Paper 04–10). FAO.
3. FAO-SARD Initiative. (2004). *Press Release*. FAO.
4. Hurst, P. (1999). *The Global Pesticide Industry's "Safe Use and Handling" Training Project in Guatemala*. International Labor Organization.
5. ILO ACTRAV. (1999). *Trade Union Actions to Promote Environmentally Sustainable Development*. International Labor Organization.
6. IUF Geneva. (2002). *Land and Freedom Project Report*. IUF.
7. IUF. (2003). *Progress Report on the Roving Safety Representative Project in South Africa*. IUF.
8. Lovendal, C., Knowles, M., & Horii, N. (2004). *Understanding Vulnerability to Food Insecurity: Lessons Learnt in Vulnerable Group Profiling* (ESA Working Paper 04–18). FAO.
9. Rugalema, G. (1999). *HIV/AIDS and the Commercial Agricultural Sector of Kenya*. FAO.
10. Tomei, M. (1995). *Indigenous and Tribal Peoples and Trade Unions: Recent Initiatives* (Labor Education 98–1995/1, Special Issue: Rural workers, p. 9). International Labor Organization.
11. UNAIDS. (2004). *2004 Report on the Global AIDS Epidemic*. UNAIDS.

CHAPTER 9

SUSTAINABLE AGRICULTURAL PRACTICES AND DECENT WORK

Contents

INTRODUCTION

Building robust food systems that can sustain the world's population while protecting the environment requires sustainable agricultural methods. These methods concentrate on preserving soil health, preserving water, encouraging biodiversity, and lowering the demand for artificial inputs such as chemical pesticides and fertilizers. By generating equitable, secure, and fulfilling job opportunities, sustainable agriculture also makes a substantial contribution to the idea of Decent Work in addition to its positive effects on the environment. By addressing concerns about worker health, job security, wage equity, and labor rights, integrating sustainability into agriculture can serve as the cornerstone of a just and sustainable agricultural economy. Developing farming systems that are both environmentally conscious and productive is one of the main objectives of sustainable agricultural practices. To improve soil fertility, stop erosion, and lower pollution, methods like crop rotation, organic farming, agroforestry, integrated pest management, and water conservation are used. These practices contribute to the preservation of natural ecosystems and resources in addition to ensuring long-term agricultural productivity. Farmers can grow crops more sustainably, lessen their environmental impact, and improve the planet's general health by implementing such practices.

Sustainable farming methods enhance working conditions and guarantee equitable treatment for employees, which supports the idea of decent work from a labor standpoint. The International Labour Organization (ILO) defines decent work as employment opportunities that offer social protection, fair wages, safe working conditions, and respect for workers' rights. Workers frequently endure difficult working conditions, low pay, and little job security in many traditional agricultural systems. To address these problems, sustainable agriculture places a strong emphasis on fair trade, human rights, and moral labor practices. Ensuring safe and healthy working conditions is a crucial component of decent work. Workers in conventional farming may be exposed to dangerous chemicals like fertilizers and pesticides, which can seriously endanger their health. Sustainable farming methods reduce or do away with the use of artificial chemicals, which lowers worker exposure to harmful substances. One essential component of decent work is a safer workplace, which is achieved as a result. Furthermore, sustainable practices frequently entail more manual labor and less mechanization, which can result in safer and more interesting job opportunities for employees. Fair compensation is another essential element of sustainable agriculture. Many agricultural workers receive low pay that is out of proportion to the value of their labor, especially in developing nations. Fair trade principles are frequently upheld by sustainable agricultural systems, guaranteeing that workers are paid a livable wage. By allowing workers to spend more money locally, this boosts local economies in addition to the financial well-being of agricultural workers. A sustainable and inclusive economy depends on social stability and poverty reduction, both of which are facilitated by fair wages and equitable pay structures.

The significance of job stability and security is also emphasized by sustainable agricultural practices. A large number of traditional agricultural jobs are informal, seasonal, and precarious, making workers susceptible to financial instability and exploitation. Sustainable agriculture can generate more consistent job opportunities all year long by encouraging varied and integrated farming systems. By reducing the risks brought on by market volatility and climate change, farmers can ensure a steadier demand for labor through techniques like crop diversification, agroforestry, and mixed farming systems. Employees gain from this stability since it gives them steady income and long-term employment opportunities.

9.1. CONNECTION BETWEEN SUSTAINABILITY AND DECENT WORK IN AGRICULTURE

In emerging and developing nations, concerns about climate change are progressively taking center stage in discussions about development. The primary worry of developing nations is the likely effects of climate adaptation and mitigation strategies on their social and economic advancement. In this context, the type and number of jobs created and destroyed by these measures are receiving a lot of attention. This viewpoint explains recent accusations at the Copenhagen COP. It is obvious that a much better job of tying the climate and development together will be necessary to advance the conversation. In this discussion, job creation and sustainability will play a significant role. The International Institute of Labour Studies (2010) found that since 38% of jobs worldwide are in carbon-intensive industries, reducing CO_2 emissions will present a significant employment challenge. But the shift to a low-carbon economy also brings with it significant new job opportunities. Certain groups may be disproportionately affected by the adjustment process when taken as a whole, and individual nations may encounter competitive limitations when acting alone.

The nature of new jobs in terms of pay, working conditions, and other respectable work parameters is also a major concern. The number and caliber of newly created jobs in the green economy will come into sharp focus as more fossil fuel-based jobs are eliminated. Green jobs are jobs in manufacturing, research, and development (R&D), agriculture, administration, and services that significantly improve or preserve environmental quality, according to the United Nations Environment Program (UNEP) and the International Labour Organization (ILO). This includes, but is not limited to, jobs that contribute to the preservation of ecosystems and biodiversity; the decarbonization of the economy; the reduction of energy, material, and water consumption through high-efficiency strategies; and the reduction or complete avoidance of waste and pollution generation in all its forms. A green job is one that contributes to and sustains the shift to ecologically friendly production and consumption methods. All green jobs, though, cannot be categorized as respectable jobs. The following elements define decent work: a) productive labor; b) rights protection; c) fair compensation; and d) social coverage. Additionally, a fifth crucial component—the tripartite approach and social dialogue—has been added.

Four strategic goals are outlined for the ILO program for 2000–2001, which must be met to guarantee decent work. These include:

- Advancing workers' rights;
- Advancing employment;
- Providing social protection for those in vulnerable situations; and
- Advancing social discourse.

According to the ILO Director General, decent work is also defined as productive labor where employees' rights are upheld, they are given security and protection, and they have the opportunity to participate in all decisions that may have an impact on their lives.

Nevertheless, it is not a given that green and decent gobs will converge. The destruction of current jobs and subpar work are possible outcomes of some adaptation and mitigation strategies. New jobs created in the green economy might not be enough to make up for these job losses. Governments, particularly those in developing nations, will have fewer incentives to implement progressive climate policies and be less able to influence public opinion if this occurs on a large scale.

To identify the best policy framework for promoting decent work in the green economy, it is crucial to have a thorough understanding of this aspect of climate policy.

Given that agriculture continues to be the primary source of income in developing nations and a significant contributor to CO_2 emissions, it is the most suitable industry to research to comprehend the connection between decent work and climate adaptation strategies, as well as the additional complexity of the food security issue. Globally, the continued emphasis on sustainable agriculture to address climate issues and guarantee food security makes this industry the most suitable choice. India, the world's second most populous nation, has been chosen as a representative example to illustrate these issues because agriculture is the primary source of income and the ongoing shift to an industrial economy highlights climate change concerns.

9.1.1. Green and Decent Jobs in Agriculture- A Conceptual Understanding

RENNER et al. (2008) have presented a classification of green and decent jobs. They have proposed a scheme of green/decent Jobs in four quadrants matrix illustrating linkages between green and decent jobs, i.e.: (i) green and decent; (ii) green but not decent; (iii) decent but not green; and (iv) neither green nor decent. A country can adopt the model of the green economy for development if entire jobs domain converges to green and decent. The matrix can be adapted to agriculture sector and can help in understanding the relationship between green and decent jobs in the context of agriculture.

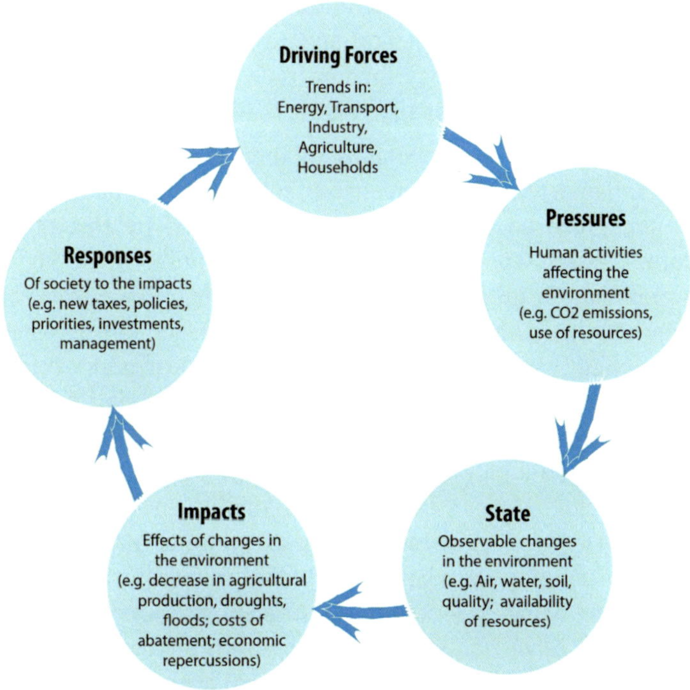

Figure 9.1. *An overview of sustainability in agriculture.*

Source: Suresh A, Krishnan P, Jha GK, Reddy AA. Agricultural Sustainability and Its Trends in India: A Macro-Level Index-Based Empirical Evaluation. Sustainability. *2022; 14(5):2540. https://doi.org/10.3390/su14052540.*

Agriculture's role in creating jobs has been extensively documented in the literature and is covered in more detail later in this essay. The agriculture industry has four main direct stakeholders in terms of employment. Agribusiness and food processing workers are among them, as are: (i) agricultural workers who work directly in the fields; (ii) transport workers and other middlemen between farms and markets; and (iii) agricultural scientists, researchers, and advisors. These stakeholders will be impacted by the shift to green agriculture in some similar ways and in some different ways. In terms of being respectable and environmentally friendly, the ideal situation from the standpoint of creating a low-carbon green economy is the quadrant that represents green and decent jobs. Those who work in agribusiness using products from sustainable agriculture, agriculture scientists and advisors working on sustainable agriculture, and well-paid sustainable coffee plantation workers certified as fair-trade producers are a few examples of these types of jobs natural juices). However, it will take a significant shift in consumer, governmental, and producer attitudes to replicate these examples throughout the agricultural economy. For example, raising consumer awareness of environmental and working conditions concerns and increasing their willingness to pay for fair trade goods are crucial if fair trade practices are to gain traction globally (Figure 9.1).

The fair distribution of the benefits of the green economy is threatened by the quadrant that represents green but subpar jobs. The working class's support for the green economy could be jeopardized by this phenomenon, and governing bodies will encounter strong opposition when implementing legislative measures to green the agriculture industry. Examples of exploited biofuel plantation day laborers, workers, and subsistence farmers operating on "Organic by default" small farms, as well as unpaid or poorly compensated employees in commercial sustainable farms that primarily produce for export markets, all illustrate this trend. In these conditions, putting too much emphasis on green job creation could backfire and distort the labor market in the agriculture sector. The promotion of sustainable and green agriculture is seriously threatened by respectable but non-green jobs. As a result, there is a pressure group of employees who have a stake in maintaining the traditional methods of working in the agricultural industry. Scientists and advisors creating chemical-based and other non-sustainable new farming technologies, workers in the food processing industry based on traditional chemical-based agriculture, and well-paid transport workers delivering agricultural produce to markets. If they are not persuaded of the stability of their employment in the new paradigm of sustainable agriculture, genetically modified food could backfire on pertinent policies and grassroots initiatives.

Jobs that fall into neither the decent nor the green categories offer an opportunity area because workers and society at large will welcome the replacement of these jobs. These measures will be fully supported by migrant workers in chemical-based agriculture, subsistence farmers who use chemical fertilizers and pesticides, and casual workers who crush sugarcane. Such change must, however, overcome numerous other obstacles, such as social discrimination, feudal systems, and a lack of effective governance. The issue of productivity, which is closely linked to food security, must be considered when discussing green and decent jobs in agriculture. The agriculture sector is distinct in the economies of developing nations, not only because it has given the vast majority of people jobs and a means of subsistence, but also because it is the main source of survival for the impoverished masses. It is necessary for food. Green agriculture has little chance of gaining the support of society and decision-makers if it cannot offer the general public a healthy diet.

9.1.2. Decent Work Opportunities in Sustainable Agriculture

By definition, sustainable agriculture involves a lot of labor and needs workers all year round. An essential component of the production process is labor costs. According to numerous studies, labor can be a significant barrier to the uptake of organic farming. According to Lampkin and Padel (1994), labor costs on organic farms are high in many European nations, even though some of those costs cover marketing and processing. Meanwhile, in Australia, Wynen (1994) discovered that labor requirements for organic and non-organic establishments were the same in the dairy and cereal-livestock sectors. Some organic farms, particularly those on plantations, and those that employed labor-intensive techniques like composting required a lot of labor, according to projects in the UNDP study (1992). Private organic farms had lower labor and overall costs in certain instances, according to the same study. The median amount of labor employed on the seven organic farms in India was less than that of the non-organic farms, according to WERF (1993). That was by no means the case for every project in which individual farmers participated.

Organic agricultural systems seem to require more work than large-scale mechanized systems. Particularly in places with little ecological potential, this is true. Rural communities in nations with employable labor can benefit from organic agriculture by creating job opportunities. Additionally, the variety of crops grown on organic farms, along with their different planting and harvesting schedules, may lead to more job openings for women and a more evenly distributed labor demand, both of which support job stability. It is still up for debate whether organic farming, with its propensity for crop diversification, results in a more evenly distributed labor requirement. However, labor needs are probably going to be dispersed throughout the year because different crops have different planting and harvesting dates. Organic farming is probably going to produce high labor productivity, but there are a lot of other variables that affect pay. The process of raising the pay and working conditions for agricultural laborers involved in organic farming can be aided by appropriate market and policy tools. Organic farming and decent working conditions can be linked through the Fair-Trade certification process, which includes the requirement that food products be produced organically. Reasonable wage conditions are a factor that many organic certification organizations favor when evaluating a project overall. If made a requirement for all certification programs, this could significantly improve the pay and working conditions for organic farming workers. According to the UK's National Survey of Economic Farms, organic farming in the country created 32% more jobs per farm than comparable non-organic farms (Table 9.1).

Table 9.1. *Logit Model Estimation of Integrated Soil Fertility Management (ISFM) Adoption by the Workforce in a Study*

Categories of Sociodemographic, Economic, and Institutional Factors	**Coe A**	**Error St**	**Wald**	**Sig.**	**Exp(B)**	**CI for Exp(B) 95%**	
						Inferior	**Superior**
Gender	0.523	0.421	1.547	0.214	1.688	0.740	3.849
Age	−0.001	0.464	0.000	0.998	1.001	0.403	2.484
Study	−0.724	1.041	0.484	0.487	0.485	0.063	3.730
Marital status	0.031	0.479	0.004	0.949	1.031	0.404	2.634
Experience of the farmer	0.011	0.489	0.001	0.981	0.989	0.379	2.580
Mutual aid	1.406	0.618	5.183	0.023 *	4.080	1.216	13.691
Perception of innovation	0.942	0.549	2.938	0.087	0.390	0.133	1.145
Land tenure status	0.203	0.435	0.217	0.641	1.225	0.522	2.873
Size of the farm	−0.625	0.638	0.960	0.327	0.535	00.153	1.868
Daily income	0.545	0.505	1.164	0.281	1.725	0.641	4.642
Income diversification	−0.701	0.544	1.664	0.197	0.496	00.171	1.440
Membership of an association	3.861	0.700	30.435	0.000 ***	47.508	12.052	187.273
Contact with extension services	0.047	0.553	0.007	0.932	0.954	0.322	2.820
Access to training	0.626	0.657	0.907	0.341	1.870	0.516	6.776

Categories of Sociodemographic, Economic, and Institutional Factors	Coe A	Error St	Wald	Sig.	Exp(B)	CI for Exp(B) 95%	
						Inferior	Superior
Access to subsidies	0.013	0.445	0.001	0.976	1.013	0.423	2.426
Access to credit	1.365	0.671	4.140	0.042 *	3.914	1.051	14.571
Constant	−1.260	1.300	0.940	0.332	0.284		

Source: Nsele MK, Fyama JNM, Maréchal K, Dogot T. Factors Influencing the Sustained Adoption of Innovative Techniques by Urban Farmers in Lubumbashi, Democratic Republic of Congo. Agriculture. *2022; 12(8):1157. https://doi.org/10.3390/agriculture12081157.*

Number of observations = 202; −2 log likelihood = 174,475 a; R2 (Nagelkerke) = 0.506; p-value = 0.000, * Indicates the significant test at 5%, *** = Indicates a very highly significant test at 5%. CI = confidence interval.

Although it is not entirely clear that organic farms require more labor than other farms, experts agree that when value-adding activities (like processing and marketing) are developed, more labor input and a different labor distribution may be needed. An increasing amount of research indicates that organic farming requires more labor than chemical farming, even when it comes to labor needs. The University of Exeter conducted a second study on the effects of organic farming on the rural economy. The study surveyed 302 organic and 353 non-organic farms across three regions of England. The results confirmed that organic farms create more jobs than their non-organic counterparts, with 64% more jobs per farm and 39% more jobs per hectare.

9.1.3. Labor Dividend of Sustainable Agriculture

The farming workforce in India can be divided into three groups: families engaged in subsistence farming, migrant agricultural labor, and small-scale farm owners operating their own farms. Agriculture still provides for 65% of India's population, although its GDP share has steadily declined to 32% in 2008. Since the majority of agricultural workers are employed in the unorganized sector, they do not sufficiently benefit from government labor policies. The most underprivileged group in India's rural structure is the agricultural workforce. Their employment is erratic and their income is low. Underemployment of labor is a natural part of the family farming system, and seasonal unemployment is a feature of the agricultural sector. Based on the first A. L. E. C. 12, 189 days of wages were paid to adult male agricultural laborers for agricultural work, and 29 days were paid for non-agricultural work, i.e., 218 days total. For 75 days, they worked for themselves. Male casual workers only had jobs for 200 days a year, whereas attached workers had jobs for 326 days. There were 134 days in a year when women worked.

Sustainable agriculture places a strong emphasis on ecological balance, resource conservation, and treating workers fairly, in contrast to conventional farming, which frequently uses methods that can harm the environment and exploit labor. Long-term productivity increases, better livelihoods, and a healthier ecosystem are all possible outcomes of this strategy, and taken together, they can make the agriculture industry more resilient and just. Crop rotation, organic farming, agroforestry, water conservation, and integrated pest management are just a few of the methods that make up sustainable agriculture. These methods are intended to improve soil health, lessen reliance on chemical inputs, and have as little of an adverse effect as possible on the environment. Although ecological sustainability is frequently the main focus, these practices also have a major labor dividend that benefits local communities and workers. Better working conditions, equitable pay, skill development, job security, and more chances for entrepreneurship and rural development are all ways that this dividend is realized.

The enhancement of farm workers' working conditions is one of the main tenets of the labor dividend. Large-scale, traditional agriculture frequently entails hard work that can be dangerous, physically taxing, and inadequately paid. Sustainable agriculture, on the other hand, typically supports methods that are less taxing on the bodies of workers, such as fewer exposures to hazardous chemicals and less physically demanding techniques. For instance, organic farming lessens the need for artificial fertilizers and pesticides, which can be harmful to farm workers' health. The agricultural industry can establish a safer and healthier workplace, which will lower employee absenteeism, illness, and injury rates, by implementing safer and more sustainable farming methods. In sustainable agriculture, fair compensation and equitable pay are also essential components of the labor dividend concept. Wage exploitation has been linked to conventional agriculture, especially when it comes to seasonal and migrant workers who frequently receive low pay and little job security. Contrarily, fair trade principles—which seek to guarantee that workers receive a living wage commensurate with the value of their labor—are given top priority in sustainable agricultural models. This strategy helps to create a more stable and motivated workforce, which is crucial for the long-term survival of agricultural businesses, in addition to enhancing the financial well-being of employees.The empowerment and development of skills is another important component of the labor dividend. A wide range of abilities are frequently needed for sustainable agriculture, such as understanding soil management, organic pest control, water conservation, and the application of sustainable technologies. Sustainable agriculture contributes to the development of a skilled labor force that can adjust to new techniques and innovations by offering farm workers education and training. In addition to increasing productivity, this gives employees the chance to become leaders, launch their own farming businesses, or diversify their revenue streams by engaging in agritourism, beekeeping, or the production of organic goods. Initiatives to develop skills in sustainable farming can be especially empowering for women, who are frequently marginalized in traditional agricultural settings. These programs give them the opportunity to become financially independent and pursue entrepreneurial roles. There are wider community benefits associated with the labor dividend as well. By strengthening ties between farmers, consumers, and local markets, sustainable agriculture frequently boosts local economies. Direct connections between producers and consumers are made possible by programs like farmers' markets and community-supported agriculture (CSA), which can help farmers earn more consistently and predictably. By maintaining local agricultural production, these models lessen the carbon footprint associated with long-distance freight transportation while also generating jobs in local communities. Furthermore, more biodiversity, better soil fertility, and better water management are all possible outcomes of sustainable practices, and they all help create a more robust and fruitful agricultural ecosystem. By guaranteeing that the land will continue to be productive and fertile for upcoming generations, this resilience helps protect jobs.

Additionally, the advancement of social justice is part of the labor dividend of sustainable agriculture. Power disparities can result in inequality and exploitation in many agricultural systems, especially for marginalized populations like migrant workers, women, and children. Sustainable agriculture promotes moral work practices, such as ending child labor, empowering women, and treating all employees fairly. Sustainable agriculture can aid in addressing systemic problems of poverty, inequality, and exclusion that are common in rural areas by emphasizing social justice and moral hiring practices.

9.2. ROLE OF FAIR TRADE, ORGANIC FARMING, AND RESPONSIBLE SOURCING IN IMPROVING LABOR CONDITIONS

The agricultural industry has long been beset by problems like low pay, hazardous working conditions, exploitative labor practices, and a lack of social protections for employees, especially in developing nations. Initiatives like

Fair Trade, organic farming, and responsible sourcing have become effective means of addressing these issues by enhancing working conditions and encouraging more sustainable and equitable practices across the agricultural value chain. In addition to stressing the importance of environmentally friendly practices, these strategies put workers' rights and welfare first, guaranteeing them decent pay, secure working conditions, and chances for both professional and personal growth. The goal of the global movement known as "fair trade" is to improve the trading conditions between producers and consumers. By guaranteeing that farmers and laborers receive fair prices for their goods, it aims to rectify the power disparities in the supply chain and allow them to make a living wage and make investments in their communities. Standards such as fair compensation, secure working conditions, freedom of association, and the ban on child labor and forced labor must all be met to receive Fair Trade certification. By giving producers a minimum price guarantee for their goods, this system protects them from unfair trading practices and volatile market fluctuations while also promoting ethical labor practices. Additionally, Fair Trade premiums are frequently used to fund infrastructure, healthcare facilities, and educational initiatives in the communities where these workers reside. This can have a positive social and economic impact on the local communities. Fair Trade can have a life-changing effect on agricultural workers. It guarantees that they receive just compensation for their work, which lowers poverty and raises their standard of living. By assisting female farmers and business owners and promoting their involvement in cooperatives and leadership positions, Fair Trade further advances gender equality. In addition to giving women more economic power, this also aids in changing gender stereotypes that frequently restrict women's access to agricultural opportunities. Furthermore, Fair Trade certification promotes accountability and transparency, which facilitates product traceability and guarantees that workers' rights are upheld across the supply chain. Another sustainable agricultural method that is essential to bettering working conditions is organic farming. Organic farming minimizes the use of synthetic chemicals like fertilizers, pesticides, and herbicides, which can be harmful to farm workers' health, by emphasizing ecological balance, soil health, and biodiversity. Workers in conventional farming are frequently exposed to dangerous chemicals that can result in long-term health problems, such as cancer and respiratory disorders. Organic farming, on the other hand, removes or drastically lowers these risks, resulting in a safer and healthier workplace. To maintain the general health of the farming ecosystem, organic standards also place a strong emphasis on crop diversity, soil fertility management, and sustainable water use.

Fair labor practices are encouraged by organic farming, which also creates a safer working environment. To guarantee that workers receive fair treatment and a living wage, many organic certification programs incorporate labor rights requirements. More manual labor is frequently needed on organic farms than on conventional ones, which can lead to more job opportunities, especially in rural areas where jobs may be hard to come by. The focus on sustainable practices can result in better working conditions overall, even though these jobs can be physically taxing. This is because organic farming usually entails a wider variety of tasks and less exposure to hazardous chemicals. A tactic used by businesses to make sure the goods they buy and sell are made in a sustainable and socially conscious way is called "responsible sourcing." Using this method, supply chains are audited to ensure that suppliers follow safety regulations, sustainable environmental practices, and ethical labor practices. Businesses that use responsible sourcing are dedicated to accountability and openness, collaborating closely with suppliers to guarantee fair labor practices, safe working conditions, and adherence to labor laws. Businesses can contribute to bettering working conditions for agricultural laborers and guaranteeing that the advantages of sustainable practices are distributed throughout the supply chain by holding their suppliers to higher standards. The potential for positive global change is one of the main benefits of responsible sourcing. Strict ethical standards for suppliers have the power to change entire industries and promote the broad adoption of better labor practices. This is especially crucial in industries like coffee, cocoa, tea, and textiles where subpar working conditions and labor exploitation have long been problems. Companies can reduce the risks of labor rights violations, including fines, harm to their reputation, and supply chain interruptions, by implementing responsible sourcing programs. Businesses that respond to consumer demands for ethically sourced products by implementing responsible sourcing practices can both improve their competitiveness and advance social justice. Additionally, responsible sourcing can promote local development by giving small-scale farmers greater access to markets. A lack of resources, expertise, and international market connections frequently makes it difficult for small farmers to compete with large agribusinesses. Companies are able to build direct trade relationships with these farmers by offering them long-term contracts and fair prices through responsible sourcing initiatives. Farmers are able to enhance their production quality, invest in sustainable farming methods, and eventually raise their incomes thanks to this stability. In rural areas, responsible sourcing promotes economic growth and reduces poverty by empowering farmers and laborers.

9.2.1. Fair Trade Certification on Labor Rights and Economic Empowerment in Developing Countries

The yin-and-yang-like Fair Trade mark is the most well-known of all the ethical and sustainable certifications in the world. For nearly 20 years, this internationally recognized certification has distinguished companies that prioritize social responsibility and ethical business practices, empowering companies of all sizes to make a positive impact in the fashion industry and beyond. Fair Trade is more than just a label. It's a movement that supports eco-friendly production methods, safe working conditions, and living wages while enabling manufacturers to create high-quality goods. The numerous advantages of the Fair-Trade certification for bag manufacturers who are dedicated to ethical production will be discussed below, along with the certification's significance and demonstrated advantages. Safe working conditions, fair wages, and ethical production are guaranteed by Fair Trade certification. It is a widely accepted representation of these standards. Designed to promote international trade across a range of sectors, it contributes to the development of a more sustainable system by assisting in the negotiation of improved pay, working conditions, and general business dealings for farmers and laborers in developing nations. This mutually beneficial certification empowers farmers and workers to have more rights while encouraging manufacturers and businesses to make positive changes within their industries. Because the Fair-Trade label denotes a commitment to ethical labor practices, sustainable sourcing, and community development, it also assists customers in making well-informed decisions about what to buy that are consistent with their values. By securing a Fair-Trade Certification for every design, bag manufacturers can show their commitment to honesty and equity.

9.2.1.1. Standards and Requirements

Manufacturers must demonstrate adherence to the strict standards and requirements set forth by the Fair-Trade organization to receive certification. Fair Trade provides a range of standards suited to various business sizes and industries. Each one is purposefully made to give ethical justice and sustainable development top priority. They provide a comprehensive approach to fair trade between suppliers and manufacturers by combining social, economic, and environmental factors.

Bag manufacturers should look to the standards in the textile category, which includes the following set of requirements:

- Providing a Living Wage: Manufacturers are required to collaborate with suppliers who provide workers with a livable wage, as established by regional legislation, local unions, or recognized practices like the Anker method. Suppliers who begin the Fair-Trade certification process will have six years to implement pay increases.
- Offering Fair Employment Conditions: Providing reasonable working hours, overtime policies, and breaks within any employment contract.

- Encouraging Worker Empowerment: To guarantee job security, employees must have the freedom to take actions like forming or joining unions. Additionally, they must be permitted to attend training on workers' rights and file complaints if they are dissatisfied with any processes or problems. .
- Ensuring Occupational Health & Safety: Stringent health and safety measures should be implemented to protect workers, including using protective clothing, building safety, and training on handling hazardous materials.
- No Child or Forced Labor: Child and forced labor laws should always be enforced to ensure ethical and safe workplaces.
- Reducing Negative Environmental Impact: Employers must actively lessen their environmental impact by using chemicals responsibly and give priority to environmentally sustainable practices. This entails regulating chemical use and getting rid of harmful and cancer-causing materials.

9.2.1.2. Benefits and Significance

For bag manufacturers, Fair Trade certification is a highly beneficial badge of honor. It immediately improves a brand's reputation, particularly in regards to sustainability and ethical labor practices.

Your products will appeal more to eco-aware consumers and retailers if they bear the Fair-Trade mark, which may lead to new business, collaborations, and whole markets. Customers are willing to pay more for products that support sustainable and ethical practices, so your products will increase in value.

Every level of stakeholders will seek out companies that share their values, and obtaining Fair Trade certification will make it simple to demonstrate the breadth of your ethical standards.

On an internal level, Fair Trade certification promotes better business practices and raises employee morale. The movement places a strong emphasis on worker empowerment because it increases job satisfaction and retention and produces a workforce that is more skilled and stable. This promotes kindness, encouragement, and support, which has a cascading effect on the community. Morale may also be raised by the environmentally friendly and sustainable practices required for Fair Trade certification. Every team member will be aware of their ability to positively impact an industry that is frequently linked to environmental harm.

9.2.2. Organic Farming: Enhancing Worker Health and Safety through Sustainable Agricultural Practices

For small-scale growers, organic farming has the potential to provide better livelihood opportunities and greater socioeconomic benefits. In Southeast Asia, organic farming has transformed and served as a strategy for peasant growth and economic deficit reduction. India is an example of a progressive country where rural development depends heavily on sustainable agriculture. Agricultural wastes, including those from wheat, sugarcane, maize, rice, wood, and cotton, can be converted into clean fuels and/or petrochemical substitutes through pyrolysis, or they can be used to make board, binder-less board, and paper. Chemical hydrolysis of organic waste can also produce sugars, which can then ferment to produce bioethanol, which can then be used in the automotive industry. Organic farming is a way of thinking and doing that has a positive effect on the environment, animal welfare, high-quality food, physical fitness, efficient use of resources, and an egalitarian society. It also uses the market to achieve its goals and compensate for externalities that are internalized.

The research's practical significance lies in ensuring that society as a whole benefit both mentally and physically from sustainable organic farming production and consumption. Currently, society consumes chemicals in conjunction with food, which is negatively impacting human health. The United Nations adopted 17 sustainable development goals (SDGs) in 2015 intending to eradicate extreme poverty, hunger, and malnutrition by 2030 while also addressing environmental issues, including the global climate. This calls for the sustainable expansion of current agricultural lands through innovative technologies and collaboration among various sectors. Sustainable Development Goals of the United Nations, including Goal No. 1 no poverty), (Aim no. 3 Well-being and health), (Goal no. 4Education of quality), (Goal no. 5 Equality of gender), (Goal no. (Goal no. 11), sustainable cities and communities. 12 Conscientious production and consumption), and (Goal no. 13 Climate action) are all in line with the goal and parameters of the present study.

To improve food safety and farmers' livelihoods, sustainable intensification is a major concern, particularly in the hills. It is believed that growing organic agriculture has the potential to improve the standard of living for small peasants and advance countries' financial stability and physical and mental well-being in the future. Longitudinal studies showed noteworthy positive outcomes when high natural consumption was linked to a lower incidence of birth defects, allergic sensitization, middle ear infections, pre-eclampsia, metabolic syndrome, high body mass index, and human infertility. However, there is one more element to it that has yet to be identified, i.e., cognitive well-being of an individual as a result of eating organic food; however, because it offers superior nutrition to traditional food, the traditional eating pattern continues to play a significant role in the eating habits of the village communities today. Farmers have been eating organic food for generations, and their optimistic outlook on life is a reflection of this. Furthermore, it appears that organic farming may pave the way for the acceptance of dietary diversification as opposed to dietary simplification. In this case, farmers are crucial as the main growers, consumers, and adopters of organic produce. The economically disadvantaged segment and sustenance farming culture, as well as the lack of alternative sources of income, are the reasons why farmers who put in long hours of work do not receive the fair price and appreciation for their goods.

9.2.2.1. Pivotal Approach in Agriculture, Emphasizing Labor Conditions

A key strategy in agriculture, organic farming emphasizes sustainable methods that improve working conditions for agricultural workers as well as the environment. Organic farming's dedication to minimizing the use of dangerous chemicals, which greatly enhances working conditions, is one of its main benefits. Conventional farming methods frequently use a lot of artificial fertilizers and pesticides, exposing employees to dangerous chemicals that can cause both short-term and long-term health problems. Organic farming, on the other hand, stays away from these chemicals and uses natural substitutes that pose fewer health hazards. Because of the safer working conditions created by this decrease in chemical exposure, farm workers can perform their jobs without constantly worrying about the negative health effects of toxic substances. For example, crop rotation, composting, and the use of natural pest control techniques are common practices in organic farms. These practices not only preserve soil health but also improve worker safety. Furthermore, there is an inherent connection between organic farming methods and wider environmental advantages, which in turn lead to better working conditions. Organic farming contributes to the preservation of ecosystems that support sustainable agricultural practices by boosting biodiversity, improving soil health, and conserving water. More stable farming operations are a result of healthier ecosystems, which lessen the volatility that comes with monoculture farming, which can result in abrupt drops in crop yields and, ultimately, worker job security. Additionally, as organic farming becomes more popular around the world, there is a greater demand for organic products, which leads to job opportunities in rural areas where these farms are frequently found. Because organic farming can boost local economies by creating new job opportunities in farming and related industries like processing, distribution, and retail, this trend is especially important in areas that have experienced economic difficulties.

Organic farming contributes significantly to the creation of jobs, especially in rural areas where job opportunities may be scarce. Communities can boost local economies and promote the expansion of related enterprises and services by establishing organic farms. This all-encompassing strategy enhances job quality in addition to expanding the number of open positions. Compared to conventional agriculture, organic farming typically places a greater emphasis on fair labor practices, which may include better pay, benefits, and working conditions. A more equal workplace is produced by the certification standards that many organic farms follow, which support moral labor practices. Organic farming places a strong focus on community and collaboration, which frequently results in a culture that values employee welfare and raises job satisfaction and retention rates. Furthermore, it is impossible to ignore organic farming's educational value. Demand for sustainable agriculture-related knowledge and expertise is rising as organic farming methods become more and more popular. This need has prompted the creation of educational initiatives and training programs meant to give employees the know-how they need to prosper in an organic farming setting. In addition to improving individual skill sets, these programs enable employees to assume more important roles in the agricultural industry, encouraging innovation and leadership in organic practices. Organic farming contributes to the creation of a more skilled labor force by investing in workforce development, which raises job satisfaction and productivity. Furthermore, by emphasizing local food systems, organic farming helps to build stronger community bonds and a sense of pride among its employees. Employees who work on nearby organic farms frequently grow invested in the prosperity of their communities and the sustainability of their farming methods. In the end, this relationship may result in cooperative efforts to address regional concerns like environmental preservation and food security, strengthening the bonds within the community. The appeal of organic farming is further increased by the incorporation of social responsibility, which not only enhances working conditions but also fosters a shared commitment to the environment.

9.2.3. Responsible Sourcing: Ensuring Ethical Labor Practices Across Global Agricultural Supply Chains

The pursuit of better working conditions in international supply chains has seen a revolution thanks to responsible sourcing. Businesses can have a big impact on the lives of workers in a variety of industries by requiring suppliers to follow ethical standards and fair labor practices. It is becoming more widely acknowledged that a sustainable and just global economy depends on this move toward responsible sourcing. Fundamentally, responsible sourcing calls on companies to carefully examine their supply chains to make sure that all parties are operating in an environment that values the welfare of employees. This can involve paying fair wages, establishing safe working conditions, abstaining from child or forced labor, and adhering to labor laws. Businesses that uphold these values create a precedent that pushes suppliers to improve their labor standards, which eventually results in significant advancements in the treatment and working conditions of employees. Accountability and transparency are essential elements of responsible sourcing because they allow businesses to monitor their suppliers' labor practices and hold them responsible. Businesses can proactively detect and resolve problems in their supply chains by putting in place stringent auditing procedures and developing mechanisms for reporting labor violations. In addition to reducing the risks associated with unethical behavior, transparency fosters consumer trust, as more and more customers seek products sourced ethically. Customers are more inclined to support businesses that exhibit a sincere commitment to responsible sourcing when they are informed about the sourcing methods used in their purchases. Pressure from customers may also encourage businesses to uphold high labor standards across their supply chains, improving working conditions for employees at all levels.

Furthermore, in the context of responsible sourcing, the advantages of developing direct trade relationships with small-scale farmers cannot be emphasized. These connections frequently give farmers more power, enabling them to get fair prices for their produce and promoting a feeling of solidarity and community. Businesses can guarantee that small-scale farmers follow moral labor standards and give them the tools they need to enhance their operations by interacting directly with these farmers. This could entail providing training courses on cooperative business models, financial literacy, and sustainable farming practices that improve farmers' ability to prosper. A larger portion of the profits can return to the farmers as a result of direct trade relationships' reduction in the number of middlemen in the supply chain. Because it allows farmers to invest in better working conditions, equipment, and ultimately the welfare of their employees, economic empowerment plays a crucial role in improving labor conditions. Responsible sourcing has an effect that goes beyond specific businesses and farmers; it may lead to changes in labor standards across the board. A collective standard that raises the bar for labor conditions across entire sectors is created when several businesses embrace responsible sourcing practices. By working together, industry-wide frameworks and certifications that guarantee all participants follow moral labor standards may be established. Additionally, businesses aiming to improve their supply chain procedures can benefit greatly from the resources, tools, and assistance offered by groups and projects that support responsible sourcing. Businesses can increase their influence and establish an industry-wide culture of accountability and responsibility by banding together. Additionally, systemic problems like poverty and limited educational opportunities that lead to subpar working conditions can be addressed with the aid of responsible sourcing. Businesses can support the economic growth of the communities they source from by placing a high priority on ethical sourcing. Workers and their families can obtain healthcare, education, and other necessities with the help of fair pay and better working conditions, creating a positive feedback loop that benefits entire communities. In addition to improving people's lives, this all-encompassing strategy builds stronger communities and a more just and sustainable global economy. Responsible sourcing can have major business benefits in addition to ethical requirements. Businesses that place a high priority on ethical labor practices frequently discover that their dedication to social responsibility is well received by customers, which boosts brand loyalty and helps them stand out in the marketplace. Businesses that can show that they are committed to ethical practices can improve their reputations and increase sales at a time when consumers are pickier about what they buy. Additionally, by making investments in ethical sourcing, businesses can reduce the risk of labor violations and reputational harm, guaranteeing their long-term viability and profitability.

9.2.4. Desperately Seeking Decent Employment in Main Policy Answers

Despite the modernization paradigm's contestation, the mainstream policy firmly adheres to its fundamental goals, which are economic profitability and productivity. In that regard, decent employment and agricultural working conditions are discussed in the policy debate but are unquestionably not at the forefront of the agenda. Except a few specific projects, they are part of a set of guiding principles, general goals, and guidelines that could be likened to a declaration of rights at the beginning of a modernization plan. Reviewing the justification for current policy solutions and the features of the associated policy toolkit is crucial to understanding the discrepancy between principles and operational practices. Because of liberalization policies that were put in place starting in the 1980s, governments now only have the authority to correct market failures, primarily involving public goods, and to facilitate the role of the private sector in promoting ongoing modernization. But it's crucial to remember that, in contrast to the modernization paradigm that has been pushed since the end of the 19th century, this private-led approach to agricultural development is relatively new. Public policies have long had a significant impact on the sector's development, especially in wealthy nations like today. In connection with the development of public agricultural research and the rise of agronomic science, governments backed the first technical innovations, which were initially the product of private initiatives. They then sought to gradually assist all kinds of farms, particularly through price support, extension, and subsidized technical packages (e.g., marketing boards), rules, and safeguards (primarily through tariffs); the European Union's (EU) Common Agricultural Policy is a prime example.

To prevent market distortions, the World Trade Organization (WTO) currently forbids and denies this kind of assistance to developing nations. Nonetheless, in many nations, farmer support remains crucial (e.g., the EU or Japan), even if they are officially separated from production in accordance with WTO regulations and focus on food safety and environmental services. The alignment of donors and international organizations, formalized by significant alliances and development programs—such as the CAADP in Africa—consolidates the market-led agenda that has emerged over the last three decades and provides a single, cohesive vision for governments with limited resources. Increased productivity and better market accessibility are the foundation of its modernization goals, which are meant to boost farmers' profitability.

9.2.5. The Market-Driven Policy Architecture

Regarding the institutional and market environments, the current policy toolkit can be broadly categorized into two groups. Public goods, which include the rule of law, education, training, information, and research, are included in the first category.

The construction of roads that allow rural areas to open is often a necessary step, but the consolidation of a favorable legal framework for smallholders receives a lot of attention. It relates to first land rights because customary arrangements typically govern agricultural land in many areas, especially SSA. Securing land access and usage rights based on best practices and widespread recognition is the primary goal. The Voluntary Guidelines on the Responsible Governance of Tenure of Land, Fisheries, and Forests in the Context of National Food Security (VGGT) are one example of a common framework that has been adopted to support it. Addressing the lack of legal status for smallholders and family farming is the second area of focus. This is because, in many nations, young people—both men and women—cannot access land tenure and farm management, which are under the authority of elders. This is a serious problem that has existed and continues to exist worldwide. This hinders projects and technological advancements that would be easier for youth to embrace. Young family workers may find it easier to access collective assets and transfer farm assets between generations if family farms are granted legal status. This agenda includes regulations for decent work as well as the rights and status of family workers, especially women and young people.

Another focus is on enhancing research, education, and innovation systems by combining strategies that support "knowledge triangles" (research, education, and innovation). These triangles are based on multi-stakeholder knowledge platforms and innovation hubs, which are thought to be more effective than conventional linear top-down research systems. The improvement of various market types (agricultural products, inputs, credit, extension, and insurance) is the focus of the second category of instruments. These markets are frequently underdeveloped and imperfect because of supply shortages, monopolies, and oligopolies, poor regulation, and missing information. The modernization package, which includes high-yield selected seeds, pesticides, fertilizers, and herbicides, is restricted in access, and the high transaction costs have an adverse effect on profitability.

Business solutions are the best options for making inputs, tools, services, and information more accessible. Governments, regional organizations, agencies, and civil society organizations have all endorsed a few soft law tools offered by the general approach to encourage adherence to common standards. Among these are the establishment of clear regulations, as well as the proper consultation and involvement of stakeholders and right holders, all of which are mentioned in policy discussions and conversations with the private sector. Apart from the previously mentioned VGGT, which was approved by the Committee on World Food Security (CFS) in 2012, the Principles for Responsible Investment in Agriculture and Food Systems (also known as RAI), which was approved in 2014, focus on inclusive and sustainable economic development, gender equality and the empowerment of women and youth, honoring tenure and access to natural resources, and preserving cultural heritage and traditional knowledge.

9.3. THE IMPACT OF SUSTAINABLE PRACTICES ON WORKER RIGHTS AND LIVELIHOODS

The sustainable livelihoods approach is a way of thinking about the objectives, scope, and priorities for development activities. It is based on evolving thinking about the way the poor and vulnerable live their lives and the importance of policies and institutions. It helps formulate development activities that are (Figure 9.2):

- People-centered;
- Responsive and participatory;
- Multilevel;
- Conducted in partnership with the public and private sectors;
- Dynamic; and
- Sustainable.

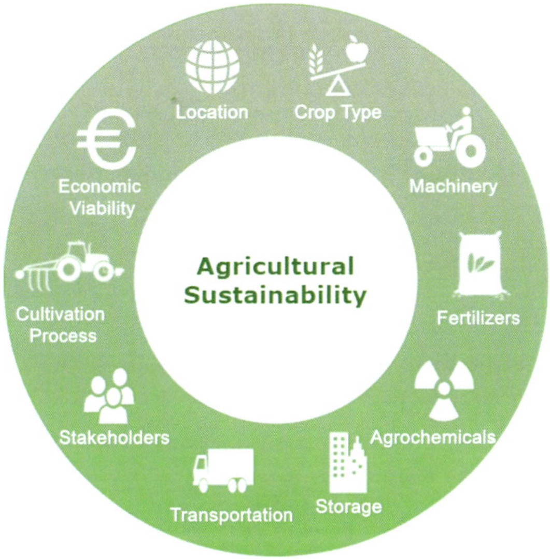

Figure 9.2. *The sustainable livelihoods framework.*

Source: Lampridi MG, Sørensen CG, Bochtis D. Agricultural Sustainability: A Review of Concepts and Methods. Sustainability. *2019; 11(18):5120. https://doi.org/10.3390/su11185120.*

Although it is not a cure-all, the sustainable livelihoods approach makes it easier to identify and prioritize actions that are grounded in the interests and opinions of those involved. Other instruments like sector-wide strategies, integrated rural development, and participatory development are not replaced by it.

The link between individuals and the broader enabling environment, which affects the results of livelihood strategies, is made, nevertheless. People's innate potential in terms of their abilities, social networks, access to material and monetary resources, and capacity to influence key institutions is brought to light. Originally created as a tool to help the industry steer clear of negative approaches to problem-solving, appreciative inquiry expands on this positive perspective. An inclusive process that emphasizes the positive rather than the negative, appreciative inquiry involves a community taking charge of information generation and collection before formulating plans based on the best past experiences (Duncan and Fridell, 2019).

9.3.1. Capital Assets

The sustainable livelihoods framework illustrates the relationships between the variables that limit or improve livelihood opportunities. One of the main ideas of the sustainable livelihood approach is that different households have varying access to livelihood assets. The livelihood assets, which the poor must often make trade-offs and choices about, comprise:

- **Human Capital:** e.g., health, nutrition, education, knowledge, and skills, capacity to work, capacity to adapt;
- **Social Capital:** e.g., networks and connections (patronage, neighborhoods, kinship), relations of trust and mutual understanding and support, formal and informal groups, shared values and behaviors, common rules and sanctions, collective representation, mechanisms for participation in decision-making, leadership;
- **Natural Capital:** e.g., land and produce, water and aquatic resources, trees, and forest products, wildlife, wild foods and fibers, biodiversity, environmental services;
- **Physical Capital:** e.g., infrastructure (transport, roads, vehicles, secure shelter and buildings, water supply and sanitation, energy, communications), tools, and technology (tools and equipment for production, seed, fertilizer, pesticides, traditional technology); and
- **Financial Capital:** e.g., savings, credit, and debt (formal, informal), remittances, pensions, wages.

9.3.2. Vulnerability Context

The state of insecurity regarding the welfare of people, households, and communities in the face of changes in their external environment is known as vulnerability. Unlike poverty line measurements, the concept of vulnerability better captures the processes of change as people enter and exit poverty. The external side of vulnerability is caused by shocks, seasonality, and critical trends; the internal side is defenselessness brought on by a lack of resources and the ability to deal with these. The vulnerability context includes:

- **Shocks:** e.g., conflict, illnesses, floods, storms, droughts, pests, diseases;
- **Seasonality:** e.g., prices and employment opportunities; and
- **Critical Trends:** e.g., demographic, environmental, economic, governance, and technological trends.

9.3.3. Policies and Institutions

The context of vulnerability and access to capital assets are not the only factors that affect livelihood strategies and results; the environment of structures and processes also has an impact. Structures are the public and private sector entities that establish and carry out laws and policies, provide services, make purchases, engage in trade, and carry out a wide range of other tasks that have an impact on people's quality of life. The laws, rules, policies, agreements, operational arrangements, societal norms, and practices that ultimately dictate how structures function are all considered processes. Without the right organizations and procedures in place to carry out policies, policy-determining structures cannot function effectively. All facets of livelihoods depend on processes. They offer rewards that encourage people to make wiser decisions. They either allow or prohibit access to resources. They make it possible for people to use markets to change one kind of asset into another. They significantly impact interpersonal relationships. One of the biggest issues facing the poor and vulnerable is that unless the government enacts pro-poor policies, which then trickle down to laws and even less formal procedures, the processes that frame their livelihoods may systematically restrict them (Duncan and Fridell, 2019).

9.3.4. Livelihood Strategies and Outcomes

The goal of livelihood strategies is to attain livelihood results. Various factors, including intensification versus diversification, short-term versus long-term outcomes, migration, and remittances, pensions and grants, natural resource-based activities, non-natural resource-based and off-farm activities, and some of which may compete, may be considered when making decisions about livelihood strategies. Projects and programs that benefit some at the expense of others are among the numerous development issues. More income, greater well-being, decreased vulnerability, better food security, more sustainable use of the natural resource base, and restored human dignity are all possible livelihood outcomes, though there may be conflict between them.

9.3.5. Implications

Thinking creatively is encouraged by the sustainable livelihoods approach. It releases development professionals from traditional methods that are frequently limited to problem identification and solution finding. To make development activities more process-oriented, it asks them to consider relationships and contexts. It forces them to search for several points of entry and to transcend a limited sectoral perspective and a uniform "community" view. It signifies a significant departure from the emphasis on project inputs, outputs, and the presumed mechanical connections between them. The sustainable livelihoods approach, in particular, emphasizes the significance of comprehending institutions by mapping the institutional framework and connecting the formal and informal, as well as the micro and macro. As a result, it advocates for a new approach to policy appraisal that shifts away from general guidelines and toward context-specific methods that let different, local viewpoints emerge in the policy framework (Duncan and Fridell, 2019).

9.4. SUSTAINABLE PRACTICES FOR PLANT UTILIZATION

The concepts of plant utilization and sustainable agriculture are closely related. The use of plants by humans encompasses a wide range of applications, such as food, medicine, fuel, building materials, and more. Contrarily, sustainable agriculture is a farming method intended to be profitable, socially and environmentally responsible, and environmentally benign. Since they enable us to maximize the natural resources at our disposal while simultaneously preserving the environment for coming generations, both ideas are crucial to building a more sustainable future. Sustainable agriculture and plant utilization are closely related because the latter relies on the wise use of plant resources to produce food sustainably and productively. In addition to providing food and other resources for human needs, sustainable agriculture can contribute to the preservation of the natural environment by using crops that are appropriate for the local environment and by utilizing practices that promote soil health and biodiversity. Apart from promoting sustainable agriculture, the use of plants can also be a significant strategy for tackling contemporary issues like poverty and climate change. For instance, using plant-based biofuels can help cut greenhouse gas emissions, and creating new plant-based medications can help treat illnesses that impact millions of people worldwide.

9.4.1. Plant Utilization

The term "plant utilization" describes the diverse ways in which people use plants for a range of objectives. Plants have been used by humans for 1000s of years; they have been used for food, medicine, clothing, shelter, and other purposes. Plant use is still a vital component of human society today, and there are numerous ways that plants help us in our day-to-day activities that are:

- **Food:** Food production is among the most significant applications of plants. Numerous nutrients that are vital to human health, such as proteins, carbohydrates, vitamins, and minerals, are found in plants. Plants are the source of many staple foods consumed worldwide, such as potatoes, rice, wheat, and maize. In addition to being significant nutrient sources, a variety of fruits, vegetables, and herbs are also frequently eaten for their therapeutic qualities.
- **Medication**: Many of the medications that are used today are made from plant compounds, making plants an important source of medicine as well. For instance, salicylic acid, which is present in willow tree bark, is the raw material used to make aspirin. Numerous other medications, such as quinine, codeine, and morphine, are also derived from plants. Furthermore, plant-based treatments are a

major component of many traditional medical systems, including Ayurveda and Traditional Chinese Medicine, which treat a wide range of illnesses.

- **Textiles:** A vast range of textiles, such as hemp, cotton, and linen, are also made from plants. One of the most popular plant fibers, cotton is used to create textiles such as clothing and bedding. Another significant plant fiber is linen, which is frequently used to create fine textiles like tablecloths and napkins. In addition to being a significant source of fiber, hemp is used to make a wide range of goods, such as paper, rope, and clothing.
- **Fuel:** Additionally, plants are a significant source of fuel, both as biofuels and as wood. For 1000s of years, people have used wood as fuel, and it remains a significant energy source in many regions of the world today. Additionally, biofuels—which are derived from plant matter—are growing in importance as an energy source. A range of plant materials, such as corn, sugarcane, and soybeans, can be converted into biofuels, which can then be used to heat buildings and power cars.
- **Construction Supplies:** Additionally, a vast range of building materials, such as bamboo, thatch, and lumber, are made from plants. Furniture and homes alike are made from wood, one of the most popular building materials in the world. Another crucial building material is bamboo, which is frequently utilized in Asian construction. In tropical areas, thatch—which is composed of dried plant material—is frequently used for roofing.
- **Other Applications:** In addition, there are many other applications for plants, such as industrial processes, landscaping, and ornamental plants. For instance, plants are utilized as ingredients in cosmetics and other consumer goods, and they are also used to make paper, chemicals, and medications. Plants are also utilized in religious ceremonies and cultural customs, and they are used to produce alcoholic beverages like wine and beer.
- **Possibilities and Difficulties:** Although there are many advantages to using plants, there are also many drawbacks. Making long-term, sustainable use of plant resources is one of the most difficult tasks. Habitat loss or overharvesting poses a threat to many plant species, potentially resulting in their extinction. In addition, the environment and the health of those who use them may suffer from the use of pesticides and other chemicals in agriculture.

9.4.2. The History of Plant Utilization

Humans have been using plants for 1000s of years. Hunter-gatherers, the first human societies relied on wild plants for sustenance, medicine, and other purposes. Humans started growing plants for particular uses as societies and agriculture advanced, which sparked the creation of new methods and tools for making use of plant resources. Consuming wild nuts and berries was one of the first documented human uses of plants. According to archeological findings, people were collecting and eating wild plants as early as 30,000 years ago. Humans developed the ability to grow plants over time, choosing and breeding cultivars with desired characteristics like bigger fruits, higher yields, and disease and pest resistance. Cultivating crops like wheat, barley, and rice, early civilizations in Egypt, Mesopotamia, and China were among the first to adopt agricultural practices. These societies also used plants for other things, like medicine to treat illnesses and fiber to make paper and textiles. The use of plants for therapeutic purposes advanced in ancient Greece and Rome. The theory of the four humors, which maintained that the body was made up of four fluids that needed to be in balance for optimal health, is attributed to the physician Hippocrates. To reestablish this equilibrium, numerous plant-based treatments were created, and the application of therapeutic plants became a significant component of ancient Greek and Roman medicine. With numerous monasteries growing therapeutic herbs and creating novel cures, the use of plants for medical purposes continued to advance during the Middle Ages. Increased knowledge and experimentation resulted from the printing press's development in the 15th century, which made it possible for information about plant-based medications and other applications of plants to be widely disseminated. The study of botany and plant physiology started to take shape as a scientific discipline in the 18th and 19th centuries. Scientists were able to examine plant structures and functions in greater detail thanks to new technologies like the microscope, which resulted in new findings and advancements in plant use. These days, plants are used for everything from fuel and building materials to food and medicine. New opportunities for creatively using plant resources are constantly being created by developments in genetic engineering and other technologies. Although using plants has many advantages, there may be drawbacks as well, such as the overuse of some plant species and the destruction of natural habitats. Therefore, it's critical to think carefully about how our use of plants affects the environment and to work toward sustainable practices that benefit both human health and the environment.

9.4.3. The Role of Plants in Modern Agriculture

Modern agriculture depends heavily on plants to produce food and other goods for both industrial and human consumption. Over time, agricultural methods have changed to increase crop yields and reduce the effect of diseases and pests on plant development. Plants have been the focus of extensive research in this area, leading to the creation of new technologies and methods for cultivating and using plant resources. Using pesticides to protect crops from diseases and pests is a crucial part of modern agriculture. There are many different types of pesticides, such as chemical sprays and genetically engineered crops that are pest-resistant. Pesticides can have detrimental effects on the environment and human health, even though they have helped to boost crop yields and decrease pest-related losses. As a result, there is continuous discussion regarding the use of pesticides in contemporary agriculture, and initiatives are underway to create more environmentally friendly methods of managing pests. Developing crops with desired traits like higher yield, drought tolerance, and disease and pest resistance through genetic modification is another important component of modern agriculture. A plant's DNA can be changed through genetic modification to achieve a particular desired outcome. Even though genetic modification has greatly advanced agriculture, there are still moral and environmental issues with it. As a result, the application of genetically modified crops in contemporary agriculture is still up for debate.

Plants are utilized for many industrial processes in addition to food crops. For instance, textiles are made from crops like cotton, flax, and hemp, while construction and other industries use lumber and other forest products. Interest in plant-based fuels, such as biofuels made from crops like corn and sugarcane, has increased recently. With growing concerns about environmental degradation and the effects of climate change, sustainable agricultural practices have gained importance in recent years. The goals of sustainable agriculture are to preserve the soil's long-term health, use fewer pesticides and other dangerous chemicals, and encourage the use of renewable resources like solar energy. Crop rotation and other methods to preserve soil health and stop nutrient depletion are also key components of sustainable agriculture. All things considered; plants play a crucial role in contemporary agriculture. In addition to serving as the basis for the global food supply, plants are essential for maintaining human health and well-being. Therefore, it's critical to keep creating new methods and technologies for responsibly and sustainably producing and using plant resources.

9.4.4. Medicinal Plants

Since ancient times, medicinal plants have been used to treat a wide range of illnesses, and they are still widely used in contemporary medicine. These plants are frequently used as an alternative or complementary treatment to traditional medicine because they contain naturally occurring compounds with therapeutic qualities.

Some examples of medicinal plants include:

- **Echinacea:** It commonly used to boost the immune system and treat colds and flu.
- **St. John's Wort:** It is used to treat depression and anxiety.
- **Aloe Vera:** It is used to treat burns, wounds, and other skin conditions.
- **Chamomile:** They are used to treat insomnia, anxiety, and digestive issues.
- **Turmeric:** It is used for its anti-inflammatory properties, and to treat a variety of conditions such as arthritis and cancer.
- **Garlic**: It is used to lower blood pressure and cholesterol, and to prevent heart disease.
- **Ginger:** It is used to treat nausea, vomiting, and other digestive issues.
- **Ginkgo Biloba:** It is used to improve memory and cognitive function.
- **Peppermint:** It is used to treat headaches, nausea, and digestive issues.
- **Milk Thistle:** It is used to protect the liver and treat liver diseases.

Medicinal plants are frequently used as topical preparations like creams or ointments, or as teas, tinctures, or capsules. Some medicinal plants can have negative effects or interfere with other medications, but most are safe when taken as prescribed. A healthcare provider should be consulted before using any medicinal plant as a treatment.

Apart from their traditional medical applications, medicinal plants have also been the focus of scientific investigation, with numerous studies looking into their potential uses and therapeutic qualities. As a result of this research, novel medications, and therapies based on naturally occurring substances in plants have been developed. Numerous traditional healing practices incorporate the use of medicinal plants to promote both physical and emotional well-being, demonstrating the cultural and spiritual significance of this practice. Medicinal plants provide a natural and frequently successful substitute or adjunct to traditional medicine, making them an essential part of healthcare overall. As studies into these plants and their medicinal qualities continue, their significance in contemporary medicine is only expected to grow.

9.4.5. Sustainable Forestry

To maintain the forest's long-term health and productivity, sustainable forestry involves managing forest resources in a way that strikes a balance between social, economic, and environmental factors. This method acknowledges that forests are intricate ecosystems that offer a variety of advantages, such as timber, clean water and air, biodiversity, and leisure activities.

The principles of sustainable forestry include:

1. **Regenerating Forests:** This involves replanting trees and managing forest growth to ensure a continuous supply of timber and other forest products.
2. **Conserving Biodiversity:** This involves protecting and preserving the many plant and animal species that inhabit forests, and ensuring that forest management practices do not cause harm to these species or their habitats.
3. **Protecting Soil and Water:** This involves minimizing soil erosion, preventing water pollution, and maintaining healthy soils to ensure the long-term health and productivity of the forest.
4. **Promoting Social Benefits:** This involves engaging with local communities and stakeholders to ensure that forest management practices are socially responsible and promote equitable access to forest resources.
5. **Ensuring Economic Viability:** This involves balancing economic considerations with environmental and social considerations to ensure that forest management practices are economically sustainable over the long term.

Stakeholder engagement, science, and research form the foundation of sustainable forestry practices, which can encompass a variety of management strategies like controlled burns, selective harvesting, and the application of natural pest control techniques. Certification schemes, which offer impartial confirmation that forest management techniques satisfy specific environmental, social, and financial requirements, can also be a part of sustainable forestry practices. The Forest Stewardship Council (FSC) is one example of a sustainable forestry certification program. It establishes guidelines for ethical forest management and certifies forest managers and producers who adhere to these guidelines. To make sure that forest management techniques adhere to FSC standards, an independent audit is part of the FSC certification process. For several reasons, sustainable forestry practices are crucial. Primarily, they contribute to the sustainability and productivity of forest ecosystems over the long term, which is crucial for delivering a variety of social, economic, and environmental advantages. By storing carbon and lowering greenhouse gas emissions, sustainable forestry techniques can also help to lessen the effects of climate change and encourage the wise use of forest resources. All things considered, managing forest resources through sustainable forestry is crucial to guaranteeing that forests will continue to yield the numerous advantages they provide to both humans and the environment.

9.4.6. Plant-Based Fuels

Renewable energy sources derived from organic materials like plants, algae, and agricultural waste are called plant-based fuels, or biofuels. Compared to fossil fuels, which are non-renewable and contribute to climate change, they are thought to be a more sustainable option. Bioethanol and biodiesel are the two primary categories of plant-based fuels. The fermentation of sugars and starches present in crops like corn, sugarcane, and wheat yields bioethanol. In contrast, biodiesel is made by removing the oils from plant sources like palm, canola, and soybeans. Compared to fossil fuels, plant-based fuels have many advantages. They can be made from a range of plant-based sources and are renewable, to start. Second, they lessen the effects of climate change by emitting fewer greenhouse gases than fossil fuels. Thirdly, they can lessen reliance on fossil fuels, which are frequently imported and whose prices are volatile.

Plant-based fuels have financial advantages in addition to these environmental ones. In addition to helping to generate new jobs in the biofuels sector, they can open up new markets for farmers and rural communities. Plant-based fuels do, however, present certain difficulties. The competition for resources and land is one of the primary obstacles. The production of food and other land uses may be hampered by the high land, water, and fertilizer requirements of plant-based fuels. Furthermore, deforestation and other environmental problems have been linked to certain plant-based fuels, like palm oil. Plant-based fuels must be produced responsibly and sustainably to meet these challenges. This includes producing biofuel from non-food crops and agricultural waste, as well as employing sustainable farming methods to reduce the negative environmental effects of biofuel production. All things considered, plant-based fuels present a viable substitute for fossil fuels, with the potential to produce sustainable and renewable energy. Despite the difficulties in producing them, the necessity of sustainable and ethical biofuel production methods that take social, economic, and environmental factors into account is becoming more widely acknowledged.

9.4.7. Plant-Based Foods

Foods that are made from plants and don't include any animal products are known as plant-based foods. They comprise a broad variety of foods, including grains, legumes, nuts, seeds, fruits, and vegetables. Growing worries about the effects of animal agriculture on the environment and human health, as well as moral dilemmas surrounding animal welfare, have led to a recent surge in the popularity of plant-based diets. A plant-based diet is linked to numerous health advantages. In comparison to animal products, plant-based foods are typically lower in calories and saturated fats and higher in nutrients like fiber, vitamins, and minerals. According to studies, eating a plant-based diet can lower your risk of developing chronic illnesses like heart disease, type 2 diabetes, and some types of cancer. Additionally, compared to animal products, plant-based foods have a smaller environmental impact. Deforestation, water pollution, and greenhouse gas emissions are all significantly impacted by animal husbandry. People can lessen the effects of climate change and their carbon footprint by eating more plant-based foods.

Vegetarian and vegan diets are just two of the many varieties of plant-based diets. Vegan diets forgo all animal products, whereas vegetarian diets forgo meat but may include dairy and eggs. As long as plant-based diets are varied and well-balanced, they can be wholesome and nourishing. Additionally, plant-based foods are increasingly offered in supermarkets and dining establishments. Nowadays, a lot of businesses sell plant-based alternatives to conventional meat and dairy products, like plant-based burgers and milk substitutes. Choosing plant-based foods has ethical implications in addition to health and environmental advantages. Animal welfare issues, such as factory farming methods and the use of hormones and antibiotics in animal feed, are linked to animal agriculture. All things considered, plant-based diets provide a healthy, ethical, and sustainable substitute for animal products. People can support more ethical food systems, lessen their environmental impact, and enhance their health by increasing the amount of plant-based foods in their diets.

9.4.8. Plant Conservation

The practice of safeguarding and conserving plant species and their environments is known as plant conservation. Because plants are essential for providing ecosystem services like soil formation, carbon sequestration, and oxygen production, it is a crucial part of biodiversity conservation. Plant species and their habitats face numerous threats, such as overexploitation, invasive species, habitat loss, and climate change. The goal of plant conservation initiatives is to counter these dangers and stop additional plant diversity loss. Habitat conservation is one of the main tactics for plant conservation. To guarantee that they can sustain robust plant populations, this entails preserving and managing natural habitats like forests, wetlands, and grasslands. Land acquisition, habitat restoration, and invasive species management are a few examples of habitat conservation strategies. *Ex situ* conservation is another crucial plant conservation tactic. This entails the preservation of plant species in settings other than their native habitats, like living collections, seed banks, and botanical gardens. For plant species that are endangered by habitat loss or other factors, *ex situ* conservation can act as a safety net.

To gain a better understanding of plant species and their habitats, plant conservation also entails research and monitoring in addition to habitat and *ex situ* conservation. In addition to monitoring plant populations and habitats to document changes over time, this can involve research on plant ecology, genetics, and population dynamics. Collaboration between a variety of stakeholders, such as governmental organizations, non-governmental organizations, educational institutions, and local communities, is also necessary for plant conservation initiatives. Partnerships that combine diverse backgrounds and viewpoints to create and carry out efficient conservation plans are frequently a component of successful plant conservation projects. One of the most important aspects of initiatives to safeguard and conserve biodiversity is plant conservation. We can contribute to the resilience and health of ecosystems and ensure that we continue to enjoy the many benefits that plants offer by preserving plant species and their habitats.

9.4.9. The Future of Plant Utilization

The future of plant utilization is likely to be shaped by a combination of factors, including advances in technology, changes in consumer preferences, and global trends related to sustainability and climate change. Here are a few potential developments that could impact the future of plant utilization:

- **Plant-Based Alternatives to Meat and Dairy Products:** It is anticipated that the demand for plant-based foods will keep rising as more people look for sustainable, healthier substitutes for conventional meat and dairy products. To produce more realistic plant-based meat and dairy alternatives that can satisfy consumer expectations for taste, texture, and nutrition, businesses are spending money on research and development.
- **Sustainable Agriculture:** Future farmers and agricultural businesses looking to lessen their environmental impact and adjust to the effects of climate change are likely to adopt more sustainable agricultural practices. Precision agriculture, which employs technology to maximize crop yields and minimize waste, and regenerative agriculture, which emphasizes enhancing soil health and biodiversity, are two examples of such practices.
- **Biotechnology:** Biotechnology developments are probably going to have a big impact on how plants are used in the future. This might entail creating new plant cultivars with enhanced characteristics like resistance to disease, drought, and increased yields. New plant-based products and materials, like bioplastics and biofuels, could also be made via biotechnology.

- **Medicinal Plants:** The use of medicinal plants is probably going to keep increasing as more people look for natural remedies instead of prescription drugs. New species of medicinal plants and more potent treatments may be discovered as a result of technological and scientific advancements.
- **Plant Conservation:** Conservation efforts will become more crucial as habitat loss and the effects of climate change continue to threaten plant diversity. More protected areas, *ex situ* conservation initiatives, and attempts to repair damaged habitats are a few examples of what this could entail.

Technology, consumer preferences, and sustainability considerations will probably all play a role in how plants are used in the future. As our world becomes more ecologically conscious and sustainable, plants will probably become more and more crucial in supplying us with food, medicine, and other necessities (Duncan and Fridell, 2019; Salgotra and Chauhan, 2023).

REFERENCES

1. Baldi, I., Gruber, A., Rondeau, V., Lebailly, P., Brochard, P., & Fabrigoule, C. (2011). Neurobehavioral effects of long-term exposure to pesticides: Results from the 4-year follow-up of the PHYTONER study. *Occupational and Environmental Medicine, 68*(2), 108–115. https://doi.org/10.1136/oem.2009.047811.
2. Duncan, J. M., & Fridell, G. (2019). *Fair Trade and the Future of Sustainable Agriculture: A Study of Smallholder Farmers in the Global South*. University of Toronto Press.
3. Gonzalez, C. A., & Osorio, J. E. (2018). Sustainable agriculture and decent work: A systematic review of the literature. *Sustainability, 10*(2), 400. https://doi.org/10.3390/su10020400.
4. Käkönen, M. (2020). Sustainable agriculture and its implications for decent work in rural areas. *International Journal of Agricultural Sustainability, 18*(1), 55–70. https://doi.org/10.1080/14735903.2019.1673729.
5. Renner, M., Sweeney, S., & Kubit, J. (2008). *Green Jobs: Working for People and the Environment* (World Watch Report 177). Worldwatch Institute.
6. Ronchi, L. (2002). The impact of fair trade on producers and their organizations: A case study with Coocafe in Costa Rica (PRUS Working Paper No. 11). University of Sussex, Poverty Research Unit.
7. Taylor, P. L. (2002). Poverty alleviation through participation in fair trade coffee networks: Synthesis of case study research question findings. (Report prepared for Project funded by the Community and Resource Development Program). The Ford Foundation.
8. Salgotra, R. K., & Chauhan, B. S. (2023). Genetic diversity, conservation, and utilization of plant genetic resources. *Genes, 14*(1), 174. https://doi.org/10.3390/genes14010174.

CHAPTER 10

POLICY FRAMEWORKS AND GLOBAL INITIATIVES FOR DECENT WORK

Contents

INTRODUCTION

Social justice and sustainable development depend on the advancement of decent work, which calls for strong legislative frameworks and international programs that give equal opportunity, labor rights, and employment standards a top priority. Fair pay, secure working conditions, and the ability for employees to exercise their rights are all components of decent work, which enhances economic stability and quality of life. Through its Conventions and Recommendations, which offer a fundamental framework for policy that nations can adopt and implement, the ILO has been at the forefront of creating international labor standards. The SDGs of the UN, especially Goal 8, which seeks to advance full and productive employment, sustained, inclusive, and sustainable economic growth, and decent work for all, are among the most important international initiatives addressing decent work. This objective highlights the necessity of laws that support entrepreneurship, create jobs, and end human trafficking and forced labor. The international community recognizes that social inclusion and environmental sustainability are essential components of economic growth, making decent work a comprehensive approach to development.

This is demonstrated by its inclusion in the SDG framework. Apart from the SDGs, many regional programs and accords seek to promote decent work. The European Union, for example, has put into effect the European Pillar of Social Rights, which delineates fundamental rights and principles for social protection, fair labor practices, and social inclusion. Member states are encouraged by this framework to create policies that support vulnerable groups in the labor market, advance gender equality, and improve the quality of jobs. These regional frameworks support international initiatives by enabling localized approaches that take into account particular social and economic circumstances.

To promote decent work, social discourse is also essential. Governments, employers, and workers' representatives can more easily develop policies that meet the needs and goals of all parties involved when they collaborate and communicate effectively. For example, workers can bargain collectively for fair pay and improved working conditions, and employers can learn how to develop a motivated and effective workforce. In addition to empowering employees, enhancing social dialogue tools encourages a sense of pride and dedication to common objectives. Promoting decent work faces both opportunities and challenges as a result of technological advancements and the evolving nature of work.

The emergence of digital labor platforms and the gig economy calls for flexible laws that uphold workers' rights and guarantee fair access to benefits and safeguards. To develop legislative frameworks that strike a balance between the need for labor standards and innovation, governments and international organizations must work together. To ensure inclusive and resilient economic reconstruction, initiatives such as the ILO's "Global Call to Action for a Human-Centered Recovery from the COVID-19 Crisis" highlight how crucial it is to give decent work a top priority in recovery efforts.

Nonetheless, there is still a long way to go before decent work is achieved globally, especially in light of social and economic injustices. To establish an inclusive labor market, policy frameworks must address systemic disparities, such as those based on gender, ethnicity, and immigration status. This calls for targeted interventions that give marginalized groups support, training, and access to resources in addition to legislation. The advancement of decent work as a fundamental human right depends on strong legislative frameworks and international initiatives. Stakeholders can work together to create a future where everyone has access to decent work by incorporating labor standards into initiatives for sustainable development, encouraging social discourse, and adjusting to changing workplaces. This dedication is essential for promoting resilient economies and prosperous societies globally, in addition to the well-being of individuals.

10.1. OVERVIEW OF INTERNATIONAL LABOR STANDARDS (ILO CONVENTIONS, SDGS)

With billions of people around the world depending on it for their livelihood, agriculture is a major contributor to the economies of many nations, especially developing ones. It is, nevertheless, also marked by serious labor issues, such as poor pay, hazardous working conditions, and a dearth of social safeguards. Promoting decent work, which is based on the notion that all workers should have the chance to obtain respect for their rights while working in a decent and productive environment, requires addressing these issues. Policies and practices that support decent work in agriculture are greatly influenced by international labor standards, especially those established by the Sustainable Development Goals (SDGs) and the International Labor Organization (ILO).

10.1.1. ILO Conventions

The ILO has developed numerous conventions aimed at establishing minimum labor standards, many of which are directly applicable to the agriculture sector. These conventions provide a framework for protecting workers' rights and improving their working conditions that are:

- **C188 - Work in Fishing Convention:** This convention aims to ensure decent working and living conditions for fishers. It covers various aspects of labor standards, including minimum age, working hours, and health and safety provisions. Although primarily focused on fishing, its principles can inform practices in aquaculture and agricultural sectors reliant on water resources.

- **C129 - Labor Inspection (Agriculture) Convention:** This convention mandates the establishment of effective labor inspection systems in agriculture to ensure compliance with labor laws and standards. It highlights the importance of monitoring working conditions, protecting workers' rights, and preventing abuses.
- **C184 - Safety and Health in Agriculture Convention:** This convention emphasizes the importance of occupational safety and health in the agriculture sector. It calls for measures to prevent accidents and illnesses related to agricultural work, addressing specific risks associated with pesticide exposure, machinery use, and ergonomic hazards.
- **C87 - Freedom of Association and Protection of the Right to Organize Convention:** This convention recognizes the right of workers in agriculture to form and join trade unions. Strengthening workers' rights to organize is essential for collective bargaining and improving working conditions.
- **C98 - Right to Organize and Collective Bargaining Convention:** This convention complements C87 by providing protections for workers engaged in collective bargaining, ensuring that their rights to negotiate fair wages and conditions are upheld.
- **C100 - Equal Remuneration Convention:** This convention emphasizes the principle of equal pay for work of equal value, which is particularly important in agriculture, where gender-based wage disparities are common. Implementing this standard is essential for achieving gender equality in the sector.
- **C111 - Discrimination (Employment and Occupation) Convention:** This convention addresses discrimination in employment and occupation, promoting equal opportunity and treatment for all workers. In agriculture, this is particularly relevant for addressing inequalities faced by women, migrant workers, and other marginalized groups.
- **C138 - Minimum Age Convention:** This convention sets the minimum age for admission to work in agriculture, aiming to eliminate child labor and promote education. Ensuring that children are not exploited in agricultural work is vital for their development and well-being.

10.1.2. Sustainable Development Goals (SDGs)

The United Nations' SDGs provide a comprehensive framework for promoting decent work and sustainable development globally. Several goals directly relate to labor standards and practices in the agriculture sector are:

- **Goal 1: No Poverty:** Ending poverty in all its forms everywhere is essential for enhancing decent work in agriculture. By promoting fair wages and access to resources, farmers and agricultural workers can achieve greater economic stability.
- **Goal 2: Zero Hunger:** Ensuring food security and promoting sustainable agriculture are critical components of decent work. This goal emphasizes the need for sustainable practices that support smallholder farmers and improve their livelihoods.
- **Goal 5: Gender Equality:** Achieving gender equality and empowering all women and girls is crucial in agriculture, where women often face significant barriers to employment and advancement. Promoting equal opportunities and addressing discrimination in the sector aligns with ILO standards on gender equality.
- **Goal 8: Decent Work and Economic Growth:** This goal explicitly focuses on promoting inclusive and sustainable economic growth, employment, and decent work for all. It highlights the need for policies that create jobs, promote entrepreneurship, and ensure safe working environments.
- **Goal 10: Reduced Inequalities:** Reducing inequalities within and among countries is vital for ensuring that all agricultural workers benefit from decent work opportunities. This includes addressing systemic discrimination and providing support for marginalized groups.
- **Goal 12: Responsible Consumption and Production:** This goal advocates for sustainable production practices, including in agriculture. It encourages responsible management of natural resources and sustainable practices that promote decent work while protecting the environment.
- **Goal 13: Climate Action:** The agriculture sector is highly vulnerable to climate change impacts. Implementing policies that promote climate resilience and sustainable practices is essential for safeguarding the livelihoods of agricultural workers.
- **Goal 17: Partnerships for the Goals:** Achieving the SDGs requires multi-stakeholder partnerships. Collaborations among governments,

civil society, the private sector, and international organizations are essential for implementing labor standards and promoting decent work in agriculture.

10.1.3. Challenges and Opportunities

Despite the existence of ILO conventions and the SDGs, numerous challenges persist in promoting decent work in the agriculture sector. These challenges include:

- **Informal Employment:** A significant portion of agricultural workers is employed informally, lacking access to legal protections and social benefits. Addressing informality is crucial for ensuring that workers can benefit from established labor standards.
- **Migration and Exploitation:** Many agricultural workers are migrants who face exploitation and poor working conditions. Strengthening protections for migrant workers is essential for ensuring decent work opportunities.
- **Climate Change:** The agriculture sector is significantly impacted by climate change, leading to unpredictable weather patterns and challenges in food production. Promoting sustainable practices and climate resilience is critical for protecting workers' livelihoods.
- **Access to Resources:** Smallholder farmers often struggle to access resources, markets, and fair prices for their products. Supporting smallholders through capacity building and access to credit is vital for enhancing their economic stability.
- **Technological Changes:** While technological advancements can enhance productivity, they may also lead to job displacement. Implementing policies that promote skills development and training for workers is essential for adapting to technological changes.
- **Cultural Norms and Gender Discrimination:** Deep-rooted cultural norms often perpetuate gender discrimination in agriculture, limiting women's access to resources and decision-making. Addressing these norms is crucial for promoting gender equality.

10.1.4. Implementation Strategies

To effectively promote decent work in the agriculture sector, a multi-faceted approach is necessary:

Strengthening Labor Laws: Governments should ratify and implement ILO conventions relevant to agriculture, ensuring that labor laws reflect international standards are:

- **Enhancing Social Protection:** Expanding social protection systems, including health care, pensions, and unemployment benefits, is essential for supporting agricultural workers, especially in times of crisis.
- **Promoting Collective Bargaining:** Encouraging the formation of trade unions and supporting collective bargaining efforts can empower agricultural workers to negotiate better wages and working conditions.
- **Investing in Education and Training:** Providing access to education and vocational training for agricultural workers can enhance their skills and improve their employment prospects.
- **Fostering Gender Equality:** Implementing policies that promote gender equality in agriculture, including equal pay and access to resources, is crucial for empowering women and addressing disparities.
- **Encouraging Sustainable Practices:** Supporting sustainable agricultural practices can enhance productivity while ensuring environmental protection. This includes promoting agroecology, organic farming, and responsible resource management.
- **Building Partnerships:** Collaboration among governments, NGOs, the private sector, and international organizations is essential for implementing strategies that promote decent work in agriculture.

Encouraging decent work in the agriculture sector is a complex issue that calls for combining global initiatives like the SDGs with international labor standards like ILO conventions. Together, stakeholders can address the systemic problems that give rise to inequality and labor exploitation and create a work environment where agricultural workers are respected for their rights, paid fairly, and have safe working conditions. Reaching these goals benefits communities and countries around the world by promoting social stability and sustainable economic growth in addition to improving individual well-being. Reaffirming the commitment to decent work principles is crucial to ensuring that the agriculture sector can prosper in a way that is sustainable and equitable, especially as the global landscape changes.

10.2. NATIONAL POLICIES PROMOTING DECENT WORK IN AGRICULTURE

Encouraging decent work in agriculture is crucial to raising the standard of living for millions of workers worldwide, especially in developing nations where agriculture serves as the main employer. To create frameworks that safeguard workers' rights, guarantee fair compensation, and provide safe working conditions, national policies are essential. The particular difficulties encountered in the agriculture industry, such as seasonal work, informality, and the vulnerability of marginalized groups like women and migrant workers, must be addressed by effective policies.

10.2.1. Legal Frameworks

Promoting decent work in agriculture requires a strong legal framework. Relevant international labor standards, especially those set forth by the ILO, ought to be ratified and implemented by national governments. This includes labor rights conventions like Occupational Safety and Health, Equal Remuneration, and the Right to Organize. These legal documents guarantee that agricultural workers have rights and protections consistent with international norms by offering guidelines that countries can modify to fit their unique circumstances. Furthermore, in light of the particulars of the agricultural industry, national labor laws ought to specifically address agricultural labor. This covers clauses pertaining to minimum pay, working hours, safety, and health standards, and dispute resolution procedures. Given that seasonal and migrant workers frequently experience exploitation and unstable working conditions, it is especially important to establish clear regulations regarding their treatment.

10.2.1.1. Social Protection Systems

To support agricultural workers, comprehensive social protection systems must be put in place. Healthcare, unemployment benefits, and pensions—all of which are frequently deficient in the agricultural industry, particularly for informal laborers—should be accessible through these systems. By incorporating agricultural workers into national social security programs, governments can strengthen social safety nets, lessening their susceptibility to economic shocks and guaranteeing improved living standards.

By introducing the Bolsa Família program, which offers financial aid to low-income families, including those involved in agriculture, nations like Brazil have made progress in this area. In addition to reducing poverty, this kind of policy promotes health and education spending, which raises the standard of living for agricultural laborers.

10.2.1.2. Access to Training and Education

Enhancing the skills of agricultural laborers through education and training is essential for fostering decent work. Vocational training programs that give workers access to contemporary farming methods, business acumen, and information about sustainable practices ought to be the main focus of national policies. In addition to boosting output, this investment gives employees more power so they can bargain for better working conditions. Initiatives in nations like Kenya and India, for instance, have effectively set up farmer training programs that emphasize market access and sustainable farming methods. These initiatives have the potential to greatly raise wages and produce a workforce with greater skill, which will boost the economy as a whole.

10.2.1.3. Gender Equality and Empowerment

Promoting decent work in agriculture requires addressing gender disparities. Despite making up a sizable share of the agricultural workforce, women frequently experience discrimination and unequal access to resources. To eradicate gender-based disparities, national policies should support women's access to training, credit, and land.

Gender equality can be promoted by laws that guarantee women's representation in decision-making bodies and offer incentives to women-owned cooperatives. By putting policies in place that encourage women's involvement in agriculture, nations like Rwanda have made impressive strides toward increased productivity and economic empowerment.

10.2.2. Promoting Sustainable Agricultural Practices

To ensure the sector's long-term viability, national policies should also promote sustainable agricultural practices. In addition to safeguarding the environment, sustainable practices increase agricultural systems' resilience, which benefits laborers. By offering subsidies, training, and access to sustainable technology, governments can facilitate this shift. Policies that support agroecological or organic farming, for example, can improve smallholder farmers' access to markets and create healthier working conditions. Adopting such practices often results in increased food security and better living conditions for agricultural laborers in the adopting countries.

10.2.2.1. Strengthening Cooperatives and Organizations

Another successful tactic for advancing decent work in agriculture is to support the establishment of worker organizations and cooperatives. By offering a forum for collective bargaining, these groups can help employees bargain for improved pay and working conditions. They can also act as a support network, providing tools and training to increase output.

Agricultural cooperatives in nations like Spain and Italy are successful examples of cooperative movements that have empowered farmers, increased their bargaining power, and improved their general economic circumstances.

A more equitable agricultural sector can be achieved by encouraging the formation of such cooperatives through advantageous legal frameworks and financial incentives.

10.2.2.2. Ensuring Access to Markets

A key component of encouraging decent work for agricultural workers is market accessibility. Fair market conditions that allow smallholder farmers to effectively compete should be the main goal of national policies. This entails removing obstacles to market entry, supporting the infrastructure, and guaranteeing just pricing practices.

Government programs that create local markets and increase supply chain transparency, for example, can greatly increase small farmers' negotiating leverage. By creating agricultural marketing cooperatives that link farmers and consumers directly, nations like Ethiopia have advanced in this field, boosting farmers' incomes and lowering middlemen's exploitation.

10.2.2.3. Monitoring and Enforcement Mechanisms

Strong enforcement and monitoring systems are necessary for the efficient application of policies. To guarantee adherence to labor regulations and standards in the agricultural industry, governments ought to set up labor inspection systems. The three main objectives of these inspections should be to promote occupational safety, expose exploitative practices, and protect workers' rights. Finland and other nations with effective labor inspection systems serve as helpful models. Frequent inspections ensure that agricultural workers receive fair treatment and that their rights are respected by holding employers accountable.

10.2.3. Multi-Stakeholder Collaboration

Various stakeholders, including government agencies, employers, labor unions, and civil society organizations, must work together to promote decent work in agriculture. To identify problems and create all-encompassing solutions, national policies should encourage communication between these groups. Initiatives for decent work can also be better implemented through partnerships with NGOs. For example, partnerships that emphasize community involvement, advocacy, and training can result in more successful policy outcomes, especially when it comes to reaching underrepresented groups.

Ensuring the well-being of millions of agricultural workers worldwide requires national policies that support decent work in agriculture. Governments can foster a favorable environment for agricultural workers by enacting strong legal frameworks, improving social protection, funding education, addressing gender inequality, and encouraging sustainable practices. The successful implementation of these policies and the realization of concrete benefits for workers will depend on stakeholder collaboration as well as efficient monitoring and enforcement. In the end, a dedication to decent work in agriculture promotes sustainable and equitable growth in the industry while also enhancing individual livelihoods and advancing larger economic and social development objectives.

10.3. ROLE OF GLOBAL PARTNERSHIPS AND MULTI-STAKEHOLDER INITIATIVES: AN EXAMPLE FROM INDONESIA

Addressing complicated global issues that cut across national boundaries, like social inequality, public health emergencies, and climate change, requires international collaborations and multi-stakeholder projects. Governments, businesses, civil society groups, and international organizations come together under these cooperative frameworks to share resources, exchange knowledge, and promote creative solutions.

These programs improve the efficacy and sustainability of responses to urgent global issues by utilizing the distinct strengths and viewpoints of various stakeholders. Facilitating knowledge transfer and best practices is one of the main benefits of international partnerships. A more thorough grasp of the issues is made possible by the unique skills and perspectives of the various stakeholders. For instance, a collaboration between governments might concentrate on creating policies, while private businesses might offer technological advancements. By offering grassroots viewpoints, civil society organizations can guarantee that underrepresented communities' voices are heard. In addition to improving the problem-solving process, this cooperative approach increases stakeholder trust and promotes a feeling of shared accountability for results.

Furthermore, international collaborations frequently mobilize financial resources more successfully than individual initiatives. Participants in multi-stakeholder initiatives can combine funds to support large-scale projects, which lessens the workload for individual organizations. Initiatives are able to take on challenging objectives that would be impossible for a single actor to accomplish thanks to this group's financial support. One example of how concerted financial efforts can result in notable health improvements in low-income countries is the Global Fund to Fight AIDS, Tuberculosis, and Malaria.

These collaborations improve the scalability and reproducibility of effective interventions in addition to mobilizing resources. When stakeholders work together, they can find strategies that work and modify them to fit various situations. This flexibility is essential in a world that is becoming more interconnected, where solutions must take into account global lessons while also being sensitive to local needs. The United Nations' SDGs, which encourage nations to cooperate toward shared goals while adjusting their strategies to suit unique local conditions, provide a framework for these kinds of partnerships.

However, fair participation and openness are critical to the success of international collaborations. Some stakeholders may be marginalized as a result of power disparities, especially those from underrepresented communities or developing nations. Building trust and encouraging true collaboration require making sure that all opinions are heard during decision-making processes. Furthermore, it is imperative to establish unambiguous communication and accountability protocols to monitor advancements and tackle obstacles as they emerge.

To address the complex issues of our day, international collaborations and multi-stakeholder projects are essential. These frameworks facilitate knowledge exchange, improve resource mobilization, and encourage creative solutions by encouraging cooperation amongst various actors. Although equity and transparency present certain difficulties, inclusive partnerships have enormous potential advantages. These initiatives are more important than ever because of the growing need for collaborative efforts to address global issues as the world grows more interconnected. Stakeholders can cooperate to build a more sustainable and just future for everybody if they maintain their commitment and cooperation.

10.3.1. Multi-Stakeholder Partnerships: A Tangible Instrument to Support the Implementation of the 2030 Agenda at the Local Level

Two primary factors—the ongoing "wicked problems" of sustainable urban development and the evolving landscape of local governance—make partnerships crucial for accomplishing global goals at the local level. Rittel and Weber described "wicked problems" as ones that are challenging to resolve because of their intricacy, conflicting concerns, intricate interdependencies, and quickly shifting surroundings. SDG 11 of the 2030 Agenda primarily aims to address urban challenges. The wicked problems at the nexus of economics, society, technology, environment, and governance are how to make cities and human settlements inclusive, safe, resilient, and sustainable.

The complexity of urban issues is further increased by the fact that cities are a nested arena where development is influenced by actors from multiple sectors and levels. Therefore, knowing how to create strong, inclusive partnerships is essential to achieving the SDGs in urban areas. Over the past 20 years, partnerships have progressively gained popularity and become an important part of development policy, primarily through international partnerships for development and more recently in the fields of climate change and sustainable development. Nonetheless, global, multilateral, and intergovernmental domains have been the subject of the majority of partnership literature to date. The North-South viewpoint has dominated the conversation, with less attention paid to emerging economies. If they have, they primarily deal with national partnerships, which deviates from the significance of addressing local multi-stakeholder partnerships.

We examine the function of locally contextualized multi-stakeholder partnerships in implementing the 2030 Agenda framework in light of this gap. By concentrating on "the state," our conversation aims to support multi-stakeholder partnerships, which are essential for realizing the revolutionary change in governance brought about by the 2030 Agenda and making a positive impact on people's well-being. This study drew on policy scholarship that documents group initiatives and collaborations. The primary data for this section came from focus groups and interviews with stakeholders and policymakers at the local and national levels. We reviewed the key takeaways from the creation of the Indonesian Multi-stakeholder Partnerships Guidelines.

The format of this article is as follows. To support policymaking and implementation, we first discuss the connections between multi-stakeholder partnerships and sustainable development, taking stock of the continued need for and advancements in these partnerships. The role of nonstate actors and important elements influencing the implementation of multi-stakeholder partnerships are then discussed. The functions of multi-stakeholder partnerships as a governance mechanism in different policy stages are covered in this section. These partnerships have the ability to address resources, guarantee knowledge production and peer-learning, and keep an eye on any gaps in the 2030 Agenda's implementation. After that, we provide three examples of SDG partnerships where nonstate actors are actively guiding the implementation process in Indonesian cities.

10.3.2. Multi-Stakeholder Partnerships for the Implementation of the 2030 Agenda

"A manifestation of inclusive development emphasizing the leave no one behind principle by engaging most vulnerable people in the development process equitably" is one of the most recent definitions of multi-stakeholder partnerships in the context of the 2030 Agenda (POST 2015/UNU IAS, 2014). "Voluntary initiatives undertaken by governments, intergovernmental organizations, major groups, and other stakeholders, which efforts are contributing to the implementation of intergovernmental, agreed on development goals and commitments," according to another report, characterize multi-stakeholder partnerships for sustainable development. The 2030 Agenda resolution text states that multi-stakeholder partnerships should "reach the furthest behind first" and be inclusive, leaving no one behind. These definitions show how the idea of partnerships has changed over time, moving from a way to help government-led initiatives achieve development goals to a way to implement them through inclusive engagement. This entails changing the focus of international collaborations from aid to development efficacy, where candid communication between various stakeholders on an equal basis is actively sought.

Multi-stakeholder partnerships, which involve a wide range of institutional arrangements between the government, private sector, civil society, the United Nations, and other multilateral agencies serving a variety of purposes to address development challenges, are frequently used interchangeably with cross-sector partnerships (Hazlewood, 2015). As multi-stakeholder partnerships offer an alternative to conventional state-led, top-down development approaches, they gradually emerge as a crucial paradigm in sustainable development. The unprecedented cost of funding the 2030 Agenda and the goal of leaving no one behind makes it more urgent than ever to overcome the limitations of traditional, state-led development, which calls for the active involvement of nonstate actors in development initiatives. The interdependence of partnerships, a healthy economy and society, and environmental protection is specifically acknowledged in the 2030 Agenda. Therefore, all development actors are acknowledged by the 2030 Agenda as essential to achieving global ambition through broad and inclusive partnerships.

In addition to addressing the regulatory gaps in the development context, multi-stakeholder partnerships may be able to close a participation gap and close the funding gap for sustainable development. The 2030 Agenda's anticipated widespread involvement of multi-stakeholder partnerships across sectors shapes the implementation response by bringing together local resources, interests, and needs. To maximize the impact of available resources and to form collective efforts on the global partnerships for SDGs, as reflected in SDG17, the SDGs also call for development aid to be more effective and catalytic. Based on critical voices that demand that the 2030 Agenda be delivered with inclusivity, credibility, impact, and quality, these partnership models are derived from the traditional literature on partnerships. Their guiding principles for collaborating on the SDGs are defined in the literature. The most widely accepted of these are transformative, inclusive, and accountable. It implies that addressing structural and systemic deficiencies should be the goal of multi-stakeholder partnerships.

The conceptual and implementation challenges of creating and managing multi-stakeholder partnerships may be rooted in the challenges associated with the range and scope of stakeholders needed, given the scope and ambition of the SDGs as well as the intricate nature of interactions involved in partnerships, coalitions, and networks. Conflicts between investment priorities, political disparities across scales, and particular socio-cultural and socio-technical aspects of program and policy conceptualization are challenges facing multi-stakeholder partnerships (Oburu and Yoshikawa, 2018). National governments can form partnerships in a variety of ways, such as through the National Statistics Agency, the leading ministry, or inter-ministry commissions. The emphasis placed on the private sector's role in the multi-stakeholder partnership of the SDGs presents another difficulty, as it is currently expected to follow a more sustainable course. According to Scheyvens, Banks, and Hughes (2016), business interest groups have frequently applied a rather limited interpretation of sustainable development principles in practice, which has resulted in a managerial focus through win-win scenarios and little interest in social sustainability. Therefore, to acknowledge the importance of the private sector in pursuing the 2030 Agenda and vice versa, a new paradigm in development thinking is required (UN Global Compact, 2014). The next query is whether multi-stakeholder partnerships could serve as a venue for companies driven by profit to significantly contribute to the SDGs' accomplishment.

10.3.3. Multi-Stakeholder Partnerships and the Role of Non-state Actors in Supporting the Achievement of SDGs in Indonesia

Multi-stakeholder collaborations that help Indonesia achieve the SDGs are still in their infancy. The challenge of creating inclusive and significant partnerships that endure at international partnerships for sustainable development is reflected in it. At both the national and local levels, the intricacy of managing cross-sector and cross-actors—which are marked by resource, regulatory, and participation gaps—presents itself. Some of the regulatory gaps have been filled, and the expectations for the contribution of multi-stakeholder partnerships have increased with the enactment of the Presidential Decree and ministerial regulations outlining the role of nonstate actors in supporting the implementation of the 2030 Agenda. As the Government of Indonesia supports the multi-stakeholder partnerships approach in the most recent National Mid-Term Development Plan 2020–2024, the expectations are further reinforced in the most recent policy development. It is anticipated that multi-stakeholder partnerships will strengthen public finances, spur local innovation, and improve the quality of public services within the framework of decentralization and local autonomy. Nonstate actors in particular have enthusiastically embraced this development. Since the 2002 Johannesburg World Summit on Sustainable Development (WSSD), there has been a global push for "Type II partnerships" between state and nonstate actors, such as the UN system and development partners, to advance global actions on local issues. There are, however, few practices at various phases of policy implementation and few insights into how the multi-stakeholder setting could support national and local policy implementation processes.

Key players in Indonesia's localization of the SDGs include academia, the private sector, philanthropy, civil society organizations, and national and local governments. These important players are given distinct roles within the governance framework established by the Presidential Decree. The implementation of the 2030 Agenda is coordinated by the national government, which also serves as its secretariat and steering committee. Promoting, advancing, and facilitating implementation at the local level is the responsibility of local governments. This includes creating baseline evaluations of the state of development in their area and the financial outlook. Nonstate actors are assigned supporting roles in consultation, policymaking, and implementation, monitoring, and awareness-raising for the SDGs, in contrast to the Millennium Development Goals (MDGs). Despite growing interest in using nonstate resources to increase domestic financial capacity and knowledge sharing from academia and civil society to improve accountability, the development provision roles are not clearly defined beyond active consulting. In any case, the majority of our interviewees concurred that, as a result of the 2030 Agenda, civil society organizations had assumed more significant roles in the policymaking process. It might be explained by the government's critical approach, which it adopted after learning from the experience of enlisting nonstate actors to deliver particular community-related development policies. Government interest has been drawn to partnerships with civil society organizations because of their capacity to promote many development areas, including: (i) health; (ii) gender equity; (iii) decreasing economic inequality; (iv) bolstering the rights of individuals with disabilities; (v) environmental resilience; and (vi) enhancing the standard of governance, including non-discrimination and transparent governance. The fact that civil society organizations represent marginalized and vulnerable groups and provide policy inputs to local governments through bottom-up approaches has also been widely acknowledged by policymakers as offering alternative solutions in the policymaking process.

Increasing the number of stakeholders involved may have a good impact on how the 2030 Agenda is implemented locally. For instance, the creation of an SDGs Center involving research institutes, universities, and other higher education institutions may encourage the participation of various stakeholders in policymaking and implementation by providing information and big data, which could result in more science-based policies. These R&D centers may serve as breeding grounds for novel technologies, cutting edge policy and data analysis, and training grounds for present and future generations of leaders. State universities in Jember and Bengkulu are just two of many that have helped local governments implement the 2030 Agenda by offering policy support. The SDGs could be effectively localized at the local level with the help of this kind of technical assistance.

10.3.4. Exploring Multi-Stakeholder Partnerships at Different Stages of Policy Implementation

From the standpoint of urban planning and governance, attaining the SDGs locally necessitates a thorough plan and budget to guarantee integrated implementation, careful monitoring, and ongoing assessment of advancement. At every stage, inclusive and significant partnerships should be sought to guarantee that the application is inclusive and responsive to the principle of leaving no one behind. Therefore, it is important to look for multi-stakeholder partnerships that involve all local stakeholders in implementing the 2030 Agenda, including when deciding on goals, targets, indicators, and implementation strategies. There are at least five primary stages that should be followed, according to case studies in Indonesia that have been observed and compared to the partnership phases suggested by the available multi-stakeholder partnership guidelines.

10.3.4.1. The Initiation Stage

To come to agreements and create plans and strategies for the partnerships, all parties involved must first build relationships with one another. The fact that cooperation between state and nonstate actors is essential to achieving the SDGs must be made widely known. As a result, the "leave no one behind" principle is a basic standard that will be covered at the beginning. The readiness to exchange information to establish mutuality, trust, and efficient channels of communication is another important component. Putting a formal agreement in place, like a memorandum of understanding or shared commitment could be pursued to guarantee future involvement of stakeholders. Despite the existence of individual interests, it will be simpler to establish shared goals and objectives when there are dedicated stakeholders. At this point, all stakeholders must work together to create and agree upon a set of partnership rules and norms, regardless of whether they want to formally formalize their relationships or decide to keep them informal. Commonly accepted guidelines and standards serve as a foundation for inclusivity and are crucial for overseeing and preserving collaborations.

10.3.4.2. The Formation Stage

Focusing on activities to develop and agree upon strategies and managing the partnership is necessary at this point. This phase is crucial because it calls for addressing and creating strategic, efficient, and responsive planning—that is, anticipating and taking into account potential risks as well as the likelihood of success. To maintain the voluntary cooperation of stakeholders, strategic planning necessitates accountability and transparency among them. To establish multi-stakeholder partnerships, it is necessary to emphasize integrated procedural and institutional arrangements, which include unambiguous rules, structures, and protocols that are acceptable to all parties as well as clear roles and responsibilities for each stakeholder. Mutual trust and a common understanding of the SDGs will grow as a result of these processes.

10.3.4.3. The Implementation Stage

The agreed-upon partnerships' action stage is the implementation stage, during which time resources are gathered and turned into plans and initiatives. Managing the relationships with all stakeholders is essential at this point. If all partners and stakeholders are transparent and have clear expectations about the indicators and targets that could or might not be met during implementation, it could be well-managed. Establishing mechanisms for risk and conflict mitigation as well as developing monitoring systems based on results and outputs provided within the framework of a multi-stakeholder partnership are also crucial.

10.3.4.4. The Monitoring and Learning Stage

At this point, all parties involved should concentrate on assessing the partnerships' performance and accomplishments by looking at the results of SDGs implemented through multi-stakeholder partnerships, among other things. Furthermore, transparency and accountability are elements that must be assessed to enhance collaboration. We'll review other factors that have an impact, like gender equality, inclusivity, and the application of leaving no one behind, particularly the marginalized and vulnerable groups. The review's findings are incorporated into the partnerships' changing strategies as lessons learned to strengthen them.

10.3.4.5. The Expansion Stage

For multi-stakeholder partnerships to be more robust and sustainable, they have reached an expansion stage where wider coverage is required to allow stakeholders to find solutions for the various challenges in the partnership. At this stage, two actions must be taken: (i) making sure that monitoring and evaluation results are available as references for input and feedback; and (ii) acting to strengthen and improve the partnerships to make them more effective and efficient and to achieve a wider impact by diversifying issues (i.e., targets for the SDGs), target audiences, funding sources, and involvement of new stakeholders.

10.3.5. Exploring Different Mechanisms of Multi-Stakeholder Partnerships for the SDGs in Indonesia

There is a lot of potential for Indonesia to use multi-stakeholder partnerships as a tool to help achieve the SDGs. Within the framework of local development, there are many problems that might be solved with cooperation. The following two examples demonstrate how multi-stakeholder partnerships assist communities and local governments in addressing poverty reduction, boosting school attendance, and enhancing the standard of education for the underprivileged through scholarships. An advocacy campaign for the creation of facilities and accessibility for people with disabilities is depicted in another example. These three case studies are explained using various multi-stakeholder partnership mechanisms.

The identifying factors are based on dimensions that determine the mechanisms of participation: who participates, how participants communicate and collaborate to make decisions, and how discussions are connected to public actions or policy. We also employ methods of cooperation. Our elements include partnership drivers, leadership, collaboration styles, and degree of formalization, results, and new value that partnerships add. These elements make up a space where the multi-stakeholder partnership mechanism may be found. Based on the research on inter-organizational collaboration, we also determined the advantages and disadvantages of multi-stakeholder partnerships. These expenses and advantages include relationship-building, service enhancement, and institutional cost benefits.

10.3.5.1. Local Government-Driven Partnerships

The most common type of partnership for service delivery in regions where the government is viewed as the exclusive supplier of public services is a government-driven partnership. The notion that the government ought to be smaller lies at the heart of this. In Indonesia, public financing plays a pivotal role in driving local development and public services. Governments function as regulators, providers, managers, or any combination of these. Since the local government serves as the partnership's initiator and primary stakeholder, public-private partnerships frequently fall under this heading. It is anticipated that the SDGs will lead to a rise in these kinds of partnerships in Indonesia, as local governments look to nonstate actors, especially the private sector, for assistance in implementing programs related to the SDGs. Partnerships led by local governments are frequently motivated by policies.

It indicates that local governments typically look to form alliances in areas of policy where resources are scarce and pressures are greatest. Social, economic, political, or environmental problems could be examples of constraints that exacerbate the pressures. Trenggalek, one of the poorest cities in East Java, has multi-stakeholder partnerships for SDG1, which were started by the local government. The partnership's two goals are to find alternate funding sources and enhance the current poverty reduction initiatives by guaranteeing the accuracy of the poor's data. Initiation of the partnership phase began in 2016. The local government started this process in response to social media criticism of the execution of programs aimed at reducing poverty.

The government's first action was to verify the data on the impoverished by holding meetings and conversations with local communities, village chiefs, and heads of districts. Involving the local communities, who expressed disapproval of the government's performance, was their second action. The so-called "Pink Force" was tasked with educating the impoverished about the program, helping them to complete the necessary paperwork to obtain government assistance, and verifying the poor's data. This community recruitment initiative is a selective recruitment mechanism, with participation processes designed to address civil disputes throughout. By offering incentives to those who are less likely to participate, selective recruitment can take place passively. Active recruitment is what we saw in Trenggalek, where particular communities are invited based on how urgent the issue is to them. This kind of participatory mechanism is an illustration of Fung's mini-publics, in which people are consciously invited to specific groups to deliberate on issues of public interest.

After the Gertak Program was redesigned and re-planned, a new partnership was formed with the National Zakat Agency (also known as BAZNAS), an external actor, to further strengthen the relationship between state and nonstate actors. Through the partnership, Islamic tax also referred to as zakat, will be used to fund initiatives to improve public services and combat poverty. For the impoverished, it covers housing, education, and health care. One noteworthy aspect of the Gertak Program's implementation is how the participation mechanism improves justice by giving citizens more say in decision-making. Accountable and transparent management is one of the benefits. The program is open to the public and is subject to independent auditors' monitoring and assessment. Additionally, the local government has set up a GERTAK center with digital and computerized systems that allow them to provide precise poverty data for prompt decision-making. In this integrated center for poverty alleviation, representatives from various government agencies collaborate.

One effective collaboration between state and nonstate actors in creating an inclusive policy for reducing poverty is the GERTAK program. Its participatory mechanism, which involves communities and nonstate actors working alongside local government to create plans and strategies for public action on poverty reduction, had a direct impact on the government. Through a more open and accountable co-governing partnership, it has replaced traditional forms of governance with new ones. One instance of collaboration between the government, local communities, and nonprofit organizations to address social needs is the GERTAK program.

We find several advantages for the local government and the communities, even though partnerships don't always benefit both parties equally (Peters, 1998). Benefits for the local government include: resolving conflicts through partnership by addressing community criticism; reducing costs by utilizing community resources, such as improved government access to volunteers; gaining more substantial financial support and stability through the involvement of third-party financing; promoting greater accountability; improving data quality; improving program coverage; enhancing government capability through experience working on strategic alliances; and boosting political capital through meeting public expectations and improving citizen satisfaction in government. Communities benefit from: (i) better poverty program delivery; (ii) social needs are addressed and met; (iii) social capital is increased through a stronger sense of community and capacity-building to contribute to the program; (iv) community access to the program is improved; (v) participation in group problem-solving; and (vi) political capital is increased through improved government-community relations, which gives the community a say in local poverty eradication policymaking.

An inventive governance solution in the context of development financing is also demonstrated by the collaboration with BAZNAS for the collection and administration of the zakat funds. The local government has been able to obtain outside funding outside of its public budget (APBD) through this partnership program. The local government has established a chance to secure outside funding for thc program, building on the initial collaboration with communities. This suggests that collaborations may inspire a desire to increase services beyond their current scope. After the scaling-up effort in 2017, the initiative's monthly total reached IDR 200 million. At first, it was generating about IDR 10 million. In 2018, the monthly amount rose once more to IDR 250 million, and it is anticipated that it will eventually reach IDR 800 million. It is consistent with the idea that resource dependence and organizational behavior are related, and that organizations with a significant financial interest would aim to maximize the objectives maintaining the relationship.

10.3.5.2. Private Sector–Driven Partnerships

Private sector partnerships existed in various forms for a long time before "Type II partnerships" were recognized by the 2002 WSSD. However, private sector-driven partnerships for sustainable development have gotten less recognition and are frequently grouped together under public-private partnerships, sometimes being referred to as social alliances or as an inventive tool for international organizations. The literature on social alliances distinguishes between two conventional goals of the private sector when seeking collaborations with nonprofit actors: economic purposes that incorporate a marketing objective and noneconomic goals that concentrate on enhancing social welfare. A large portion of the scholarships in this field concentrate on the procedure for exchanging resources and obtaining resource support from the private sector. There hasn't been much knowledge of the organizational and collective interactions between businesses and nonprofit actors.

It entails going beyond social responsibility to form strategic alliances that combine corporate strategy with social responsibility through the exchange of assets, skills, and knowledge. The evolving political, managerial, and intellectual landscape typically leads to the development of a public-private partnership. By including more parties and giving the private sector a bigger role, research saw the goal in Indonesia to broaden public-private partnerships beyond corporate social responsibility. Though the number of private sector-driven partnerships for the SDGs in Indonesia is currently increasing, most of them fall under the traditional social responsibility framework of philanthropy, in which the government or civil society organizations look for and receive corporate donations. The research, which looks at private sector-driven collaborations in the provinces of Bangka Belitung and Riau Islands, aims to comprehend how the business interacts with other stakeholders outside of the process of acquiring and sharing resources.

10.3.5.3. Civil Society-Driven Partnerships

To address the wicked failure of legitimacy, justice, and effectiveness in representative and bureaucratic institutions, citizens could offer local knowledge, wisdom, dedication, and authority if they were deployed effectively. The 2030 Agenda's implementation anticipates notable advancements in development cooperation's engagement of civil society organizations. According to Arhin (2016), civil society organizations play three primary roles in development: advocacy or catalyst, service delivery or implementers, and facilitation or brokering.

The literature also made clear how important it is for civil society organizations to be involved in the implementation of the SDGs and how eager they are to contribute to their success. Nevertheless, it is not always clear how the actions of civil society organizations relate to the SDGs. Additionally, it's unclear if their efforts have an impact on the formulation of national policy. Yogyakarta's civil society-driven partnership serves as an illustration of how nonstate actors use the participatory mechanism to communicate with the local government. The local government in Yogyakarta is encouraged to implement participatory policymaking by the civil society organization. With assistance from the Sepuluh Nopember Institute of Technology's Department of Architecture, Ohana worked with other civil society organizations to create a feasibility study following the plans of there provincial and local governments to create an inclusive Yogyakarta by 2024. The study found serious issues with the creation of public policies about accessibility and facilities for the disabled, such as the absence of participation of pertinent parties in decision-making procedures.

Nonetheless, the project's implementation and the creation of local regulations are seen as exclusive processes. To advocate for policy, Ohana and its partners created a feasibility study. The study emphasized that accessibility for people with disabilities should be taken into account when agencies at the local and provincial levels communicate and coordinate. To enable a crucial role in comprehending the idea of inclusive, nondiscriminatory, and sustainable urban development, public officials' and pertinent stakeholders' ability to disseminate information will also be enhanced. The study also noted obstacles and suggested policies for efforts to create an inclusive Yogyakarta by 2024. Initiating multi-stakeholder partnerships is crucial when designing and constructing accessibility for people with disabilities, according to one of the 11 recommendations.

The study concluded by offering legal and policy references for the application of inclusive design and universal access. In this collaboration, Ohana serves as a facilitator by bridging the gap between the government and local needs through policymaking, and as an advocate or catalyst by sharing knowledge.

In this instance, the local government has been successfully persuaded to use a participatory policymaking process by the civil society organization. Its announcement of a partnership could serve as a model for implementing the NUA, the 2030 Agenda, and the Universal Declaration of Human Rights, emphasizing that cities should be places where everyone is welcome. We discovered that to inform plans and strategies for the partnership, a multi-stakeholder partnership must consider social needs at the outset. To achieve this, leadership has been essential in bridging the gap between the global agenda and local needs as well as in identifying the transformative leadership elements that contribute to transformative capacity. Through its founder, Ohana is a member of the Committee on the Rights of Persons with Disabilities, an impartial panel of experts that oversees how the States Parties are implementing the Convention.

10.4. GLOBAL AGRICULTURAL MONITORING AND GLOBAL POLICY FRAMEWORKS

The G20 Action Plan on Food Price Volatility and Markets launched the GEOGLAM (Group on Earth Observations Global Agricultural Monitoring) Initiative in 2011 to enhance crop production outlooks and early warning systems through the use of satellite data. Regardless of their internal remote sensing knowledge or experience, empowering agricultural policy and program decisions is at the heart of GEOGLAM's mission. Our efforts are motivated by these needs for decision support, and we bring together a varied community to transform data into metrics, information, and knowledge before reaching long-term decisions that have real policy implications.

10.4.1. The GEO Global Agricultural Monitoring (GEOGLAM) Initiative

GEOGLAM is a global initiative supported by the G20 that aims to improve the use of Earth observations (EO) to support policy, action, and decision-making in the areas of sustainable agriculture and food security. NASA Harvest provides funding for the co-chairpersonship of GEOGLAM as a significant contribution (C.O. justice), the GEOGLAM Crop Monitor's leadership and coordination, as well as co-leadership of the Essential Agriculture Variables for GEOGLAM working group and leadership of the EO Data Thematic Coordination Team (Whitcraft).

10.4.1.1. How Satellites Make This Work

The Group on Earth Observations Global Agricultural Monitoring Initiative is known by the acronym GEOGLAM. The G20 Action Plan on Food Price Volatility, which also includes the Agricultural Market Information System (AMIS), includes the GEOGLAM initiative, which was started in June 2011 by the Group of 20 (G20) Agriculture Ministers and reiterated in January 2017. GEOGLAM is a cooperative project with a synthetic orientation by design. It makes use of the expertise and experience of global agricultural specialists who concentrate on food production and security at the national, regional, and international levels.

10.4.1.2. Developing and Implementing a Satellite Data Acquisition Strategy in support of GEOGLAM: Collaboration with CEOS

Coordination of Earth observations, including satellite-based data from both commercial and civil space agencies, is at the heart of GEOGLAM's operations. In the latter instance, a close partnership with the CEOS enables this to be achieved. Dr. Alyssa Whitcraft is spearheading a project to "reboot" the requirements to take into consideration developments in agricultural monitoring methods, approaches, and research findings derived from studies conducted within the framework of GEOGLAM. She still collaborates with CEOS to define "analysis ready" data for optical and SAR systems and to convert requirements into precise, actionable acquisition requests for site, sub-national, national, regional, and worldwide scope.

10.4.2. Project Background Mission and Goals

To improve future cooperation, the U. S. National Aeronautics and Space Administration (NASA) and the Department of Agriculture (USDA) recently inked a Memorandum of Understanding (MOU). A new project to integrate NASA's Moderate Resolution Imaging Spectroradiometer (MODIS) data and products into an existing decision support system (DSS) run by the International Production Assessment Division (IPAD) of the USDA Foreign Agricultural Service (FAS) was jointly funded by the two agencies in support of this partnership. The project is putting in place an intuitive system that will enable the integration and analysis of MODIS data products in IPAD's DSS, building on NASA's investment in the MODIS Science Team.

10.4.2.1. FAS Mission & Tasks

0FAS supports U.S. stability and security. S. food supply, enhances U.S. access to international markets. S. agricultural goods, provides information on global food security, and counsels the U. S. government on the need for international food aid. USDA's foreign operations, including market development, international trade agreements and negotiations, and the gathering and evaluation of market data and statistics, are primarily the responsibility of FAS. Additionally, it manages USDA's food assistance and export credit guarantee programs and, by leveraging expertise for agriculturally led economic growth, contributes to raising incomes and food availability in developing countries.

10.4.2.2. IPAD's Scope

The FAS provides agricultural data for global food security through IPAD. It generates timely, unbiased, and consistent evaluations of the state of food security and the outlook for agricultural production worldwide. IPAD is in charge of estimating the production and yield of grains, oilseeds, and cotton as well as conducting assessments of crop conditions worldwide. The monthly crop assessments published by USDA's World Agricultural Outlook Board, a major global source of agricultural information, include IPAD assessments as a crucial component.

10.4.3. The Application of NASA EOS MODIS Data to FAS

10.4.3.1. Agricultural Assessment and Forecasting

To achieve its goals, FAS/IPAD uses satellite data and data products to track and monitor natural disasters like floods, droughts, and persistent snow cover that affect agricultural productivity, as well as to monitor agriculture globally. In the United States, non-military sector, FAS is the biggest user of satellite imagery. S. the government. FAS has been tracking crop conditions and reporting on sporadic events using a combination of Landsat and NOAA-AVHRR satellite data for the past 20 years.

Through an information delivery system for MODIS data and derived products, FAS is improving and modernizing the satellite component of its IPAD decision support system. NASA's MODIS, which is installed on two Earth Observing System (EOS) platforms, was created in part to track land cover, vegetation production, and subtle vegetation responses to stress with regional to global coverage. Because of this, FAS can better characterize land surface conditions at the regional level and monitor changes in the major agricultural areas of FAS focus regions more quickly and at a higher resolution than was previously feasible with NOAA-AVHRR data thanks to the integration of MODIS data and derived products into the IPAD FAS DSS.

10.4.3.2. MODIS: An Operational Prototype

Despite being a NASA experimental mission, the Visible Infrared Imager Radiometer Suite (VIIRS), which will be launched in 2008 as a component of the National Polar Orbiting Environmental Satellite System (NPOESS) and fully operational in 2009, will expand the capabilities of MODIS. As a result, this research project's techniques and system can be transformed into a fully functional domain.

10.4.3.3. Project Components

Delivery and integration of MODIS Rapid Response data into the FAS monitoring system to facilitate improved monitoring of the impact of climate hazards, such as drought, large-scale flooding, and snow storms, on agricultural production. Development and delivery of a long-term database of MODIS composite Vegetation Index (VI) time series including analysis tools and a graphic user interface that provides mosaicking, projection capabilities, and easy access to the moderate-resolution image archive.

Establishment of the relationship between MODIS VI data and the long-term archives from the AVHRR and SPOT-VEGETATION used by FAS/IPAD.

Development of enhanced MODIS cropland products including a crop mask, a crop type map, new band combination products, and a crop stress index.

10.4.3.4. MODIS Rapid Response

The MODIS Rapid Response (RR) system provides rapid access to MODIS data collected twice daily from the Terra satellite in the morning (10:30 am) and the Aqua satellite in the afternoon (2:30 pm).

The RR system provides FAS analysts with access to georeferenced, calibrated, mosaicked daily global MODIS imagery for FAS regions of interest within 2–4 hours of satellite acquisition. RR data are available to FAS analysts at spatial resolutions of 250m, 500m and 1km in different band combinations.

Such rapidly accessible data allows the FAS analysts to evaluate, and asses, in near real-time, the effect of disaster events on crops.

10.4.4. Long-Term Database of MODIS Composite Vegetation Index Time Series

The evolution of the growing season concerning reference long-term mean conditions is tracked using remotely sensed vegetation index data when crop conditions are monitored for a particular region. Similar to the current archive of NOAA-AVHRR-derived NDVI, a global normalized difference vegetation index (NDVI) is created from MODIS data and is known as the continuity index. A 16-day compositing period was used to create a global NDVI time-series database with a 250-meter spatial resolution, enabling interannual comparisons of growing season dynamics. To fit the FAS regions of interest, this MODIS NDVI dataset is automatically reprojected and mosaicked.

The time-series data are accessible to FAS analysts through a powerful web interface and analysis tool. From this stand-alone interface, the analysts can query these data by pre-defined sub-regions, by interactive regional sub-setting, and by implementing crop/water masks to:

- Plot time-series graphs over the crop growing seasons to quickly assess crop conditions and anomalies;
- Monitor current conditions;
- View spatially, NDVI anomalies comparing current conditions to previous year, or historical mean; and
- Plot histograms of current and historical NDVI data.

These data and utilities are fundamental for crop yield forecasts and can serve as an early warning system for regions suffering from crop loss and food shortages.

10.4.5. Long-Term Monitoring: An inter-sensor Calibration

USDA/FAS receives an international data stream of NDVI spanning more than 20 years (1981-present) from the Global Inventory Monitoring and Modeling Studies (GIMMS) group at NASA Goddard Space Flight Center (NASA/GSFC). The Global Area Coverage (GAC) data from the National Atmospheric Oceanic Administration's (NOAA) polar orbiting series of satellites and measurements from the Advanced Very High-Resolution Radiometer (AVHRR) are used to calculate the GIMMS NDVI. To create and preserve a consistent NDVI archive, GIMMS inter-calibrated the data from the NOAA-AVHRR satellite series and carried out atmospheric correction to reduce the impact of volcanic aerosols.

The historical database for tracking how vegetation reacts to climate conditions is provided by the GIMMS NDVI archive. To provide a consistent multi-source long-term data record for agricultural monitoring, it is essential to link the MODIS data to the long-term GIMMS AVHRR/NDVI, archive, and SPOT Vegetation sensor data. This makes it possible for FAS analysts to contrast the spatial extent and intensity of NDVI anomalies linked to crop failures, droughts, and floods with the data that is currently available (Figure 10.1).

Figure 10.1. *Long-term monitoring: An inter-sensor calibration.*

Source: By Robert Markowitz - NASAhttp://www.nasa.gov/mission_pages/station/multimedia/ISS_FCR.htmlhttp://www.nasa.gov/images/content/160446main_jsc2006e43860_high.jpgTransferred from en.wikipedia to Commons. Transfer was stated to be made by User:TheDJ., Public Domain, https://commons.wikimedia.org/w/index.php?curid=3172773.

VIs from different sensors are related by examining their temporal and spatial behavior (means, ranges, and anomalies) at target FAS agricultural regions throughout the globe and interpreting these statistical patterns concerning the growing seasons for various crops. This project component aims to build a consistent time series that best represents the vegetation dynamics over these agricultural regions.

10.4.5.1. Enhanced Cropland Products using MODIS: A Dynamic Continuous Cropland Mask for use with MODIS Time-Series Web Interface

The spatial distribution of croplands must be understood to effectively monitor agricultural regions around the world and provide accurate assessments of agricultural production.

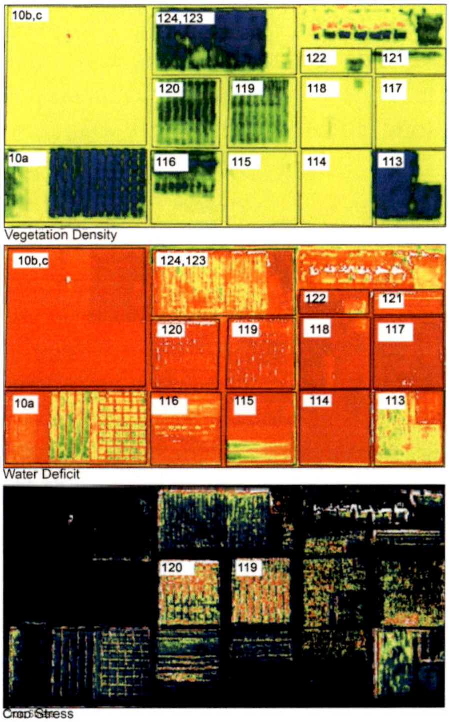

Figure 10.2. *Successfully monitoring worldwide agricultural regions and provide accurate agricultural production assessments.*

Source: By Susan Moran, Landsat 7 Science Team and USDA Agricultural Research Service (U.S. Government work) - earthobservatory.nasa.gov (NASA Earth Observatory, Image of the Day 2001-01-30 "Precision Farming"), Public Domain, https://commons.wikimedia.org/w/index.php?curid=10747279.

To identify every location used for crop production, GLAM has created a global croplands mask. Croplands exhibit significant temporal and spatial variability. Croplands differ greatly throughout the world in terms of traits like cropping intensity and field size, and they fluctuate from year to year as a result of occurrences like drought and fallow periods. To illustrate these diverse aspects of global crop cover, a flexible crop likelihood mask is employed (Figure 10.2).

Four years' worth of MODIS data were analyzed to create this adaptable crop lands mask. With the help of this dynamic mask, FAS analysts can use the MODIS time series web interface to set cropland membership thresholds based on their requirements and areas of interest.

In contrast to less intensively farmed regions in parts of Sub-Saharan Africa where cropland identification is partially confused with natural background vegetation phenologies, regions with intensive agro-industrial farming practices, like the Maize Triangle in South Africa, will have higher confidence values in the crop mask. Thus, regions with different cropping intensities can be examined using a customized threshold.

10.5. A COMPARISON OF GLOBAL AGRICULTURAL MONITORING SYSTEMS

SDG 2, which aims to "End hunger, achieve food security and improved nutrition, and promote sustainable agriculture," places achieving food security at the top of the Sustainable Development Goals list. Sustainable global agricultural production has come a long way, but there are still many obstacles to overcome. These include the intricate interactions between various factors like extreme weather patterns, population growth, increasing wealth, water scarcity, rising energy costs, and civil unrest. The price of major commodities, including food, increased significantly between 2007 and 2011, for maize, wheat, and rice, which caused social and political upheaval globally. Several factors were proposed as potential causes, including the shift in many developing nations toward a meat-based diet, rising oil prices that affected fertilizer prices, financial speculation, and the main agricultural producing nations' inadequate supplies of major grain commodities. Timely and accurate information on current and projected global food production is required to better prepare for disruptions in the food supply and global crop market price fluctuations of the kinds that have been observed over the last 10 years. Global food insecurity will be lessened with improved monitoring since it will allow for more precise commodity price forecasting and a better comprehension of the main threats to the food supply.

National and regional agricultural monitoring has existed for many years, e.g., the Food and Agriculture Organization's (FAO) Global Information and Early Warning System (GIEWS), the European Commission's (EC) Monitoring

Agriculture with Remote Sensing (MARS) system, the United States Agency for International Development's (USAID) Famine Early Warning Systems Network (FEWS NET), and CropWatch in China. These systems have a history of functioning largely independently and sharing little information, with an emphasis on either global food production or food security for developing nations. The first system to offer globally comprehensive data on crop production and condition was the Foreign Agricultural Service (USDA-FAS) of the US Department of Agriculture. These estimates of crop production are also utilized as economic indicators, for early warning alerts, for trade policy and exporter assistance, for commercial market trends and analysis, and for foreign aid assessments for food import needs. However, it was acknowledged that more coordination and information sharing are required because national and regional monitoring systems are unable to adequately monitor agriculture at all scales. A worldwide system of systems for monitoring agriculture. To bring together important players in the global agricultural monitoring community, share information internationally, and produce two regular bulletins (one for the Agricultural Market Information System (AMIS) and one for Early Warning covering roughly 95% of the world's croplands) that represent the consensus of the current situation globally, the Group on Earth Observations (GEO) became the perfect body to launch the flagship initiative GEOGLAM (GEO Global Agricultural Monitoring). Other programs that assist GEOGLAM have also been started, such as beginning in 2017, the Anomaly Hot Spots of Agricultural Production (ASAP) system.

Eight major operational systems that either contribute to GEOGLAM on an international level or are crucial to regional or global agricultural monitoring were identified to better understand the current state of global agricultural monitoring. Additionally, this section offers a thorough update to Atzberger's earlier review of agricultural monitoring systems. A brief description of each of these systems is provided, along with the findings from the two questionnaires that were used to collect data on them. In addition to general inquiries about geographical scope, stakeholders, interactions with other systems, etc., the first one concentrated on many questions about the inputs and outputs of the system. The significance of (i) various data inputs—which are then described along with cutting-edge advancements from the literature—and (ii) perceived gaps in methodologies—which are also described—were assessed using a second questionnaire.

10.5.1. Description of the Main Global and Regional Scale Agricultural Monitoring Systems

There are currently eight operational agricultural monitoring systems at the regional and global levels that give stakeholders information to help them make evidence-based decisions. Below is a synopsis of each system, arranged chronologically by date of establishment. A few of these systems are a component of larger monitoring programs for food security (e.g., FEWS NET and GIEWS), but this section's main objective is to compare the elements of global crop production.

10.5.1.1. Global Information and Early Warning System (GIEWS)

One of the first important worldwide sources of data on food production and food security within FAO was GIEWS, which was founded in the early 1970s. The system offers more focused regional reports based on intelligence from FAO's regional and country offices in addition to regular bulletins about food crop production and markets worldwide. 115 governments, 61 non-governmental organizations (NGOs), and a large number of trade, research, and media organizations are all part of the GIEWS network. To identify weather-related issues that might affect food security in member nations, the GIEWS team uses geospatial data as an auxiliary variable to continuously monitor the global food supply and demand situation. GIEWS employs the Agricultural Stress Index (ASI), which was adopted in 2013 and was created to close an information gap in the current early warning system, in addition to rainfall estimates and the Normalized Difference Vegetation Index (NDVI). The ASI is an indicator for early identification of agricultural areas that may be impacted by droughts or dry spells. The ASIS (ASI System) creates a map every 10 days that highlights hotspots where crops experience water stress during the growing season. These hotspots are then confirmed by data from government agencies or by agrometeorological models that use data from national meteorological networks, which eventually demonstrate indicator convergence.

10.5.2. Famine Early Warning Systems Network (FEWS NET)

USAID launched the FEWS NET system2 in 1985 to support decision-making for relief organizations and food assistance programs. Although it doesn't directly track production, FEWS NET makes an effort to measure changes in both planted area and crop yield (Brown, 2008). In addition to publishing specialized monthly reports on current and projected food security, the system, which currently covers 36 of the world's most food-insecure nations, also promptly alerts users to new crises. To accomplish its objectives, FEWS NET employs a convergence of evidence approach. To create scenarios, perform livelihood analysis, and generate data for efficient decision support, information is gathered from field assessments, agro-climatology, market/price monitoring, nutrition surveillance, and conflicts. As new data becomes available, FEWS

NET updates scenarios every month in addition to its quarterly outlook reports. For its analysis of food security, FEWS NET primarily uses anomaly analysis and heavily relies on agro-climatology data.

10.5.2.1. MARS Crop Yield Forecasting System (MCYFS)

In 1992, the MARS program of the JRC developed the operational MCYFS to fill the need for operational estimates of area, yield and production at the pan-European level for EU member states. It is run in accordance with European Regulation No. 1306/2013 (Articles. 6 and 22). To manage agricultural markets, this regulation calls for forecasts of production and yield as well as an agricultural monitoring system. By tracking crop growth and predicting yields, the MCYFS serves as a decision support system that offers unbiased, fact-based information on the state of annual crops in the EU and its neighboring nations. Three primary data sources are acquired and processed in near real-time by the MCYFS: crop model simulations, weather data (observations and forecasts), and biophysical parameters obtained from satellite remote sensing to track crop health. A statistical yield forecasting process makes use of all of these data in addition to a time series of historical area and yield statistics. The monthly MARS bulletins, which include a pasture analysis, country-specific analyses, and yield forecasts for cereals, oilseeds, and tuber crops, give a summary of the progress of the major crops and areas of concern. The JRC MARS Explorer3 can be used to visualize historical and near real-time data on crop growth and weather conditions. There are maps for many crop and weather indicators, and the data is updated three times a month.

10.5.2.2. CropWatch

CropWatch assesses crop production at the national and international levels and is run by the Chinese Academy of Sciences' Institute of Remote Sensing and Digital Earth. To plan crop imports, exports, and prices and guarantee national food security, this system was established in 1998 to provide timely, accurate, and independent predictions of crop conditions and production, both domestically and internationally. CropWatch has been publishing bulletins globally since 2013. There are four spatial levels taken into consideration: sub-national (for the nine largest countries) and global, regional, and national (for 31 important countries, including China). More than 80% of maize, rice, soybean, and wheat production and exports come from these 31 nations. Indicators for rainfall, temperature, photosynthetically active radiation (PAR), and potential biomass are used to examine global trends in growing conditions. Crop conditions, farming intensity, and stress are described at the regional level using additional indicators like the Vegetation Health Index (VHI) and the Vegetation Condition

Index (VCI). To estimate food production, CropWatch also conducts thorough crop condition analyses at the national and sub-national levels using a wide range of variables and indicators.

10.5.3. United States Department of Agriculture-Foreign Agricultural Service (USDA-FAS)

The Honduran Ministry of Agriculture required almost real-time agricultural data during the reconstruction phase following Hurricane Mitch's 1998 devastation of the country. To help agricultural economists and researchers forecast global crop production, the USDA and FAS launched the Crop Explorer service in 2001. Every 10 days, the system publishes data visualization products by automatically extracting and processing agro-meteorological indicators from the Jason-1, TOPEX/Poseidon, and MODIS satellites. FAS is also in charge of providing market intelligence for all major commodities in all foreign nations in the form of timely, objective, unclassified, worldwide crop conditions and production estimates. The US Office of Management and Budget (OMB) uses these estimates as a key component of the World Agricultural Production and World Agricultural Supply and Demand figures as economic indicators. FAS gathers data from its global network of agricultural economists, marketing specialists, meteorologists, and remote sensing scientists to complete this task. Remote sensing aids in the verification of crop production data obtained from field visits and reports from foreign governments.

The USDA-FAS's operational capabilities to track and predict global crop production have been greatly enhanced by many data sets combined with crop models and meteorological data. These data sets originate from the Global Agricultural Monitoring (GLAM) project, which was started in 2002 and is a joint research effort between South Dakota State University (SDSU), the University of Maryland (UMD), USDA-FAS, and the GIMMS at NASA. The integration and analysis of MODIS data products to feed the USDA-FAS decision support system is the main goal of the GLAM project. The project integrates many satellite data sources, such as MODIS and GIMMS's NDVI. Semi-automated classification algorithms and remote sensing data are used in USDA-FAS to calculate changes in crop area estimates and mid-season dominant crop masks. Regression and analog year algorithms are used to generate yield estimates and maps from mid-season to end-of-season based on MODIS NDVI data.

10.5.3.1. GEOGLAM

As stated in the introduction, GEOGLAM is a flagship project of GEO that was approved by the G20 in 2011 to give the AMIS an evaluation of crop status, growing conditions, and agroclimatic factors that could affect the world's production of rice, soy, wheat, and maize. Operating since September 2013, these assessments are documented in the Crop Monitor bulletin. To guarantee timely information for the Crop Monitor, assessments are carried out in the last 10 days of every month. Each month, a group of global experts participates in a conference call to review and discuss these assessments, which were produced by many independent but complementary sources, and to provide a chance to agree on any differences. Following that, a report containing the consensus data is created and iteratively reviewed by the partners.

10.5.4. World Food Programme Seasonal Monitor

The World Food Programme (WFP) Seasonal Monitor system7 has been in use since 2014, mainly to track the status of the growing season and provide early warning of conditions that could harm crop and pasture production in WFP regions of interest. The need for more detailed information than other systems could provide and the capability to generate customized outputs for internal clients were the driving forces behind the creation of this system. The system uses NDVI (MODIS) and near real-time rainfall estimates (CHIRPS) data to generate indicators. A variety of outputs are generated, primarily summaries of rainfall amounts over different time periods, dates of the growing season and vegetation index onset, and anomalies of the majority of parameters. Region-specific reports that describe the current growing season conditions and offer an outlook for the upcoming months based on available seasonal forecast information are the system's primary outputs, and they are produced at a roughly monthly frequency. Dataviz, a WFP visualization platform, offers an alternative viewpoint on growing season monitoring. It incorporates data from the Seasonal Explorer, which allows users to view charts of time series of seasonal rainfall and NDVI values and anomalies for administrative regions, including administrative Level 0 (country) and Level 1 (state, province, region, governorate, etc.). As well as information from the Economic Explorer, which displays price and price-forecast data, and Level 2 (district, locality, and county). Either the entire administrative unit or just the areas covered by pasture or cropland (based on a land cover layer) can receive the data from the Seasonal Monitor.

10.5.5. ASAP (Anomaly Hot Spots of Agricultural Production)

The last system is the ASAP, which was introduced in June 2017 by the European Commission's Joint Research Centre (JRC). This information system was created to support multi-agency information systems like GEOGLAM as well as for use by EC. To inform EU-supported food security assessments like the Integrated Food Security Classification and the Global Report on Food Crises, ASAP focuses on identifying regions where unfavorable growing conditions for both crops and rangelands may represent a potential food security issue.

This system consists of two components. In the first section, warnings about possible issues with crop and rangeland production worldwide are automatically generated at the first sub-national level using rainfall estimates and the NDVI obtained from remote sensing. Every 10 days, when fresh rainfall and NDVI data are available, these early warnings are sent out. For about 80 countries with a high risk of food insecurity, the second component of the system uses agricultural monitoring experts to confirm these "hotspots" of possible food security issues; these assessments are updated monthly. The primary benefit of the ASAP system is that it uses the available weather and EO data to create brief warning messages that decision makers can use without any prior knowledge of geospatial data analysis. The ASAP, which covers food insecure regions outside of Europe, supplements the data made available by the MCYFS.

10.5.6. Gaps in Agricultural Monitoring Needs

The first step was to identify the existing data gaps in agricultural monitoring through conversations with each of the monitoring systems. The significance of various data inputs to the system, such as cropland maps, crop calendars, maps of cropping intensity and crop types, crop management data sets, meteorological data, and statistical and in-situ data on agricultural production, yield, and area, was assessed using a second questionnaire to formalize this gap analysis.

10.5.6.1. Cropland Maps

Global cropland maps were deemed essential or extremely important to seven of the eight systems. While there is significant disagreement among the major land cover products regarding cropland, crops are included in one or more land cover types in global land cover maps. There have also been specific cropland maps created, but each one has its own set of uncertainties related to cropland definitions, methods, spectral separation problems, especially when seasonal effects are present, cloud cover in specific areas when using optical data, etc. Higher-resolution photos will be necessary to achieve the accuracy requirements for smallholder farmers' highly fragmented landscapes. There is still room for improvement as the most recent 30 m cropland map has an overall accuracy of 94% (the percentage of correctly classified sample locations) but a user's accuracy of 68.5%, which is a class-specific accuracy for cropland that shows how frequently this class will actually be present on the ground. Although this is not yet operationally available on a global scale, annual updates are preferred, particularly in areas like the Sahel where cropland can be highly dynamic.

Another method that combines several cropland or land cover products to provide the best description of cropland or land cover at a specific location is the Unified Cropland Layer, the GLC-SHARE hybrid land cover product, and the global hybrid cropland map created by IIASA and IFPRI. For the IIASA-IFPRI product to be utilized in global models that call for official statistics, it has also been calibrated using FAO statistics. The ASAP cropland and rangeland maps were created using a similar methodology, in which the best land cover data sets for agricultural monitoring were chosen by comparing six global and 16 regional land cover data sets at the national level using multi-criteria decision analysis. We have compiled existing cropland maps to highlight priority areas for improving cropland mapping because it is evident that more work is needed to accurately map the world's cropland.

10.5.6.2. Crop Calendars

Though the MCYFS system only uses European ones, all eight systems considered global crop calendars to be critical to extremely critical, much like they did with global cropland maps. The planting and harvesting dates of various crop varieties for a given area or region are listed on crop calendars. Among other uses, these are helpful for forecasting and estimating crop yield, crop condition monitoring, and crop type area estimation. Accurate harvest timing information would also help with agricultural policies, food aid mobilization, and commodity transportation to markets.Additionally, this data may help achieve UN SDG goal 12.3, which is to decrease post-harvest losses. Crop calendars for a wide range of crops are provided by FAO for 44 African nations according to their agro-ecological zones. Household surveys and national censuses are the traditional sources of this kind of data, which takes a lot of time to compile and preserve. Furthermore, regional variations are not captured by the use of agro-ecological zones.

10.5.6.3. Maps of Cropping Intensity (CI)

The number of crops grown in a year is known as CI, and it can reveal important details about food security. Information about CI is considered extremely important by both FEWS NET and WFP's Seasonal Monitor, while it is considered critical by four other systems. They suggested using data from multiple sensors after discovering that, when using only MODIS data, they underestimated the number of cropping cycles because of missing data, especially during cloud-covered monsoon periods. The significance of including additional data types, such as crop model outputs or agricultural statistics, has been demonstrated by other studies. The CI data are not documented, although the FAO's agricultural statistics subtly include information on multiple cropping in the figures for total production and area harvested. To facilitate the creation of global maps of CI, it is crucial to compile this data into a global database at both the national and subnational levels.

10.5.6.4. Maps of Crop Type

Maps of crop type were viewed as very critically important by half of the systems and critical by two others. Neither the ASIS system of GIEWS nor the MCYFS of MARS uses crop type information; in the case of MARS, agricultural statistics are used instead. Spatially explicit crop-type maps have been produced by downscaling national and sub-national statistics using different methodologies. The M3 Cropping System Model consists of harvested area and yield downscaled for 175 crops including tree crops and managed grasslands. Using remote sensing, cropland, and pasture maps are constructed at a 5 arc-minute resolution and national and sub-national agricultural data are then downscaled using a regression approach.

10.5.6.5. Crop Management Data Sets

Crop management data sets were rated as critical by five systems, but less critical by the other systems. Crop growth models, which can simulate yields under various management practices, require spatially explicit data sets, but there is a glaring lack of information on crop management practiccs, which are important factors influencing agricultural productivity. The absence of information on nitrogen fertilization and tillage was noted in an inventory of the primary data sets available for various land management techniques. While there are globally available, spatially explicit data sets on nitrogen fertilization, there are currently no data on tillage, despite the development of promising methods using Sentinel 1. However, these data sets are based on models, and there are significant differences between them. This data set has several uncertainties as estimates may be further uncertain because the M3-crop distribution data set was used for crop types, and it does not account for grassland fertilization, which could cause it to overestimate fertilizer use in some places.

The environment may also be negatively impacted by crop management techniques, e.g., contamination, eutrophication, and salinization of the regions around agricultural land, common environmental issues associated with agriculture include rivers, groundwater, and populations of natural pollinators, all of which are impacted by excessive pesticide use. Researchers typically need to piece together data on fertilizers, pesticides, and their dosages and frequency to comprehend their impact on the environment. Building better spatially explicit products of crop management practices and assessing their effects on the environment, including soil health, water quality, and biological activity surrounding and within crop fields, would benefit from a clear registry and characterization of these and other environmentally related practices from cropping areas worldwide.

10.5.6.6. Meteorological Data

Rainfall data is the most crucial meteorological input, and six of the eight systems considered meteorological data to be extremely important for system performance. All eight monitoring systems use rainfall estimates (Table 10.1). Even though rain gauges provide the most accurate rainfall data, there are gaps in the spatial data across many areas due to the sparseness of the station network, especially in areas with the highest risks to food security. Generally speaking, the network of global weather stations has dwindled and deteriorated in certain developing nations, especially those in Central Asia, and many stations are unable to be maintained because of inadequate national capacity and underfunding. Satellite observations and atmospheric circulation models can provide spatially explicit measurements to supplement ground-based meteorological data, especially rainfall. An extensive overview of the available gauge and satellite rainfall data for Africa is provided by precipitation forecasts from the European Centre for Medium-Range Weather Forecasts (ECMWF), rainfall estimates from the Climate Hazards Group InfraRed Precipitation with Station data (CHIRPS) of the USGS, rainfall from the Tropical Rainfall Measuring Mission (TRMM), and rainfall from the Tropical Applications of Meteorology using SATellite (TAMSAT). These forecasts compare seven different rainfall products and demonstrate discrepancies between them and with rain gauge data. Products with shorter historical records but incorporating rain gauge data clearly have a trade-off with those with longer historical records but not using rain gauge data. As a result, these products are surrounded by uncertainties, which are further exacerbated in crop models that rely on rainfall estimates from satellites. A promising line of current research that could eventually help agricultural monitoring is rainfall estimation using radio interference in cellular networks. Furthermore, there are current plans to use cell phone towers as locations for weather stations and to deploy inexpensive sensors to increase the density of meteorological observations.

10.5.6.7. Data on Agricultural Production, Area, and Yield

This information is considered critical or very critical in six of the eight systems. From 1961 until the most recent year available, the FAO has been gathering agriculturally relevant statistics for 245 countries and 35 regions in its FAOSTAT system. The data sets are derived from questionnaire-based surveys, agricultural samples, and censuses. Despite this extensive historical record, there are many disadvantages to depending solely on the caliber and correctness of these national reports because no efforts have been made to standardize data collection techniques across the various nations. Numerous nations produce national agricultural statistics, but there are wide variations in their timeliness, accuracy, consistency over time, and accessibility. Some nations use national sampling frames, surveys, or censuses to gather administrative data. An established statistical technique for gathering agricultural data—like actual crop yield—that cannot be directly acquired by satellite is called area frame sampling. However, there are no internationally accepted standards for the collection of in-situ or survey data for agricultural statistics, and other nations lack a formal data collection procedure. FAOSTAT only offers a limited amount of agricultural information because FAO statistics are reported nationally, especially when spatially explicit data are needed. One issue with crop forecasting systems and remote sensing has been identified as the absence of sub-national agricultural statistics. Nonetheless, there are programs like FAO's CountrySTAT and Agro-MAPS (Mapping of Agricultural Production Systems) that gather data on crop yield, area, and production at the subnational level. To create and validate crop models that are used to forecast production or as inputs to statistical models for the prediction of yield anomalies using remote sensing, more in-situ yield data is required. In-situ data was considered essential to extremely essential for the functioning of six of the eight systems. In certain EU nations, the Land Parcel Information System (LPIS) is publicly accessible like the Czech Republic and the Netherlands, which might inspire other nations to make their agricultural databases publicly available in the future. Crop trials provide additional sources of in-situ yield data, such as from FAO, the Global Yip Atlas, the International Maize and Wheat Improvement Center (CIMMYT), and the Global Agricultural Trial Repository and Database on CCAFS (Climate Change, Agriculture, and Food Security), which is housed by the Consultative Group on International Agricultural Research (CGIAR). With the latter initiative, scientists will be encouraged to share their data with the community instead of keeping it locked away in their own institutions. The private sector is another possible source of data, and developments in mobile phone and location-based technologies are also expanding the amount of yield data being collected in-situ. Access to and sharing of these varied data sets are uncommon, though.

Soil data is another essential input for crop growth models. The Harmonized World Soil Database v1.2, which has integrated national and regional soil data sets, many of which are available from the International Soil Reference and Information Centre (ISRIC), is available at a 1 km resolution for yield forecasting for agricultural monitoring purposes. This map has drawn criticism, though, due to its coarse resolution and the fact that it combines existing soil data sets of varying quality and time periods rather than accurately depicting the state of the soil today. Several other data sets of interest are available from ISRIC, including the more recent SoilGrids250m product, gridded data sets of soil water capacity, and the Harmonized Global Soil Profile data set v3.1, which contains over 10,000 soil profiles worldwide.

10.5.6.8. Gaps in Methods

There are still gaps in the science underlying agricultural monitoring and forecasting, despite the claim that data limitations and scarcity are more significant than theoretical and methodological flaws in agricultural system models. The eight international agricultural monitoring systems were asked about their opinions of the methodological gaps that exist today to better understand these problems in the context of this section. Better yield and crop production forecasts are needed, according to three of the systems. This can be achieved using statistical models or crop growth models according to NDVI. Several recent reviews of crop models and their shortcomings were brought to light in the review of agricultural system models.

These limitations were primarily related to the data, but they also acknowledged the need for ongoing methodological advancements. Another gap that was found was improving comprehension of the variations among various input data sets. For instance, different sources produce different amounts of precipitation, and the same is true for various vegetation indices. Consensus in monitoring requires a deeper comprehension of the areas and causes of discrepancies in these data sets, and there are currently insufficient tools available for conducting an automated comparison.

10.6. GLOBAL POLICY FRAMEWORKS

Gender equality is promoted by many normative policy frameworks at the national and international levels, including FCAS. Resolutions 1325 and 1820 of the Security Council seek to protect and empower women in times of violence and instability. They combinc spccial protections (based on the differences between men and women) with general protections for women (based on the similarities between men and women).

The General Recommendation No. was adopted by the Committee on the Elimination of Discrimination Against Women (CEDAW) in 2013. 30, which requires states to stop, look into, punish, and make sure that non-state actors who commit crimes against women are held accountable. *See* Table 10.1 below for additional resources on these and other significant international frameworks and commitments. Some experts contend that some laws problematically portray women as victims and men as perpetrators, and place an excessive amount of emphasis on protecting women, even though international frameworks reflect a growing global consensus on the importance of gender in conflict. This disregards women's broader experiences and roles in armed conflict and may compromise their agency.

10.6.1. What Impact Has International Law Had?

Evaluations of how international law affects gender and conflict are scarce. An impact analysis of 12 UN peacekeeping missions across 11 nations reveals that overall gains have been small and that the implementation of UN Security Resolution 1325 has been implemented with varying degrees of success. Despite a notable increase in women's political participation, efforts to include more women in peace talks and in disarmament, demobilization, and reintegration programs have only partially succeeded. Though there are still far too few women in the security industry, the presence of female peacekeepers has encouraged women to enter the field and challenged gender stereotypes. The adoption of gender equality clauses in national constitutions and pertinent national laws has been aided by legal and judicial reforms. It is unclear, nevertheless, how far the judicial system has come in terms of increasing capacity and promoting gender parity. Even though there are stronger legal frameworks in place to address gender-based and sexual violence, it still goes unchecked in many areas affected by conflicts. According to an OSCE evaluation, there isn't a systematic exchange of best practices or lessons learned within the OSCE region, nor is there a coordinated initiative to implement UNSCR 1325. It requires states to commit funds for audits to track progress, set baselines for collecting useful statistics, and set timelines for implementing reform.

10.6.2. What Impact Have National Action Plans Had?

To implement UNSCRs 1325 and 1820, several nations, regional organizations, and international organizations have created National Action Plans (NAPs). The goals of NAPs are to protect women from gender-based violence, mainstream women's perspectives in all facets of conflict-related crime prevention and prosecution, and support women's involvement in relief and recovery. Although this does not always happen in practice, the process of creating a plan offers the chance to coordinate pertinent cross-ministry actions. The quality and content of NAPs vary, as do the approaches taken. Although there is little proof of NAPs' effectiveness, the literature addresses many implementation issues. These include a lack of commitment from important political figures and stakeholders, a lack of accountability, an absence of a designated budget, and an emphasis on measuring outcomes rather than results. Concern over the discrepancies between commitments made at the national level and their implementation at the subnational level is growing. To decentralize the implementation of NAP and involve local governance structures, strategies are being developed to monitor and evaluate at the regional and local levels. The likelihood of NAPs being implemented is higher when they were created using inclusive processes. According to recent reports, more precise progress indicators must be created.

Table 10.1. *Principles of Effective Governance for Sustainable Development*

Principles	
Effectiveness	1. Competence
	2. Sound Policymaking
	3. Collaboration
Accountability	4. Integrity
	5. Transparency
	6. Independent oversight
Inclusiveness	7. Leaving no one behind
	8. Non-discrimination
	9. Participation
	10. Subsidiarity
	11. Intergenerational equity

Source: Meuleman L. Public Administration and Governance for the SDGs: Navigating between Change and Stability. Sustainability. 2021; 13(11):5914. https://doi.org/10.3390/su13115914.

REFERENCES

1. Abshagen, M. L., Cavazzini, A., Graen, L., & Obenland, W. (2018). Highjacking the SDGs? 32. Retrieved from: www.unfairtobacco.org, www.globalpolicy.org, www.misereor.de (accessed on 17 December 2024).
2. Adam, C., & Ajakaiye, O. (2011). Causes, consequences and policy implications of global food price shocks: Introduction and overview. *Journal of African Economies, 20*(i3–i11). https://doi.org/10.1093/jae/ejr012.
3. Anderson, W., You, L., Wood, S., Wood-Sichra, U., & Wu, W. (2014). A comparative analysis of global cropping systems models and maps (IFPRI Discussion Paper 1327). International Food Policy Research Institute.
4. Andonova, L. B., & Levy, M. (2003). Franchising global governance: Making sense of the Johannesburg Type II partnerships. In O. S. Stokke & O. B. Thommessen (Eds.), *Yearbook of International Co-Operation on Environment and Development 2003/04* (pp. 19–31). Earthscan.
5. Arhin, A. (2016). Advancing post-2015 sustainable development goals in a changing development landscape: Challenges of NGOs in Ghana. *Development in Practice, 26*(5), 555–568. https://doi.org/10.1080/09614524.2016.1189513.
6. Atstaja, Z. (2017). Sentinel-1 data use for monitoring agricultural activities. Retrieved from: http://eoscience.esa.int/landtraining2017/files/posters/ATSTAJA.pdf (accessed on 17 December 2024).
7. Atzberger, C. (2013). Advances in remote sensing of agriculture: Context description, existing operational monitoring systems and major information needs. *Remote Sensing, 5*(2), 949–981. https://doi.org/10.3390/rs5020949.
8. Austin, J. E. (2000). *The Collaboration Challenge: How Nonprofits and Businesses Succeed Through Strategic Alliances*. Jossey-Bass Publishers.
9. Bäckstrand, K. (2008). Accountability of networked climate governance: The rise of transnational climate partnerships. *Global Environmental Politics, 8*(3), 74–102. https://doi.org/10.1162/glep.2008.8.3.74.
10. Bäckstrand, K., & Lövbrand, E. (2006). Planting trees to mitigate climate change: Contested discourses of ecological modernization, green governmentality and civic environmentalism. *Global Environmental Politics, 6*(1), 50–75. https://doi.org/10.1162/glep.2006.6.1.50.
11. Bahagijo, S., Santono, H., & Okitasari, M. (2019). *Panduan Kemitraan Multipihak* (pp. 7–30). Ministry of National Development and Planning/Bappenas.
12. Baruth, B., Royer, A., Klisch, A., & Genovese, G. (2008). The use of remote sensing within the MARS crop yield monitoring system of the European Commission. In *International Archives of Photogrammetry,*

Remote Sensing and Spatial Information Sciences–ISPRS Congress, Beijing XXXVII (pp. 935–940).

13. Beisheim, M., & Ellersiek, A. (2017). *Partnerships for the 2030 Agenda for Sustainable Development: Transformative, Inclusive and Accountable?* SWP Research Paper. Stiftung Wissenschaft und Politik.
14. Berger, I. E., Cunningham, P. H., & Drumwright, M. E. (2004). Social alliances: Company/nonprofit collaboration. *California Management Review, 47*(1), 58–90. https://doi.org/10.2307/41166287.
15. Biermann, F., Chan, M. S., Mert, A., & Pattberg, P. (2007). Multi-stakeholder partnerships for sustainable development: Does the promise hold? In P. Glasbergen, F. Biermann, & A. P. J. Mol (Eds.), *Partnerships, Governance and Sustainable Development–Reflections on Theory and Practice* (pp. 239–260). Edward Elgar.
16. Blowfield, M. (2012). Business and development: Making sense of business as a development agent. *Corporate Governance: The International Journal of Business in Society, 12*(4), 415. https://doi.org/10.1108/14720701211267775.
17. Bojanowski, A., Avis, T. J., Pelletier, S., & Tweddell, R. J. (2013). Management of potato dry rot. *Postharvest Biology and Technology, 84*, 99–109. https://doi.org/10.1016/j.postharvbio.2013.04.008.
18. Booth, D. (2011). Towards a theory of local governance and public goods provision. *IDS Bulletin, 42*(2), 11–21. https://doi.org/10.1111/j.1759-5436.2011.00207.x.
19. Boryan, C., Yang, Z., Mueller, R., & Craig, M. (2011). Monitoring US agriculture: The US Department of Agriculture, National Agricultural Statistics Service, Cropland Data Layer Program. *Geocarto International, 26*, 341–358. https://doi.org/10.1080/10106049.2011.562309.
20. Brown, M. E. (2008). *Famine Early Warning Systems and Remote Sensing Data*. Springer Science & Business Media.
21. Classifying multiyear agricultural land use data from Mato Grosso using time-series MODIS vegetation index data. *Remote Sensing of Environment, 130*, 39–50. https://doi.org/10.1016/j.rse.2012.11.009.
22. Dahiya, B., & Okitasari, M. (2018). *Partnering for Sustainable Development: Guidelines for Multi-Stakeholder Partnerships to Implement the 2030 Agenda in Asia and the Pacific*. United Nations University Institute for Advanced Study.
23. Ding, M., Chen, Q., Xiao, X., Xin, L., Zhang, G., & Li, L. (2016). Variation in cropping intensity in northern China from 1982 to 2012 based on GIMMS-NDVI data. *Sustainability, 8*(11), 1123. https://doi.org/10.3390/su8111123.

24. Duveiller, G. (2012). Caveats in calculating crop-specific pixel purity for agricultural monitoring using MODIS time series. In C. M. U. Neale & A. Maltese (Eds.), *Proceedings of SPIE, 85310J*. S. Fritz et al. *Agricultural Systems, 168*, 258–272.
25. Fowler, A., & Biekart, K. (2017). Multi-stakeholder initiatives for sustainable development goals: The importance of interlocutors. *Public Administration and Development, 37*(2), 81–93. https://doi.org/10.1002/pad.1795.
26. Frantzeskaki, N., Dumitru, A., Anguelovski, I., Avelino, F., Bach, M., Best, B., & Rauschmayer, F. (2016). Elucidating the changing roles of civil society in urban sustainability transitions. *Current Opinion in Environmental Sustainability, 22*, 45–50.
27. Fung, A. (2006). Varieties of participation in complex governance. *Public Administration Review, 66*(Suppl. 1), 66–75. https://doi.org/10.1111/j.1540-6210.2006.00667.x.
28. Gazley, B., & Brudney, J. L. (2007). The purpose (and perils) of government-nonprofit partnerships. *Nonprofit and Voluntary Sector Quarterly, 36*(3), 389–415. https://doi.org/10.1177/0899764006295997.
29. Hale, T. N., & Mauzerall, D. L. (2004). Thinking globally and acting locally: Can the Johannesburg partnerships coordinate action on sustainable development? *The Journal of Environment & Development, 13*(3), 220–239.
30. Hasan, S., & McWilliams, C. (2015). Civil society participation in urban development in countries of the South: The case of Syria. *International Planning Studies, 20*(3), 231–247. https://doi.org/10.1080/13563475.2014.984663.
31. Haywood, L. K., Funke, N., Audouin, M., Musvoto, C., & Nahman, A. (2019). The sustainable development goals in South Africa: Investigating the need for multi-stakeholder partnerships. *Development Southern Africa, 36*(5), 555–569. https://doi.org/10.1080/0376835X.2018.1461611.
32. Hazlewood, P. (2015). *Global Multistakeholder Partnerships–Scaling up Public Private Collective Impact for the SDGs*. Independent Research Forum.
33. Minister of Home Affairs of the Republic of Indonesia. (2018). *Ministerial Regulation of Ministry of Home Affairs No. 7/2018 on Preparation and Implementation of Strategic Environmental Assessment in Formulating the Regional Mid-Term Development Plan*. Jakarta: Ministry of Home Affairs.
34. Minister of National Development and Planning/Bappenas of the Republic of Indonesia. (2018). *Peraturan Menteri Perencanaan Pembangunan*

Nasional No 7/2018 Tentang Koordinasi, Perencanaan, Pemantauan, Evaluasi, dan Pelaporan Pelaksanaan Tujuan Pembangunan Berkelanjutan. Jakarta: Ministry of National Development and Planning.

35. Minister of National Development and Planning/Bappenas of the Republic of Indonesia. (2019). *Rancangan Rencana Pembangunan Jangka Menengah Nasional 2020–2024* (p. 91). Jakarta: Ministry of National Development and Planning.

36. Minister of Public Works and Housing of the Republic of Indonesia. (2017). *Panduan Praktis Implementasi Agenda Baru Perkotaan*. Jakarta: Ministry of Public Works and Housing Indonesia.

37. Nelson, J. (2017). *Partnerships for Sustainable Development: Collective Action by Business, Governments, and Civil Society to Achieve Scale and Transform Markets*.

38. Nilsson, M., & Persson, Å. (2017). Policy notes: Lessons from environmental policy integration for the implementation of the 2030 Agenda. *Environmental Science & Policy, 78*, 36–39. https://doi.org/10.1016/j.envsci.2017.09.003.

39. Oburu, P. O., & Yoshikawa, H. (2018). Roles of multiple stakeholder partnerships in addressing developmental and implementation challenges of sustainable development goals (p. 434). https://doi.org/10.1007/978-3-319-96592-5_24.

40. OECD. (2018). *Making Blended Finance Work for the Sustainable Development Goals*. OECD Publishing. https://doi.org/10.1787/9789264288768-en.

41. Peters, B. G. (1998). 'With a little help from our friends': Public-private partnerships as institutions and instruments. In J. Pierre (Ed.), *Partnerships in Urban Governance* (pp. 11–33). Palgrave Macmillan. https://doi.org/10.1007/978-1-349-14408-2_2.

42. Pinkse, J., & Kolk, A. (2012). Addressing the climate change—Sustainable development nexus. *Business & Society, 51*(1), 176–210. https://doi.org/10.1177/0007650311427426.

43. President of the Republic of Indonesia. (2017). *Presidential Regulation No. 59/2017 on Implementation of Achievements of Sustainable Development*. Jakarta: Secretary of Cabinet of Republic of Indonesia.

44. Rittel, H. W. J., & Webber, M. M. (1973). Dilemmas in a general theory of planning. *Policy Sciences, 4*(2), 155–169. https://doi.org/10.1007/BF01405730.

45. Samii, R., Van Wassenhove, L. N., & Bhattacharya, S. (2002). An innovative public–private partnership: New approach to development.

World Development, 30(6), 991–1008. https://doi.org/10.1016/S0305-750X(02)00015-3.

46. Scheyvens, R., Banks, G., & Hughes, E. (2016). The private sector and the SDGs: The need to move beyond 'business as usual.' *Sustainable Development, 24*(6), 373–384. https://doi.org/10.1002/sd.1623.
47. Stibbe, D., Reid, S., & Gilbert, J. (2019). *Maximizing the Impact of Partnerships for the SDGs* (1st ed.). The Partnering Initiative and United Nations Department of Economic and Social Affairs.
48. Streck, C. (2004). New partnerships in global environmental policy: The clean development mechanism. *The Journal of Environment & Development, 13*(3), 295–322. https://doi.org/10.1177/1070496504268696.
49. Sullivan, H., & Skelcher, C. (2002). *Working Across Boundaries: Collaboration in Public Services*. Palgrave Macmillan.

INDEX

B

C

D

G

H

I

N

O

P

T

U

V

W

FSC
www.fsc.org
MIX
Papier | Fördert
gute Waldnutzung
FSC® C083411

Kolibri 360 GmbH
Ferdinand-Jühlke-Straße 7
99095 Erfurt, Deutschland
produktsicherheit@kolibri360.de